Astrometry – the study of the positions of astronomical objects and how they move across the sky – is a cornerstone of modern astronomy. Pinning down the distances and motions of stars in our Galaxy is fundamental to an understanding of the origin of these stars and how they affect the evolution of our Galaxy. Similarly, measuring the motions of planets, their satellites and asteroids is crucial to unravelling the origin and evolution of our Solar System (as well as to the planning of space missions to explore them).

A workshop held jointly by the Royal Greenwich Observatory and the Institute of Astronomy in Cambridge, offered an unusual opportunity to bring together astronomers from the fields of Galactic and Solar System astrometry. Gathered in this volume are the articles they presented. Together, these provide an auspicious review of optical astrometry and our understanding of the evolution of the Galaxy and Solar System for graduate students and researchers.

Galactic and Solar System
Optical Astrometry

Galactic and Solar System Optical Astrometry

Proceedings of the Royal Greenwich Observatory
and the Institute of Astronomy Workshop,
held in Cambridge, June 21–24, 1993

Edited by
L. V. MORRISON
Royal Greenwich Observatory

and

G. F. GILMORE
*Institute of Astronomy
University of Cambridge*

Published by the Press Syndicate of the University of Cambridge
The Pitt Building, Trumpington Street, Cambridge CB2 1RP
40 West 20th Street, New York, NY 10011–4211, USA
10 Stamford Road, Oakleigh, Melbourne 3166, Australia

© Cambridge University Press 1994

First published 1994

Printed in Great Britain at the University Press, Cambridge

A catalogue record for this book is available from the British Library

ISBN 0 521 46240 1 hardback

Contents

List of participants xiii

Preface and acknowledgements xvii

Dedication xviii

GALACTIC ASTROMETRY: OBSERVATION & APPLICATION

Proper motion catalogues - I (Chair: *C.A. Murray*)

Estimates of proper motions and of spherical coordinates obtained by the
overlapping-plates method 1
   H. Eichhorn

Proper motions of the PPM 7
   S. Röser

The U.S. Naval Observatory Astrographic Catalog project 11
   S.E. Urban & T.E. Corbin

Assessing the AGK3, AGK3U, ACRS (Part 1), and PPM (North) 14
   D. Daou & L.G. Taff

Proper motion catalogues - II (*H. Eichhorn*)

Lick NPM program: NPM1 Catalog and its applications 20
   A.R. Klemola, R.B. Hanson & B.F. Jones

Southern Proper Motion Program: progress, scope and accuracy 26
   W.F. van Altena, I. Platais, T.M. Girard & C.E. López

Proper motions from fourfold astrographic coverage of the northern hemisphere:
first results 31
   G.A. Ivanov, V.S. Kislyuk, L.K. Pakulyak, T.P. Sergeeva & A.I. Yatsenko

Improving proper motions in the southern hemisphere using CPC2 34
   N. Zacharias

Improving proper motions with the Carlsberg Automatic Meridian Circle 37
   C. Fabricius, L.V. Morrison & L. Helmer

Ground-based parallaxes (A.R. Upgren)

Ground-based parallaxes for nearby stars 44
    H. Jahreiss

The new edition of the Yale Parallax Catalogue: a look at systematic errors 50
    W.F. van Altena, J.T. Lee & E.D. Hoffleit

USNO parallax program: directions, results and applications 55
    C.C. Dahn

An analysis of the accuracy of the parallaxes in the 1952 Yale General Catalogue 61
    K.Aa. Strand

Parallaxes for the non-astrometrist 63
    C. Tinney, N. Reid & J. Mould

HIPPARCOS and HST astrometry (G. Westerhout)

HIPPARCOS proper motions and parallaxes 66
    L. Lindegren, J. Kovalevsky & M.A.C. Perryman

TYCHO astrometry of one million stars 71
    E. Høg, V.V. Makarov & H. Pedersen

A test of preliminary HIPPARCOS parallaxes using photometric parallaxes of
distant stars 74
    R. Wielen, H.-H. Bernstein, C. Dettbarn, R. Hering, R Jährling,
    H. Lenhardt & H.G. Walter

Status of Hubble Space Telescope Fine Guidance Sensor Astrometry.............. 80
    P.D. Hemenway, W.H. Jefferys & G.F. Benedict et al.

Galactic structure, dynamics and evolution - I (R. Wielen)

Stellar kinematics: the Galactic context.. 89
    G. Gilmore & R.F.G. Wyse

Stellar kinematics at the SGP 96
    J. Stock, R.J. Agostinho, J.A. Rose & A.R. Upgren

Galactic kinematics from the Cambridge APM Proper Motion Project............ 105
    D.W. Evans

Structure and kinematics of the Galaxy with Schmidt plates.................... 110
    N.V. Kharchenko, E. Schilbach & R.-D. Scholz

Galactic structure, dynamics and evolution - II (*D. Lynden-Bell*)

Evolution of old populations of our Galaxy 113
   P.E. Nissen

Galactic structure with the GSC plate archive – Colors and proper motions in
the main meridional section of the Galaxy: first results 119
   *M.G. Lattanzi, A. Spagna, B.M. Lasker, G. Massone, B. McLean,
B. Bucciarelli & M. Postman*

Galactic internal motions derived from proper motion surveys 126
   M. Miyamoto

Kinematics of the stellar populations from a proper motion survey 134
   C. Soubiran

Galactic structure, dynamics and evolution - III (*P. Nissen*)

Dynamical analysis of local kinematical data 141
   J.J. Binney

Absolute proper motions and tangential velocities to B=22.5 at the SGP 150
   X. Guo, T.M. Girard, Wm.F. van Altena & C.E. López

Expected phase space distribution of disc stars 156
   J.A. Sellwood

Local kinematics and star clusters - I (*A. Fridman*)

The Solar Neighborhood 165
   A.R. Upgren

Disk density from a survey of F stars at the NGP 174
   J. Knude

Proper motions of Stephenson's spectroscopically selected Red Dwarfs 182
   D.H.P. Jones & M Azzaro

Observations in the region of the Orion Association 186
   R.L. Smart

Local kinematics and star clusters - II (*M. Miyamoto*)

Stellar Clusters, Superclusters and Groups 191
    O.J. Eggen

The Hyades Cluster 204
    N. Reid

Is the Stellar Luminosity Function Universal? 211
    I. Platais

Studies of nearby OB associations 215
    P.T. de Zeeuw, A.G.A. Brown & W. Verschueren

Local kinematics and star clusters - III (*J. Knude*)

Measuring velocity dispersions in Open Clusters 223
    F. van Leeuwen

The mass-luminosity relation from proper motions in Galactic open clusters 230
    I.V. Petrovskaya

Proper motions of Galactic halo globular clusters and nearby Galactic dwarf spheroidal satellites 233
    R.-D. Scholz & M.J. Irwin

The central Galactic star cluster and the 'mini-spiral' morphology 239
    A.M. Fridman, O.V. Khoruzuii, V.V. Lyakhovich & L. Ozernoy

Magellanic Clouds and future space astrometry (*D. Polojentsev*)

On the motion of the Magellanic Clouds 243
    P. Kroupa, S. Röser & U. Bastian

ROEMER satellite project: The first high-accuracy survey of faint stars ... 246
    E. Høg & L. Lindegren

SOLAR SYSTEM ASTROMETRY: OBSERVATION & APPLICATION

Solar System - I (K. Seidelmann)

Meridian circle observations of the planets 253
    E.M. Standish

Astrometric observations of minor planets and comets: present and future needs   263
    B.G. Marsden

Astrometry and space missions to asteroids and comets 276
    D.K. Yeomans

Occultation astrometry: predictions and post-event results 286
    C.B. Olkin & J.L. Elliot

Solar System - II (D. Yeomans)

Methods for development of satellite theories 291
    P.J. Message

CCD Observations at the Bureau des Longitudes: analysis of the positions of satellites 297
    J.-E. Arlot, F. Colas, W. Thuillot & D.T. Vu

An appraisal of the USNO program for photographic astrometry of bright planetary satellites 304
    D. Pascu

CCD Observations of Saturn's satellites 312
    K. Beurle

Solar System III (B. Marsden)

Planetary satellites 318
    P.K. Seidelmann

Satellite astrometry with a long-focus astrograph 325
    A.A. Kisselev

Some results obtained during the 1991 campaign of observation of the mutual events of the Galilean satellites 329
    B. Morando & P. Descamps

Observation and analysis of mutual satellite events 335
    K. Aksnes

List of poster papers in alphabetical order of principal author 341

Participants

Name	Institute	Country
Ables, H.	US Naval Observatory	USA
Agostinho, R.	Lisbon University	Portugal
Aksnes, K.	Institute of Theoretical Astrophysics	Norway
Andrei, A.	Observatorio Nacional CNPq	Brazil
Andrews, P.	Royal Greenwich Observatory	UK
Appleby, G.	Royal Greenwich Observatory	UK
Argue, N.	Institute of Astronomy	UK
Argyle, R.	Royal Greenwich Observatory	UK
Arlot, J.-E.	Bureau des Longitudes	France
Assafin, M.	Observatorio Nacional CNPq	Brazil
Baron, N.	Bureau des Longitudes	France
Bell, S.	Royal Greenwich Observatory	UK
Beurle, K.	Queen Mary Westfield College	UK
Bienayme, O.	Observatoire de Besancon	France
Binney, J.	University of Oxford	UK
Bougeard, M.-L.	Observatoire de Paris	France
Brown, A.	Sterrewacht Leiden	The Netherlands
Bucciarelli, B.	Space Telescope Science Institute	USA
Buontempo, M.	Royal Greenwich Observatory	UK
Cannon, R.	Anglo-Australian Observatory	Australia
Cardini, D.	Instituto di Astrofisica Spaziale	Italy
Casertano, S.	University of Illinois	USA
Cepeda-Pena, W.	Observatorio Astronomico	Colombia
Colas, F.	Bureau des Longitudes	France
Colin, J.	Observatoire de Bordeaux	France
Corbin, T.	US Naval Observatory	USA
Crifo, F.	DASGAL	France
Dahn, C.	US Naval Observatory	USA
Daou, D.	Space Telescope Science Institute	USA
Dauphole-Fouillet, B.	Observatoire de Bordeaux	France
de Zeeuw, T.	Sterrewacht Leiden	The Netherlands
Debarbat, S.	Observatoire de Paris	France
Dejonghe, H.	Ghent University	Belgium
Dewhirst, D.	Institute of Astronomy	UK
Dinescu, D.	Yale University	USA

Doggett, L.	US Naval Observatory	USA
Ducourant, C.	Observatoire de Bordeaux	France
Eichhorn, H.	University of Florida	USA
Emanuele, A.	Instituto di Astrofisica Spaziale	Italy
Evans, D.	Royal Greenwich Observatory	UK
Fabricius, C.	Copenhagen University Observatory	Denmark
Figuras, F.	Universitat de Barcelona	Spain
Fridman, A.	Moscow Institute of Astronomy	Russia
Gauss, S.	US Naval Observatory	USA
Geffert, M.	Sternwarte der Universtat Bonn	Germany
Gemmo, A.	ESO, Garching	Germany
Gilmore, G.	Institute of Astronomy	UK
Girard, T.	Yale University	USA
Guo, X.	Yale University	USA
Harper, D.	Queen Mary Westfield College	UK
Helmer, L.	Copenhagen University Observatory	Denmark
Hemenway, P.	University of Texas	USA
Henden, A.	USRA/USNOFS	USA
Hilton, J.	US Naval Observatory	USA
Hindsley, R.	US Naval Observatory	USA
Høg, E.	Copenhagen University Observatory	Denmark
Hohenkerk, C.	Royal Greenwich Observatory	UK
Holdenried, E.	US Naval Observatory	USA
Ianna, P.	University of Virginia	USA
Jahreiss, H.	Astronomisches Rechen-Institut	Germany
Jones, D.	Royal Greenwich Observatory	UK
Jordi, C.	Universitat de Barcelona	Spain
Kharchenko, N.	Main Astronomical Observatory	Ukraine
Kislyuk, V.	Main Astronomical Observatory	Ukraine
Kisselev, A.	Pulkovo Observatory	Russia
Klemola, A.	Lick Observatory	USA
Knude, J.	Copenhagen University Observatory	Denmark
Kolesnik, Y.	Moscow Institute of Astronomy	Russia
Kroupa, P.	Astronomisches Rechen-Institut	Germany
Lattanzi, M.	Space Telescope Science Institute	USA
Lee, J.	Yale University	USA
Lenhardt, H.	Astronomisches Rechen-Institut	Germany
Lindegren, L.	Lund Observatory	Sweden
Loden, K.	Stockholm Observatory	Sweden
Loden, L.	Uppsala Observatory	Sweden
Lopez Garcia, A.	University of Valencia	Spain

Lutz, T.	Washington State University	USA
Lutz, J.	Washington State University	USA
Lynden-Bell, D.	Institute of Astronomy	UK
Maia, D.	University of Lisbon	Portugal
Marsden, B.	Smithsonian Astrophysical Observatory	USA
Martin, R.	Royal Greenwich Observatory	UK
Mendez, R.	Yale University Observatory	USA
Message, J.	University of Liverpool	UK
Miyamoto, M.	National Astronomical Observatory	Japan
Monet, D.	US Naval Observatory	USA
Morando, B.	Bureau des Longitudes	France
Morrison, J.	University of Florida	USA
Morrison, L.	Royal Greenwich Observatory	UK
Murray, C.A.	Eastbourne	UK
Murray, C.	Queen Mary Westfield College	UK
Nemec, J.	University of Washington	USA
Nissen, P.	University of Aarhus	Denmark
Oblak, E.	Observatoire de Besancon	France
Odenkirchen, M.	Sternwarte der Universitat Bonn	Germany
Oja, T.	Kvistaberg Observatory	Sweden
Ojha, D.	Observatoire de Besancon	France
Olkin, C.	MIT	USA
Ortiz Gil, A.	University of Valencia	Spain
Pannunzio, R.	Observatorio Pino Torinese	Italy
Pascu, D.	UC Naval Observatory	USA
Penston, M.	Royal Greenwich Observatory	UK
Perryman, M.	ESA-ESTEC	The Netherlands
Petrovskaya, I.	St. Petersburg University	Russia
Pilkington, J.	Royal Greenwich Observatory	UK
Platais, V.	Yale University	USA
Platais, I.	Yale University	USA
Polojentsev, D.	Pulkovo Observatory	Russia
Rafferty, T.	US Naval Observatory	USA
Rapaport, M.	Observatoire de Bordeaux	France
Ratnatunga, K.	John Hopkins University	USA
Reid, N.	Caltech	USA
Requieme, Y.	Obervatoire de Bordeaux	France
Richer, H.	University of British Columbia	Canada
Rohde, J.	US Naval Observatory	USA
Röser, S.	Astronomisches Rechen-Institut	Germany

Schilbach, E.	Universitat Potsdam	Germany
Schmidt, J.	Sternwarte der Universitat Bonn	Germany
Scholz, R.-D.	Universitat Potsdam	Germany
Seidelmann, P.K.	US Naval Observatory	USA
Sellwood, J.	Rutgers University	USA
Shaw, L.	ISDA Inc.	USA
Sinclair, A.	Royal Greenwich Observatory	UK
Smart, R.	University of Florida	USA
Soubiran, C.	Observatoire de Paris	France
Spagna, A.	Observatorio Pino Torinese	Italy
Standish, E.M.	Jet Propulsion Laboratory	USA
Steel, D.	Anglo-Australian Observatory	Australia
Stock, J.	CIDA	Venezuela
Stone, R.	US Naval Observatory	USA
Strand, K.	Washington, DC	USA
Taff, L.	Space Telescope Science Institute	USA
Taylor, D.	Royal Greenwich Observatory	UK
Tel'nyuk-Adamchuk, V.	Kiev University Observatory	Ukraine
Thomas, D.	Edinburgh University	UK
Tinney, C.	ESO, Garching	Germany
Torra, J.	Universitat de Barcelona	Spain
Tritton, K.	Royal Greenwich Observatory	UK
Upgren, A.	Van Vleck Observatory	USA
Urban, S.	US Naval Observatory	USA
van Altena, W.	Yale University Observatory	USA
Van Flandern, T.	Meta Research	USA
van Leeuwen, F.	Royal Greenwich Observatory	UK
Verschueren, W.	Sterrewacht Leiden	The Netherlands
Viateau, B.	Observatoire de Floirac	France
Vikki, E.	Mont St. Aignan	France
Vu, D.	Bureau des Longitudes	France
Weis, E.	Van Vleck Observatory	USA
Westerhout, G.	US Naval Observatory	USA
Wielen, R.	Astronomisches Rechen-Institut	Germany
Wilkins, G.	University of Exeter	UK
Yallop, B.	Royal Greenwich Observatory	UK
Yao, Z.-G.	US Naval Observatory	USA
Yeomans, D.	Jet Propulsion Laboratory	USA
Zacharias, N.	US Naval Observatory	USA

Preface and acknowledgements

There have been several important advances in recent years in the techniques of optical astrometry, and a growing realisation that astrometric data is relevant to topical questions concerning the formation and evolution of galaxies in a Universe dominated by Dark Matter. Not least among the technical advances are the ESA astrometric satellite HIPPARCOS and the increasing use of CCDs in ground-based measurements. These techniques have brought about a considerable improvement in accuracy. Although the HIPPARCOS data will not be released until 1996, several papers presented at the Conference show from the analyses of preliminary results that the satellite is on track to deliver the target accuracy of 2 mas. The future application of HIPPARCOS results in helping to solve specific questions of Galactic structure and evolution is addressed in several papers.

Progress in both the scientific analysis and the technical aspects of ground-based astrometry is reviewed, with attention to developments in the application of CCDs, as well as photography and photoelectric methods. Compilations of large photographic catalogues, and in particular the advent of the Astrographic Catalogues in machine-readable form, has enabled accurate proper motions to be determined for several hundred thousand stars down to a visual magnitude of about 11. The application of these and other recent results for proper motions and parallaxes are reported in papers dealing with local Galactic kinematics and the stellar luminosity function. The use of Schmidt plates for probing deep along selected Galactic directions—particlarly the Galactic poles—is addressed in several papers. A few papers stipulate what is needed by way of astrometric data in order to test specific theories of Galactic evolution and present dynamical structure, particularly considering the topical Cold Dark Matter cosmologies.

Astrometry of objects in the Solar System, including the major planets, their satellites, asteroids and comets, is reviewed. The important part played by accurate, long-term series of observations in fitting numerical integrations is emphasised, both from the point of view of future space missions and the prediction of occultations.

The papers are given in the order in which they were read at the Conference, except for the sessions on the Solar System, which have been grouped together here, instead of being interspersed between the sessions on Galactic topics as they were at the Conference. The titles and principal authors of the poster papers are listed at the end of the book.

We thank the members of the Scientific Organizing Committee for their help in arranging the programme, and in particular, for the advice proffered by C. Andrew Murray. Bob Argyle and Michael Buontempo of the RGO gave valuable assistance in the running of the meeting and preparing the contributions for publication. Anne Reynolds did a fine job as Secretary of the Conference. The Conference was sponsored by the Royal Greenwich Observatory and the Institute of Astronomy, Cambridge, and was held at Robinson College, Cambridge, on 21-24 June 1993.

The main objective of the Conference was to bring together astrometrists who measure postions, proper motions and parallaxes with astronomers who use these fundamental data in their research into the composition and evolution of our Galaxy and the Solar System. Sometimes these two groups of people lose sight of the other's objectives. Judging by the comments of the participants and the blend of papers presented here, the Conference went some way towards achieving the main objective.

L.V. Morrison	Gerard Gilmore
Royal Greenwich Observatory	Institute of Astronomy

Dedication

The proceedings of this conference are dedicated to the memory of our colleagues:

| Wilhelm Gliese |

| Robert S. Harrington |

| James A. Hughes |

| Clayton A. Smith |

| Yu. K. Zverev |

Estimates of proper motions and of spherical coordinates obtained by the overlapping-plates method

By HEINRICH EICHHORN

SSRB 211, Dept of Astronomy, University of Florida, Gainesville, Florida 32611, U.S.A.

This is an extension of the author's article [Eichhorn 1985, "paper (I)"], which developed an overlapping-plates algorithm that regards the stars' spherical coordinates as adjustment parameters under the presumption that all frames were generated at the same epoch (e.g., that all plates were exposed at the same epoch). After showing that only reference stars can influence the estimates of the frame parameters in a single-frame reduction, this article expands the algorithm to the (in practice always realized) situation that the frames were not generated contemporaneously, but at different epochs which are, in addition, sufficiently widely separated to allow one the estimation of the stars' proper motion components along with that of their spherical coordinates. A list of errors and misprints in paper (I) is appended.

1. Introduction

Paper (I) showed how the spherical coordinates of stars are estimated by the overlapping-plates method from contemporaneous frames. This is patently an idealized situation which cannot occur exactly in practice, although the overlapping-plates adjustment can still (as many examples have shown) be applied as long as the frame epochs do not vary so much that the stars' proper motions during the epoch intervals render the positions sensibly different at different epochs.

The strictly correct way to deal with this complication is to regard the stars' proper motion components as adjustment parameters (*i.e.*, unknowns) along with their spherical coordinates at a certain epoch; in principle, any other astrometric star parameters (such as parallaxes) could also be included in the set of the adjustment parameters. Eichhorn (1988) developed an algorithm to accomplish this, but only for the situation of a central overlap, *i.e.*, that in which the coordinates of an image of each star were measured on each of the frames.

The present study formulates an estimation algorithm that adds the stars' proper-motion components to the set of adjustment parameters.

2. Single-frame reductions

It is intuitively obvious that the measurement of the field stars' coordinates cannot contribute in any way to the estimation of the frame parameters in a single-frame adjustment, but no analytical proof of this proposition has yet been published. This section will provide the proof.

We follow the notation in paper (I) and note that the validity of Eq. (I,1) presumes the origin of the coordinate system in which x_ν and y_ν are reckoned to be very close to the tangential point of the frame, *i.e.*, the origin of the system with respect to which the ξ and η are reckoned. Section 2 of paper (I) implies [in the formulation of Eqs. (I,2) and (I,3)] that each star involved in the reduction was also a reference star.

In order to show that field stars *i.e.*, non-reference stars, have no influence on the value of the plate-parameter estimates [computed from Eq. (I,16a)], we regard the stars numbered $m+1, m+2, \ldots, k$ as field stars. In what follows, an f subscripted to a symbol will always indicate that the quantities represented by this symbol were generated by the field stars. These generate equations of the type (I,2), namely $\boldsymbol{F}_f = \boldsymbol{0}$, but not equations of the type (I,3). The set of condition equations is then $\boldsymbol{H} = \boldsymbol{0}$, with $\boldsymbol{H}^T = (\boldsymbol{F}^T \; \boldsymbol{F}_f^T \; \boldsymbol{G})$. Thus we still have $\left(\dfrac{\partial \boldsymbol{H}}{\partial \boldsymbol{x}}\right) = \boldsymbol{I}$, while Eq. (I,5) is replaced by†

$$\boldsymbol{\Sigma} = \begin{pmatrix} \boldsymbol{\Sigma}_x & 0 & 0 \\ 0 & \boldsymbol{\Sigma}_{fx} & 0 \\ 0 & 0 & \bar{\boldsymbol{\Sigma}}_a \end{pmatrix}. \tag{2.1}$$

$\boldsymbol{\Sigma}_x$ and $\bar{\boldsymbol{\Sigma}}_a$ are defined as in Eq. (I,6) and we now have in addition

$$\boldsymbol{\Sigma}_{fx} = \operatorname{diag}(v_{m+1}, \varphi_{m+1}, v_{m+2}, \varphi_{m+2}, \ldots, v_k, \varphi_k) \tag{2.2}$$

and

$$\underset{2 \times 2}{\boldsymbol{\Sigma}_{fx\nu}} = \begin{pmatrix} v_\nu & 0 \\ 0 & \varphi_\nu \end{pmatrix}; \qquad \nu \in \{m+1, m+2, \ldots, k\}. \tag{2.3}$$

While it was possible to write $\boldsymbol{H}_0^T = (\boldsymbol{d}^T \; \boldsymbol{0})$ in the case when each star was also a reference star (by regarding the position estimates from the reference catalogue as first approximations), we will now have to split the residual vector as follows:

$$\boldsymbol{H}_0^T = (\boldsymbol{d}^T \; \boldsymbol{d}_f^T \; \boldsymbol{0}), \tag{2.4}$$

where the components of the $(k-m)$-vector $\boldsymbol{d}_f$ are the numerical values of $\boldsymbol{F}_f$ evaluated at $\boldsymbol{a}_0 = \boldsymbol{0}$ (as before) and plausible approximations to the field stars' parameters found somehow. Eq. (I,10) is then accordingly changed to

$$\boldsymbol{A} = -\begin{pmatrix} s\boldsymbol{B} & 0 & \boldsymbol{\Xi} \\ 0 & s\boldsymbol{B}_f & \boldsymbol{\Xi}_f \\ \boldsymbol{I} & 0 & 0 \end{pmatrix}, \tag{2.5}$$

where $\boldsymbol{\Xi}$ is [as in paper (I)] defined by Eq. (I,12) and

$$\boldsymbol{\Xi}_f^T = (\boldsymbol{\Xi}_{m+1}^T \; \boldsymbol{\Xi}_{m+2}^T \; \ldots \; \boldsymbol{\Xi}_k^T). \tag{2.6}$$

It is crucial for our argument that the matrices $\underset{2 \times 2}{\boldsymbol{B}_\nu}$ ($\nu \in \{1, 2, \ldots, k\}$) are non-singular and can therefore be inverted. This has (but only in a single-frame reduction) the consequence that only the reference stars have an effect on the determination of the plate parameters $\boldsymbol{a}$.

A little algebra shows that

$$\boldsymbol{A}^T \boldsymbol{\Sigma}^{-1} = -\begin{pmatrix} s\boldsymbol{B}^T \boldsymbol{\Sigma}_x^{-1} & 0 & \bar{\boldsymbol{\Sigma}}_a^{-1} \\ 0 & s\boldsymbol{B}_f^T \boldsymbol{\Sigma}_{fx}^{-1} & 0 \\ \boldsymbol{\Xi}^T \boldsymbol{\Sigma}_x^{-1} & \boldsymbol{\Xi}_f^T \boldsymbol{\Sigma}_{fx}^{-1} & 0 \end{pmatrix} \tag{2.7}$$

and

$$\boldsymbol{A}^T \boldsymbol{\Sigma}^{-1} \boldsymbol{a} = \begin{pmatrix} s^2 \boldsymbol{B}^T \boldsymbol{\Sigma}_x^{-1} \boldsymbol{B} + \bar{\boldsymbol{\Sigma}}_a^{-1} & 0 & s\boldsymbol{B}^T \boldsymbol{\Sigma}_x^{-1} \boldsymbol{\Xi} \\ 0 & s^2 \boldsymbol{B}_f^T \boldsymbol{\Sigma}_{fx}^{-1} \boldsymbol{B}_f & s\boldsymbol{B}_f^T \boldsymbol{\Sigma}_{fx}^{-1} \boldsymbol{\Xi}_f \\ s\boldsymbol{\Xi}^T \boldsymbol{\Sigma}_x^{-1} \boldsymbol{B} & s\boldsymbol{\Xi}_f^T \boldsymbol{\Sigma}_{fx}^{-1} \boldsymbol{B}_f & \boldsymbol{\Xi}^T \boldsymbol{\Sigma}_x^{-1} \boldsymbol{\Xi} + \boldsymbol{\Xi}_f^T \boldsymbol{\Sigma}_{fx}^{-1} \boldsymbol{\Xi}_f \end{pmatrix}. \tag{2.8}$$

† Note that we are changing notation; it appears preferable to use upper case symbols for all matrices.

Instead of Eq. (I,8) we now have

$$\mathbf{A}^T \mathbf{\Sigma}^{-1} \mathbf{A} \begin{pmatrix} \beta \\ \beta_f \\ a \end{pmatrix} = \mathbf{A}^T \mathbf{\Sigma}^{-1} \begin{pmatrix} d \\ d_f \\ 0 \end{pmatrix}. \qquad (2.9)$$

From Eq. (9) we see, using Eqs. (7), (8) and (I,33), that

$$\beta = \frac{1}{s}(\mathbf{\Sigma}_a^{-1} + \mathbf{B}^T \mathbf{\Sigma}_x^{-1} \mathbf{B})^{-1} \mathbf{B} \mathbf{\Sigma}_x^{-1}(d - \mathbf{\Xi} a) \qquad (2.10)$$

and

$$\beta_f = \frac{1}{s} \mathbf{B}_f^{-1}(d_f - \mathbf{\Xi}_f a), \qquad (2.11)$$

because $\mathbf{B}_f$ is non-singular.

We find a from an equation obtained by using Eqs. (10) and (11) in the third equation of the system Eqs. (9). We notice that all symbols generated by the field stars cancel; it is in this context essential that we remember the nonsingularity of $\mathbf{B}_f$. After gathering all remaining terms in a on the left and those in d on the right-hand side we are left with

$$\begin{aligned} \mathbf{\Xi}^T [\mathbf{\Sigma}_x^{-1} - \mathbf{\Sigma}_x^{-1} \mathbf{B}(\mathbf{\Sigma}_a^{-1} + \mathbf{B}^T \mathbf{\Sigma}_x^{-1} \mathbf{B})^{-1} \mathbf{B} \mathbf{\Sigma}_x^{-1}] \mathbf{\Xi} a = \\ \mathbf{\Xi}^T [\mathbf{\Sigma}_x^{-1} - \mathbf{\Sigma}_x^{-1} \mathbf{B}(\mathbf{\Sigma}_a^{-1} + \mathbf{B}^T \mathbf{\Sigma}_x^{-1} \mathbf{B})^{-1} \mathbf{B} \mathbf{\Sigma}_x^{-1}] d, \end{aligned} \qquad (2.12)$$

which is Eq. (I,16). We see thus that the equations generated by the field stars make no contribution to a. This is, of course, heuristically and intuitively obvious.

3. The determinacy of overlapping-plates reductions

We now generalize the approach to non-contemporaneous frames. The star parameters are then no longer just the stars' spherical coordinates, but their coordinates (at a judiciously chosen, but in principle arbitrary common central epoch t_0†) and the right ascension and declination components μ and μ', respectively, of their proper motions. We now assume that there are catalogues which contain estimates of positions and possibly also of proper motions for a number of stars. In what follows, these estimates will be called reference positions and reference proper motions, respectively. For the geometry of the problem to be determinate, i.e., to yield an unambiguous result, it is only necessary that either a sufficient number of reference proper motions exists even though all reference positions refer to the same epoch, or that there exists a sufficient number of reference positions referring to different epochs.‡

To appreciate this, consider first the case that all reference positions are contemporaneous (i.e., refer to the same epoch) but that there is a number of reference proper motions (none of which need belong to a star for which a reference position is also available). The proper motions then allow one to derive proper-motion estimates (on the system of the reference proper motions) for all stars in the overlapping region of two non-contemporaneous frames which, in turn, make it possible to find positions for the reference stars at the epochs of the frames and thus to derive position estimates (on the system of the reference positions) from the frames.

In the second case (i.e., reference positions *at one—not necessarily always the same— epoch only* exist for a sufficiently numerous set of stars, but there are no reference proper

---

† In what follows, epochs written as τ_ν indicate epochs t_ν reckoned from the central epoch, i.e., $\tau_\nu = t_\nu - t_0$.
‡ The problem is indeterminate in the "pathological" cases when all reference positions are contemporaneous and there are no reference proper motions, and also when there are only reference proper motions and only one reference position.

motions), estimates of *relative* proper motions and positions can be found for all stars in the region common to two overlapping frames. These relative positions and proper motions can be made absolute by fitting them into the system of the reference positions.

These are rather general statements and illustrate the principle only; it is doubtful, though, that more specific statements would contribute substantially to elucidating the situation.

4. The algorithm for the estimation of the extended set of star parameters

It is now no longer appropriate to divide the stars just into "reference" stars and "field" stars, because reference positions at only one (not necessarily at the same) epoch for each star may be available for one set S_1 of stars, reference proper motions only for another set S_2 of stars and a third set S_3 may contain those stars for which reference positions are available at no fewer than two epochs (which may vary from one star to the next). The sets $S_1 \cap S_2$ and $S_3 \cap S_2$ are not necessarily empty.

In any case, all measured rectangular coordinates of a star on any one frame will generate a set of condition equations of the type (I,2), where a will, of course, vary from one frame to the next.

We must remember, however, that the auxiliary frame-specific standard cordinates ξ and η are now functions of (besides of the stars' spherical coordinates and those of the frame's tangential point, which we regard as given) all four of the star's parameters, *i.e.*, α_0, δ_0, μ and μ'. This is important when we set up the contributions of the equations $\boldsymbol{F} = \mathbf{o}$ to the matrix $\mathbf{A}$.

We have to modify the equations (I,3) which represent the condition equations provided by the reference material. Depending on the kind of reference data available for a star, the following types of condition equations are possible.

Reference positions for the ν-th star† generate equations of the type

$$(\alpha_{c\nu_\lambda} - \alpha_{0\nu} - \mu_\nu \tau_{\nu\lambda})\cos\delta_\nu = 0 \quad \text{and} \quad \delta_{c\nu_\lambda} - \delta_{0\nu} - \mu'_\nu \tau_{\nu\lambda} = 0; \quad \lambda \in \{1,\ldots,\lambda_\nu\} \quad (4.13)$$

where $\alpha_{0\nu}$, $\delta_{0\nu}$, μ_ν and μ'_ν are the star parameters of the ν-th star, as well as

$$(\mu_{c\nu_\kappa} - \mu_\nu)\cos\delta_\nu = 0 \quad \text{and} \quad \mu'_{c\nu_\kappa} - \mu'_\nu = 0; \quad \kappa \in \{1,\ldots,\kappa_\nu\}. \quad (4.14)$$

The factor $\cos\delta$ in the first of Eqs. (13) and (14) performs the reduction to the great circle; *cf.* paper (I).

Depending on whether (or not) reference data are available for the ν-th star, λ_ν and κ_ν will (or will not) be different from zero. λ_ν and κ_ν may be different from 1 in order to accommodate the cases when more than one reference position (and proper-motion pair) is available for the ν-th star.

A (slight) complication arises when reference positions and reference proper motions are communicated together (as, *e.g.*, in a fundamental catalogue). In these cases, the proper-motion estimates and position estimates of any one star are—as a rule—correlated at the epoch of the position estimates. This would render the covariance matrix of these data—the corresponding block in the matrix $\boldsymbol{\Sigma}_a$—nondiagonal. This correlation can be removed by using the directly given data to calculate the position estimates for the *central date* (*cf.* Eichhorn 1974, pp. 112–113), at which position estimates and proper-motion estimates will no longer be correlated with each other. Without restricting generality we may thus assume that $\boldsymbol{\Sigma}_a$ is diagonal, but remember the caution expressed in paper (I).

---

† It is relatively safe to assume that spherical coordinates are almost always available in complete pairs.

Note that we always consider the variances of $\alpha \cos \delta$ and $\mu \cos \delta$ rather than those of α and μ.

We have shown in Section 2 above, that only reference stars have an effect on the resulting values of the frame-parameter estimates in a single-frame reduction. The analogue to this statement in our multi-epoch multi-frame reduction is as follows: Non-reference stars imaged on one frame only should not be included in the calculations for the adjustment; their position estimates (without proper-motion estimates) are found from Eqs. (I,1) once the parameters of the frame on which they are imaged have been estimated. Non-reference stars imaged on two frames only have no effect on the values of the frame parameters estimates; their proper-motion estimates can also be found from Eqs. (I,1) after estimates for the frame parameters become available, and their proper-motion estimates can be computed from those, but will be reasonably precise only if there is a sufficiently large interval between the epochs of the two frames.

Measurements of star images' relative rectangular coordinates will have an effect on the estimation of any frame parameters only and exactly when there are, besides the measured coordinates of an image on at least one frame, at least two more data sets which can be either reference data (positions or proper motions) or measured coordinates on another frame.

We now regard all equations (13) and (14) as the set $\boldsymbol{G} = \boldsymbol{0}$.

Since the geometry of the problem is such, that a coordinate will always be estimated together with its corresponding time derivative (i.e., proper-motion component), we arrange the star parameters star by star, and within a star as $\alpha \cos \delta$, δ, $\mu \cos \delta$ and μ'. It is realistic to assume that both spherical coordinates of each star will always be involved in the reduction, so that now

$$\boldsymbol{\beta}^{\mathrm{T}} = (\alpha_1 \cos \delta_1 \; \delta_1 \; \mu_1 \cos \delta_1 \; \mu_1' \; \alpha_2 \cos \delta_2 \; \delta_2 \; \ldots \; \mu_m'), \tag{4.15}$$

where m [as in paper (I)] is the total number of stars involved in the adjustment.

We now reconsider Eq. (I,10') in light of the new vector $\boldsymbol{\beta}^{\mathrm{T}}$ of star parameters given by Eq. (15). First we note, that—retaining the meaning of the symbols in Eq. (I,10) and remembering Eq. (13)—we get analogously to Eq. (I,10)

$$\left(\frac{\partial \boldsymbol{F}}{\partial \boldsymbol{\beta}}\right) = -s\mathbf{B}(\mathbf{I}_2 \; \mathbf{T}) = -s \underset{2 \times 4}{\mathbf{C}} \tag{4.16}$$

with $\mathbf{T} = \tau \mathbf{I}_2$, where τ is the epoch of the frame. With these changes the discussion in Section 6 of paper (I) remains valid as long as we remember that the dimensions of the matrices $\left(\dfrac{\partial \boldsymbol{F}_\nu}{\partial \boldsymbol{\beta}_\nu}\right) = \mathbf{C}_\nu$ (which are defined analogously to the $\mathbf{B}_\nu$ and replace them) are $2m_\nu \times 4m$.

Mutatis mutandis, all formulas and statements in paper (I) retain their validity.

Errors and misprints in paper (I)

The following errors and misprints in paper (I) have been found.

p. 252 left, 3rd line from bottom: $\ldots \cos \delta da + \ldots$ should be $\ldots \cos \delta d\alpha + \ldots$
p. 252 right, Eqs. (13) and (14): Replace σ_a^{-1} with $\bar{\sigma}_a^{-1}$
p. 252 right, line after Eq. (15): Replace α_a^{-1} with $\bar{\sigma}_a^{-1}$
p. 254 left, Eq. (5') and right, Eq. (14'): Replace σ_a with $\bar{\sigma}_a$
p. 254 right, line 4 from top: Replace $2m_\nu \times 2m_\nu$ with $2m_\nu \times 2m$
p. 254 right: Eq. (40) should read:

$$\mathbf{L} = s^2(\mathbf{K}^{\mathrm{T}} \boldsymbol{\sigma}_a^{-1} \mathbf{K} + \mathbf{B}^{\mathrm{T}} \boldsymbol{\sigma}_x^{-1} \mathbf{B}).$$

p. 254 right, line 16 from bottom: Replace σ_ν with σ_x
p. 255 right: Eq. (15'), should read:

$$\beta_\mu = \frac{1}{s}\left(\delta_{\mu,\mu_n}\sigma_{a\mu_n}{}^{-1} + \sum_{\nu=1}^{n}\cdots\right)^{-1} \cdots (\&\text{c. as printed})$$

p. 255 right, line 4 from top: Replace σ^{-1} with $\sigma_a{}^{-1}$

REFERENCES

Eichhorn, H. 1974 Astronomy of Star Positions, Ungar, New York
Eichhorn, H. 1985 A&A, **150**, 251
Eichhorn, H. 1988 ApJ, **334**, 465.

The proper motions of the PPM

By SIEGFRIED RÖSER

Astronomisches Rechen-Institut, Mönchhofstraβe, 12-14, D-6900, Heidelberg 1, Germany

At present, the PPM catalogue with its 378 910 stars is the largest source of accurate proper motions. The accuracy of the PPM proper motions is tested by external comparisons. Some applications in galactic kinematics are discussed.

---

1. Accuracy of the PPM proper motions.

In 1992, the final southern part of the PPM Star Catalogue was completed. On the whole sky PPM contains the positions and proper motions of 378 910 stars. An overview on the principal properties of PPM is given by Röser and Bastian (1993a). In the context of galactic astronomy the proper motions of PPM are of particular interest. Figure 1 shows the distribution of the *rms* errors of the proper motions in right ascension and declination, respectively. Different subgroups of PPM stars can be distinguished in this figure. The small number of stars with *rms* errors around 1 mas/yr represents the FK5 stars. Stars forming the peak at 3 mas/yr are the stars from PPM South and from the High Precision Subset of PPM North. The remaining part of the distribution peaking somewhere around 4 mas/yr and the tail extending to 8 mas/yr belongs to the bulk of the stars on the northern hemisphere. For these stars AGK3 is the latest epoch. With an accurate modern epoch such as FOKAT-S in the south, all PPM stars would have proper motions of 3 mas/yr or better.

The estimates of the *rms* errors of the proper motions in PPM are calcuted from the weights attributed to the individual observation catalogues entering PPM. For more details see Röser and Bastian (1991). It is quite important to check the estimates in PPM by comparing the PPM proper motions with independent sources of proper motions with better intrinsic accuracy. The only whole-sky-catalogue suited for this purpose will be the Hipparcos Output Catalogue.

As an indirect test of the accuracy of the PPM proper motions, one can use comparisons of the present-day PPM positions with highly accurate star positions. As the average mean epoch of the northern hemisphere part of PPM is about 1930 (for HPS 1950), the accuracy of the present-day PPM positions is dominated by the proper motions. The average epoch of FOKAT-S is around 1985, so on the southern hemisphere one would essentially test the quality of FOKAT-S instead of the estimates of the *rms* errors of the PPM proper motions. As sources of accurate star positions modern meridian circle catalogues such as the CMC5-7 catalogues could be used. Alternatively, Turon et al. (1992) derived preliminary Hipparcos star positions from an analysis of 12 weeks of data spread over 1.2 years, and compared them with PPM. The *rms* difference between Hipparcos and PPM for their sample of 21 000 stars came out to be 270 mas. The expectation value for this sample of PPM stars (where HPS stars are over-represented) is about 240 mas, an over-estimate of some 10%.

In applications of global galactic kinematics for example, the systematic accuracy of the PPM proper motions is of great importance. The PPM proper motions should represent the system of the FK5 proper motions. Again, without the Hipparcos Output Catalogue an assessment of the systematic accuracy of the FK5 or PPM proper motions is not possible. However, the system of the present-day PPM positions can be tested

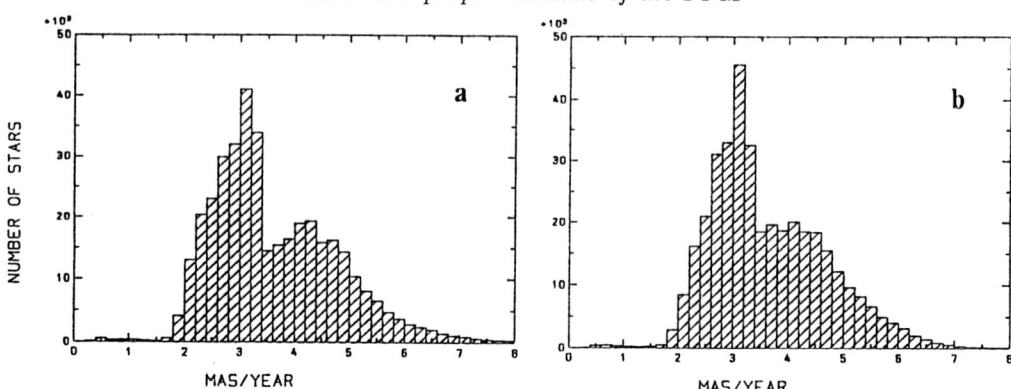

FIGURE 1. The distribution of the *rms* errors of PPM proper motions: (a) in right ascension ($\times \cos \delta$), (b) in declination.

using preliminary Hipparcos results. Lindegren (1992) carried out such comparisons for FK5 and PPM. After subtraction of a global rotation the systematic differences between Hipparcos and FK5 are generally less than 100 mas, only in some small regions on the sky 200 mas may be reached. In general, the systematic differences on the southern hemisphere are larger than on the northern. PPM shows a pattern quite similar to that of FK5, but because of the much larger number of PPM stars, the comparison extends to smaller scales. As a general result of this comparison, one can conclude that PPM is on the system of FK5 for present-day positions. As the mean epoch of the PPM in the north is relatively early, there are good indications that the PPM proper motions also represent the FK5 system. As said above, these tests are not equally relevant for southern proper motions, but the fact, that in constructing PPM on the southern hemisphere, the same methods as in the north were adopted, lead me to expect that the conclusions for the northern hemisphere also hold for the southern one.

2. Applications of PPM proper motions.

Being an all-sky catalogue PPM is not particularly tailored for investigations of galactic open clusters. However, studying open clusters with PPM is worthwhile in several respects. For clusters which are already carefully studied using small-scale relative proper motions, comparisons with PPM can be performed to estimate the accuracy of PPM proper motions locally. For clusters, studied poorly so far, membership of PPM stars can be determined via their proper motions. With the exception of only a few nearby clusters, such as the Hyades or the Pleiades, projection effects or the intrinsic velocity dispersion are so small that they are not reflected in the PPM proper motions. So, in practically all galactic open clusters, the proper motions of the member stars are internally the same, and the dispersion in proper motions reflects the mean errors of the proper motions in the star catalogues. In the case of the Pleiades one can demonstrate the improvement of PPM with respect to AGK3, and also the superiority of a special small-scale catalogue over PPM. Figure 2a-c shows plots of proper motions in an 16-square-degree area around the Pleiades in AGK3, PPM and the catalogue by Vasilevskis et al. (1979) covering the central 1.5-square-degree field. The average *rms* error of the proper motions in AGK3 is 9 mas/yr, in PPM 4 mas/yr, and in Vasilevskis et al. 1 mas/yr or better.

So far, only a few open clusters have been studied in order to examine the estimates of the accuracy of PPM proper motions. These are shown in Table 1. For each respective

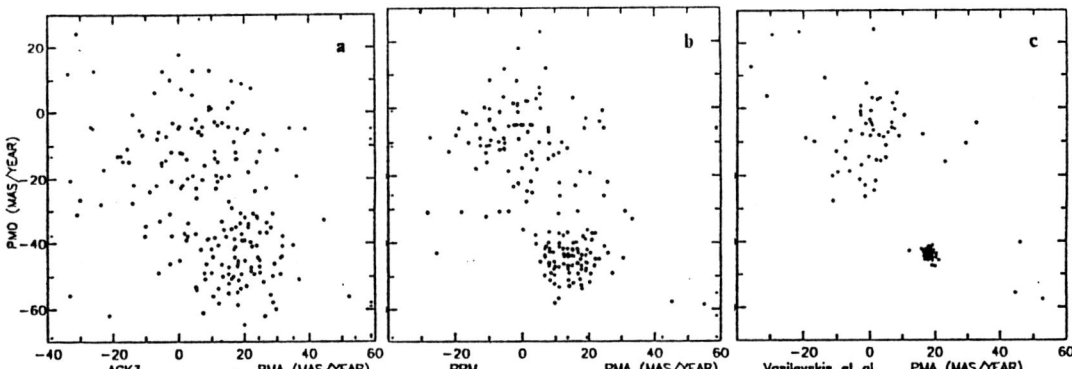

FIGURE 2. Vector diagram of the proper motions in the region of the Pleiades: (a) Stars in a 16-square degree field in AGK3, (b) the same field in PPM, and (c) the central 1.5-square degree field from Vasilevskis et al. (1979).

Name	N	expected		measured	
rms error of proper motions		α	δ	α	δ
		mas/yr		mas/yr	
Pleiades	52	3.5	3.6	3.5	4.1
Blanco 1	6	3.4	3.2	2.5	3.7
NGC 2547	17	3.2	3.2	3.3	3.3
Car OB2	16	3.0	3.1	2.7	3.7

TABLE 1. Comparison of the expected rms error of the PPM proper motion components with the measured one, as derived from the dispersion of proper motions of members in galactic open clusters.

cluster, Table 1 gives the number of stars compared and the expected and measured rms errors of the PPM proper motions in right ascension and declination. As can be seen from this table the estimates in PPM are rather reliable. In the case of the Pleiades – by comparing with Vasilevskis et al. – a systematic trend in the PPM proper motions in right ascension of −3 mas/yr/deg has been detected in addition to the rms scatter of 3.5 mas/yr in Table 1. This trend can already be seen in the PPM proper motions themselves. It extends over an angle of 1.5 degrees in right ascension, and vanishes outside. In declination, no such effect can be seen. As an explanation a small-scale systematic error in PPM is assumed, the alternative of a contraction of the Pleiades cluster in right ascension with a contraction time of 1 million years is unrealistic.

Several applications of PPM proper motions have already been published in literature, four of which are reported below. Schwan (1991) extended his study of the Hyades using FK5 stars to the larger sample of Hyades members in the PPM catalogue. In addition to the 44 FK5 stars, Schwan found 101 additional PPM stars as members of the Hyades. In the colour-magnitude diagram the PPM stars populate the fainter end of the main-sequence, and the transition from FK5 to PPM is completely smooth. This proves that the individual distances of the Hyades stars derived from PPM proper motions are consistent with those derived from FK5. So, there are no significant systematic differences between PPM and FK5 proper motions in the region of the Hyades.

Comeron et al. (1993) investigated the Cyg OB1 and OB9 regions at the edge of the Cynus superbubble. Using PPM proper motions, they found that these two regions move outwards from the centre of the superbubble, Cyg OB2.

The controversial case of the existence of NGC 2451 as an open cluster has been discussed in literature during the past decade. The reason for this controversy was the fact that only photometric data were used to prove the existence of this cluster. Röser and Bastian (1993b) showed using PPM proper motions that whereas the bright stars which optically form NGC 2451 on the sky do not form a real galactic open cluster, there is an open cluster in the direction of NGC 2451 whose existence is proven by combining photometry with astrometry.

Forty-three of the brightest stars in the Magellanic Clouds are contained in PPM. Kroupa et al. (1993) used the proper motions of these stars from PPM to determine the motion of the LMC, and, with less accuracy, of the SMC. They found the motion of the Magellanic Clouds not in contradiction with various models, which have been developed recently.

3. More proper motions

The high quality of the PPM proper motions, especially on the southern hemisphere, comes from the fact, that excellent modern photographic catalogues such as FOKAT-S and CPC2 are now available, and that the Astrographic Catalogue has been put into machine-readable form (see Nesterov et al., 1990). During the compilation of the southern part of PPM it turned out that some 100 000 stars from the CPC2 were not contained in the star list for PPM, the FOKAT-S. These stars were identified in the Astrographic Catalogue with the result that 95 per cent of them had at least two observations in the AC. For almost 90 000 stars proper motions have been successfully determined. They will form a Southern Supplement to PPM. The average rms error of a proper motion component is 4.5 mas/yr. The publication of the Southern Supplement is planned for the end of 1993.

REFERENCES

Comeron, F., Torra, J., Jordi, C. and Gomez, A.E. (1993). Anomalous proper motion in the Cygnus Superbubble region. A&AS (in press).

Kroupa, P., Röser, S. and Bastian, U. (1993). On the Motion of the Magellanic Clouds. MNRAS (in press).

Lindegren, L. (1992). Status and early results of the Hipparcos astrometry project. ESA-SP 349.

Nesterov, V.V., Kislyuk, V.S., Potter, Kh. I.(1990). In: IAU Symp. 141. Eds. J.H. Lieske and V.K. Abalakin. Kluwer Academic Press.

Röser, S. and Bastian, U. (1991). PPM Star Catalogue. Astronomisches Rechen-Institut, Heidelberg. Spektrum Akademischer Verlag.

Röser, S. and Bastian, U. (1993a). The final PPM star catalogue for both hemispheres. Bull. Inform. CDS, 42, 11.

Röser, S. and Bastian, U. (1993b). NGC 2451 - What is it ? submitted to Astronomy and Astrophysics.

Schwan, H. (1991). AA, 243, 386.

Turon, C., Arenou, F., Evans, D.W. and van Leeuwen, F. (1992). AA, 258, 125.

Vasilevskis, S., van Leeuwen, F., Nicholson, W. and Murray, C.A. (1979). AAS, 37, 333.

The U.S. Naval Observatory Astrographic Catalog project

By SEAN E. URBAN AND THOMAS E. CORBIN

U.S. Naval Observatory, 34th Massachusetts Ave. NW, Washington DC 20392, U.S.A.

For the last several years, the U.S. Naval Observatory has been committed to a new reduction of the Astrographic Catalogue (AC) plates. The result of this project will be an accurate catalog of positions of stars down to magnitude 12.5 at epochs mostly from 1890 to 1920. The first phase of this work has been the data entry of the published x-y values. The next phase is a conventional plate adjustment using the Astrographic Catalog Reference Stars (ACRS), with positions accurate to $0\rlap{.}''24$ at 1900, to derive plate parameters. This process has been applied to the Cape zone of the AC (declinations −41 to −51) which is being used to establish the procedures to be applied to the other zones. Project details and preliminary results from the Cape zone will be discussed.

---

1. Introduction

The Astrographic Catalogue (AC), with about seven million stars and a limiting magnitude of around 12.5, is widely recognized as an ideal source of highly accurate star positions. Its early epoch (most plates taken prior to 1920, with many around 1900) makes the AC an excellent source of first-epoch positions for determining proper motions. However the data remain largely unused. The main reason is that participating observatories published coordinates in x-y measures as measured on a plate, not right ascensions and declinations. Another problem with the AC is the presence of systematic errors in the plate constants used to compute equatorial coordinates. The plate constants published in the AC were, for the most part, derived from the old AG catalogs, and the systematic errors in the AG catalogs have propagated into the plate constants of the AC. Still another reason the AC has been used only sparingly is that until recently few of the data have been in machine-readable form. The U.S. Naval Observatory is currently transferring all x-y and magnitude data to magnetic tape and new reductions of the AC plates, using the Astrographic Catalog Reference Stars (ACRS), are underway.

2. Data transfer status

All x-y and magnitude data must be made machine-readable before a new reduction of the AC can occur. The U.S. Naval Observatory has been transferring the AC's published coordinates to magnetic tape for the past several years. Currently, all x-y measures and diameter values are now on tape for all zones with the exception of Greenwich, Catania, and Helsingfors. These zones are expected to be completed by the end of 1994. Other institutions have also aided in the transfer process. The French zones along with Oxford I, from declination −2 deg to +31 deg declination, were originally keypunched at Strasbourg. Help with the Cordoba and Melbourne zones was provided to the USNO by the Sternberg Institute and University of Florida respectively. The U.S. Naval Observatory thanks Jurgen Stock of CIDA for his considerable help in transferring the Potsdam and Vatican zones and his continuing work with the Greenwich data. In addition to the x-y measures and diameters, other information such as meteorological data, time and duration of exposures, measuring machine used and plate emulsion type has also been keyed when available. Quality control has been stressed; a rate of no more than one error per 10 000 records due to the transfer process is achieved for most zones.

3. ACRS

The Astrographic Catalog Reference Stars (ACRS) catalog was compiled at the U.S. Naval Observatory for the purpose of providing accurate star positions on the FK5 system that could be used to reduce the AC plates. In order to accomplish this it was felt that combining a large number of catalogs, all reduced to the FK4 with the International Reference Stars (IRS), would give the best results for individual stars and for the catalog system. A total of 167 catalogs were combined to give positions and proper motions of 320 211 stars which were then transformed to FK5. The errors of the positions average $0\rlap{.}''08$ in each coordinate at the mean catalog epoch of 1949.5 The standard deviation of the proper motions average $0\rlap{.}''47$ per century in right ascension and $0\rlap{.}''46$ per century in declination. Thus the individual ACRS positions are accurate to about $0\rlap{.}''24$ at 1900. Since the AC plates in most zones contain 20 to 30 ACRS stars, the plate solutions will be very well determined.

The processes of compiling the ACRS and the IRS are given elsewhere (Corbin & Urban, 1990; Corbin, Urban & Warren, 1991; Corbin & Warren, 1991). Both catalogs are available at data centers and are included on the Goddard CD-ROM Selected Astronomical Catalogs Volume 1. The data and reduction procedures of both catalogs have been prepared in such a way as to allow revision as new observational catalogs become available. First, however, all catalogs used to compile the IRS will have to be reduced directly to the FK5 system to give the best possible result.

4. New reduction of the Cape Zone

The first zone being reduced by the U.S. Naval Observatory is the Cape zone with declinations of plate centers located at –41 deg to –51 deg. The software developed for the reduction of the Cape zone will be used to reduce the other AC zones. The Cape zone has over 900 000 individual x-y measures for about 500 000 stars on 1512 plates. All plates were exposed between 1897 and 1911, with 97% exposed prior to 1906. The Cape zone was measured using a short eyepiece micrometer screw. Theoretically, this type of screw when used on a high quality exposure can produce an accuracy of about $0\rlap{.}''2$ per coordinate (Eichhorn, 1974).

Preliminary positions for all images were generated using the algorithms from the printed Cape volumes. These positions were used to uniquely identify stars whose images appear on different plates including those stars in common with the ACRS. Unlike most zones, the printed coordinates of Cape have preliminary plate constant terms, as well as differential refraction terms, already applied. These must be removed and a small linear scale factor applied to convert printed values to original measures in millimeters (Gill and Hough, 1913).

A version of the Hamburg Block Adjustment Program Package is being used for the weighted least-squares plate adjustment as well as much of the data analysis (Zacharias et al., 1992). Preliminary results show a small, but significant, third-order distortion term present in the Cape data. At its largest, this term can cause a $0\rlap{.}''15$ offset at the plate edges. This term is epoch dependent and changes abruptly four times during the program. No instrumental changes have been found that correspond to the changes in the distortion term. The distortion term has been applied to the x-y measures prior to further adjustment.

It was hoped that a minimal, linear model could represent the mapping. A six-constant linear plate model, shown below, was used.

$$X = Ax + By + C + Ex + Fy$$
$$Y = Ay - Bx + D - Ey + Fx \qquad (4.1)$$

In this formula, X,Y are the standard coordinates, x,y are the measured plate coordinates. A, B, C, and D are the orthogonal plate parameters; E and F are the non-orthogonal parameters. After an initial adjustment, removal of misidentified reference stars, as well as reference stars with poor positions, was performed. Plate adjustments were made for all plates, and the values for each parameter were plotted against epoch. The results show that the linear model parameters are well-behaved with no abrupt changes corresponding to epoch. Analysis was then performed on residuals of the field stars. The results show that the six-constant model may not be sufficient for the Cape zone. Currently, investigations are being performed on the tilt terms (p,q), as well as a magnitude dependent, radial scale term (coma).

It is hoped that the new reductions of the Cape zone will be completed by the end of 1993. Since much of the software written for the Cape zone can be applied to other AC zones, it is envisioned that, following the completion of Cape, reduction of several zones per year will be performed.

5. Conclusion

The USNO AC project is an important part of the current efforts in astrometry to extend the reference system of stellar positions and motions to fainter magnitudes and higher densities. The AC is unique in that there is no other available data source prior to 1950 that can be combined with modern astrometric catalogs to give well determined proper motions for stars as faint as the 12th magnitude. By using the best possible reference star catalog and the most modern reduction techniques the U.S. Naval Observatory intends to extract from the AC as much as this invaluable resource is capable of giving to modern astronomy.

REFERENCES

Corbin, T.E. and Urban, S.E., 1990 in IAU Symposium 141, Inertial Coordinate System on the Sky, ed. J. Lieske and V.K. Abalakin, Kluwer, Dordrecht p. 433

Corbin, T.E. and Warren, W., 1991, International Reference Stars, NASA, NSSDC 91-11

Corbin, T.E., Urban, S.E. and Warren, W., 1991, Astrographic Catalog Reference Stars, NASA, NSSDC 91-10

Eichhorn, H., 1974, in Astronomy of Star Positions, p. 284

Gill, D. and Hough, S.S., 1913, Cape Astrographic Zones v. 1, p. xxi.

Zacharias, N., de Vegt, C., Nicholson, W. and Penston, M.J., 1992, A&A, 254, 397

Assessing the AGK3, AGK3U, ACRS (Part 1), and PPM (North)

By D. DAOU AND L. G. TAFF

Space Telescope Science Institute, 3700 San Martin Drive, Baltimore, MD 21218, U.S.A.

We have extended our work on the systematics in fundamental catalogs to the large-scale catalogs mentioned in the title. Essential is a procedure for minimizing catalog-to-catalog differences and a reference catalog. For the former the method of infinitely overlapping circles is utilized while for the latter we have heretofore used the FK5. As the FK5 will have a small, bright overlap with these catalogs, we have turned to the GC with which we anticipated an overlap of $\sim$ 10 000 stars. Finally, because heretofore unknown problems of a systematic nature within GC might lead us to false conclusions, we have also used the AGK3RN as a test case. It shows that the northern hemisphere half of the GC is an acceptable standard. Only the AGK3U passes these tests.

1. Introduction

Two previous papers—Bucciarelli, Lattanzi, and Taff (1993) and Bucciarelli and Taff (1993)—described the results of applying a new star catalog evaluation procedure to the FK3 (Kopff 1937, 1938), the FK4 (Fricke and Kopff 1963), and the FK5 (Fricke, Schwan, and Lederle 1988) and to the GC (Boss et al. 1937), the N30 (Morgan 1952), and the FK5. Although invented to provide insight into the true utility of coordinate and proper motion error estimates, the algorithm also allows one to investigate residual systematics in a catalog. Much to our surprise, the results of this work showed that the FK sequence is inhomogeneous and ridden with rapidly varying systematics while both the GC and N30, although based on essentially the same material, were much smoother catalogs at their mean epochs of place. New work on re-compiling the Southern Reference Star catalog (Taff 1993) similarly shows that the FK problems continue into the FK5. Given the power of the new technique, we have looked for other areas of catalog astronomy to which we can apply it. The large-scale catalogs mentioned in the title have a very high overlap in their star lists in the Northern celestial hemisphere, and in their sources of material, but differ substantially in their compilation methods. Hence, an investigation of these four star catalogs seemed like the logical next step.

There are two aspects of this extension that make it problematical. First is the question of a reference catalog. Continuing to use the Basic FK5, even supplemented with the FK5 Extension (Fricke, Schwan, and Corbin 1991), would only yield a bright, and not very dense, intersection set with these four catalogs. Hence, we adopted instead the General Catalogue as a reference catalog. Utilizing the GC brings in two additional difficulties; namely the inhomogeneity between the very bright part of the GC (i.e., the *fundamental stars*) and those GC stars a couple of magnitudes fainter and the known-to-be-poor GC proper motions. We can build support for our reliance on the GC by probing the GC magnitude distribution in steps by, say, first using the Basic FK5 to test the GC, then the FK5 Extension, and then the AGK3RN (Corbin 1978) before using the large-scale catalogs. This we have done (see Table 1).

To alleviate potential problems with the GC proper motions all the comparisons have been performed at B1900.0 which is approximately the GC mean epoch of place. (The mean epoch of place is 1902.3 $\pm$ 5.1 for the right ascensions and 1899.5 $\pm$ 5.8 for the declinations). Note that by backdating our four catalogs to B1900.0 we are necessarily evaluating a combination of the positions and proper motions for the large-scale catalogs. In order to only investigate their positional systems requires updating the GC positions to the mean epochs of place of the large-scale catalog coordinates.

The second questionable aspect of this work has to do with an intrinsic difference between the fundamental compilation catalogs and the large-scale AGK-based catalogs. The essence of the difficulty is that our statistical hypotheses are framed in terms of the Central Limit theorem. While one can confidently invoke the Central Limit theorem for the coordinates or proper motion components for a fundamental catalog, doing so for a catalog which only contains 3 or 4 or even 10 observations per star strains the notion of asymptotic convergence on which the Central Limit theorem rests. Hence, this work should be regarded as experimental on two levels: we have to show that the GC is a reliable standard and we have to remember that our statistical force now comes from the number of stars in the comparison set and not on the number of observations per star.

2. Overview

We assume that a weighted least squares adjustment was performed, for each coordinate of each catalog star, to a linear polynomial in the time. For example

$$\alpha(t) = \alpha(t_0) + \mu_\alpha(t - t_0) \quad \text{and} \quad \delta(t) = \delta(t'_0) + \mu_\delta(t - t'_0).$$

Because the constant terms in such polynomials are linear in the coordinates and because the multipliers of the first-order temporal terms are the sought-for proper motion components, good estimates for these parameters should have been obtained (i.e., the hypotheses of the Gauss-Markov theorem are satisfied; see Kendall and Stuart 1973) as long as the systematic differences among the source catalogs were successfully minimized. For a fundamental catalog such as the GC, whatever the underlying probability distribution of the individual observational errors, because of the many measures of the stars' coordinates, because of the large number of separate source catalogs, and because of the weighted least squares averaging used to compute their mean positions and proper motions, this should result in statistical properties, after the systematic effects have been removed, which are Gaussian. The Central Limit theorem leaves no room for maneuver on this point.

The comparisons we describe below clearly demonstrate that the statistically predicted conclusions are not borne out for the ACRS (Part 1; Corbin and Urban 1990), nor for the AGK3 (Heckmann et al. 1975) while they are more so for the PPM (North; Röser and Bastian 1989) and clearly so for the AGK3U (Bucciarelli et al. 1992). The simplest explanations we have for these results are that the catalog-to-catalog differences for the ACRS compilation were not adequately treated and that 2 is not sufficiently close to infinity for the AGK3 (but 3 is for the AGK3U!?).

3. Method

After backdating all the catalogs to B1900.0, which we have used as an approximation for the average mean epoch of the GC stars' epochs of observation, we need to put the AGK3U, and so on, on the GC system. We use the method of infinitely overlapping circles (Taff, Bucciarelli, and Lattanzi 1990; Bucciarelli, Taff, and Lattanzi 1993) which explicitly and automatically includes "the proper motion systematic differences" because the systematic differences are at the mean epoch of observation t_{GC}. Column 2 of Table 1.1 contains the mean number of stars inside this circle for the different catalogs.

Catalog Name	Number of stars per circle	RA Offset ×cos δ (″) B1900.0	DEC Offset (″) B1900.0	Total number of stars in GC overlap	Mean mgnitude
FK5[1]	10.2 ± 3.6	-0.12 ± 0.15	0.03 ± 0.06	1534	4.7 ± 1.3
FK5 EXT[2]	11.0 ± 4.1	0.15 ± 0.16	-0.04 ± 0.08	1083	6.6 ± 0.8
AGK3RN	10.1 ± 3.1	0.12 ± 0.34	-0.31 ± 0.39	4687	7.4 ± 0.7
AGK3	11.8 ± 4.8	0.21 ± 0.34	-0.06 ± 0.29	10008	7.5 ± 0.9
ACRS[3]	11.4 ± 5.2	0.12 ± 0.18	-0.06 ± 0.12	12961	7.3 ± 0.9
PPM[4]	12.4 ± 5.3	-0.11 ± 0.28	-0.05 ± 0.21	13806	7.2 ± 1.0
AGK3U	12.1 ± 5.0	0.21 ± 0.31	-0.10 ± 0.28	8058	7.6 ± 0.8

[1] This is an all sky result.
[2] Northern hemisphere only (i.e., declination > -5 deg).
[3] Part 1 of the ACRS and northern hemisphere only (i.e., declination > -5 deg).
[4] Using the northern hemisphere half of the PPM only.

TABLE 1. Catalog comparison statistics.

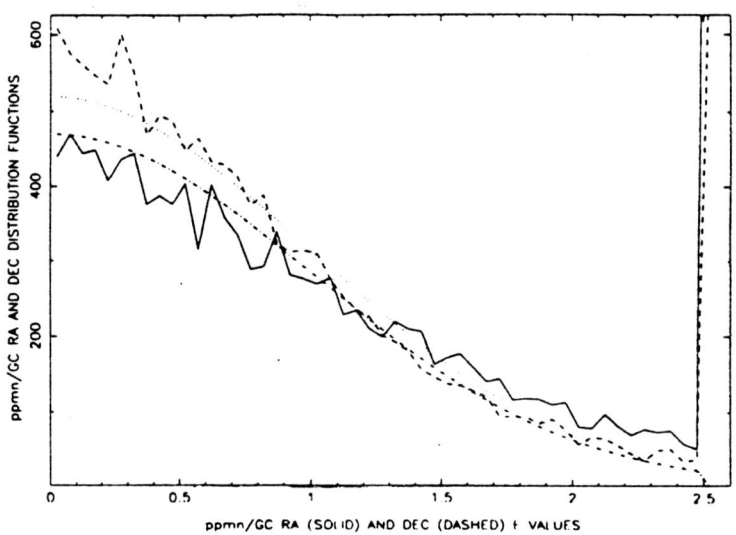

FIGURE 1. Theoretical (smooth) and empirical (jagged) distribution functions for the PPM North/GC overlap. The theoretical curves are dot–dash (RA) and dotted (DEC) while the empirical curves are solid (RA) and dashed (DEC). Outliers exceed 800 and lie beyond 2.5.

The value of the radius of the circle was about 3 deg, varying slightly among the four large-scale catalogs.

If there are no rapidly varying residual systematics within a catalog, then after applying the catalog-to-catalog difference, when we evaluate the difference between a star's right ascensions from the GC (α_{GC}) and its right ascension from the AGK3U (α_{3U}) at B1900.0 viz. $\alpha_{GC} - (\alpha_{3U} + \Delta\alpha_{3U,GC})$ we can expect this quantity to be a normally distributed random variable with zero mean. This quantity must be normalized by

$$\sigma^2_{3U,GC} = M^2(\alpha_{3U}) + m^2(\alpha_{GC}), \quad M^2(\alpha_{3U}) = m^2(\alpha_{3U}) + m^2(\mu_{3U})(t_{GC} - t_{3U})^2$$

to make into a "standardized random variable" where $t_{GC} \simeq B1900.0(t_{3U})$ is the mean epoch of the GC(AGK3U) right ascension $\alpha_{GC}(\alpha_{3U})$,

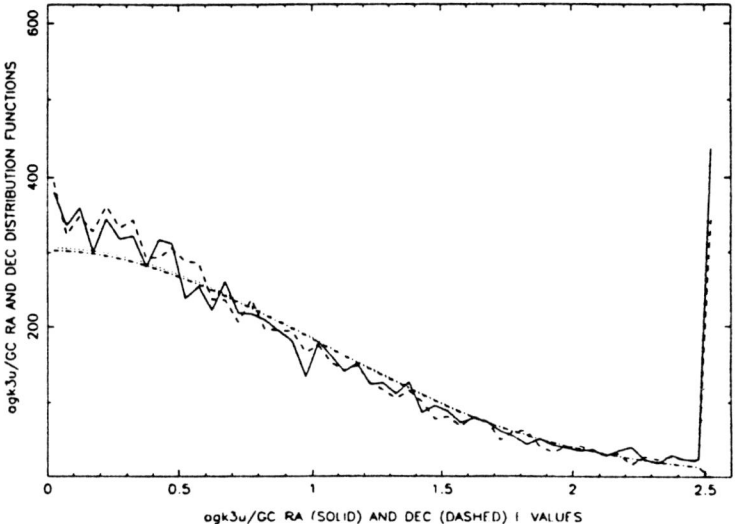

FIGURE 2. Same as Figure 1 but for the AGK3U/GC overlap set. Note the better agreement for the declination and the diminished number of outliers.

$m(\alpha_{GC})[m(\alpha_{3U})]$ is the mean error of $\alpha_{GC}(\alpha_{3U})$, and $m(\mu_{3U})$ is the mean error of the proper motion in right ascension μ_{3U}. The use of t_{GC} as the inter-comparison epoch means that the GC proper motions play no substantial role in the computation. This is important because we know that the GC proper motions, especially for the fainter stars, are relatively poor.

The statement that the right ascensions should have the same expectation refers to each star. Thus, we anticipate that catalog-wide average of $\alpha_{GC} - (\alpha_{3U} + \Delta\alpha_{3U,GC})$ will be zero. However, its absolute value will not have a vanishing mean so we prefer it as our test statistic. Our hypothesis must be that both $\alpha_{3U} + \Delta\alpha_{3U,GC}$ and α_{GC} are independently normally distributed random variables with the same (but unknown) mean value and with different standard deviations about their means; $m(\alpha_{GC})$ and $M(\alpha_{3U})$ at the epoch $t_{GC}(\simeq B1900.0)$. Therefore, our test variate is

$$e_{3U,GC} = |\alpha_{GC} - (\alpha_{3U} + \Delta\alpha_{3U,GC})|/[M^2(\alpha_{3U}) + m^2(\alpha_{GC})]^{1/2}.$$

The analytical counterpart to this problem yields $\sqrt{2/\pi}$ for the expectation of $e_{3U,GC}$. Thus, more fully (but still symbolically because we use a second order power series to incorporate the proper motions) the numerator of $e_{3U,GC}$ is given by

$$|[\alpha_{GC}(B1950.0) + \mu_{GC}(t_{GC} - B1950.0)] - [\alpha_{3U}(B1950.0)$$
$$+ \mu_{3U}(t_{GC} - B1950.0) + \Delta\alpha_{3U,GC}(\alpha_{3U}, \delta_{3U}, t_{GC})]|.$$

We have backdated α_{3U} by its proper motion to the mean epoch of observation of the GC ($= t_{GC}$) and put it onto the GC reference frame by adding the local systematic differences for the right ascension system at that place on the celestial sphere and at that instant of time. This is the quantity $M(\alpha_{3U})$ is supposed to characterize. Similarly we have backdated $\alpha_{GC}(B1950.0)$ using μ_{GC}. The local systematic differences in the motions are already embedded in $\Delta\alpha_{3U,GC}$ because it is evaluated at t_{GC}.

The distribution function for e (see the formula below) has only one parameter and is of a Gaussian type. The mean value of e^2 is predicted to be unity and the standard deviation of e about its mean (σ_e) is forecast to be $(1 - 2/\pi)^{1/2} = 0.603$. However, as the standard deviation about the mean involves the second moment (i.e., the mean of e^2), separately examining these two may prove to be beneficial. Moreover, the result that the mean of e is $\sqrt{2/\pi}$ does not tell us how sharp the distribution about the mean is. Therefore, we also tested the entire distribution function.

To do so we define an intermediate variable $w = |z|$, $z = x - y$. We know that x and y (eg. δ_{3U} at the mean epoch of observation of δ_{GC}, namely t'_{GC}, and δ_{GC} also at t'_{GC} and adjusted for all the systematic effects) share the same (but unknown) average value. There exists a joint probability distribution—namely a bivariate Gaussian—which describes the distribution function of x and y. A straightforward computation for the probability distribution of w yields $f(w) = [2/(2\pi\sigma_z^2)^{1/2}]\exp[-w^2/(2\sigma_z^2)]$ for w in $[0, \infty]$ where σ_z is the standard deviation of $z = x - y$.

Our preferred test variate is not w but its normalized version ω; $\omega = w/\sigma_z$. The distribution of ω is given by g,

$$g(\omega) = \sqrt{2/\pi}\,e^{-\omega^2/2}.$$

The advantage of ω over w is that σ_z is star dependent while the ω distribution is the same for all stars. The empirical distribution of ω (separately for right ascension and declination) furnishes us with $\sim$ 10,000 samples from which we can build up $g(\omega)$. Hence, we can simply perform a goodness-of-fit test to $g(\omega)$. If these distributions are non-normal, as measured for instance by a chi-squared test, then we have found a serious problem.

4. Results

Let us first examine the equatorial coordinate offset columns in Table 1. As can be clearly seen by looking at the FK5 and FK5 Ext rows, there is a clear difference—as Kopff (1954) had already told us—between the very bright fundamental stars in the GC and those $\sim$ 2 mag fainter. (The fact that the PPM North right ascension system offset alone agrees with the FK5 value and it is the PPM North right ascension which does poorly may be connected. The PPM North was also the only large-scale catalog provided to us on the J2000.0/FK5 system. We precessed it back to B1950.0, added the elliptic aberration terms to make the positions catalog mean place, adjusted the proper motions for the changes in the general precession parameters as well as the time dependence of the FK5/4 equinox motion, and adjusted the right ascensions for the FK5/4 equinox offset. Thus, the PPM(N)/GC results should be in accordance with the FK5/GC values because the version of the FK5 we used was similarly obtained.) Another thing to note from Table 1 is the general stability in the systematic differences as one probes ever fainter into the GC star list (beyond the fundamental stars). Also by reading the mean magnitude column we can see that there is no substantial difference among the large-scale catalog/GC intersection sets on this basis. Hence, we conclude that the GC is an acceptable standard.

Table 2 has the numerical values of the comparisons and clearly shows the failure of the AGK3 and the ACRS to conform to the predictions of statistics. For the PPM North, which still contains $\sim$ 6000 High-Precision Subset stars, the declination system does very well. The AGK3U is clearly the best of all the large-scale catalogs in this regard. Figures 1 and 2 show the empirical and theoretical distributions for e for the latter two catalogs and both equatorial coordinates. The theoretical curves have been normalized by the total number of stars in the catalog/GC intersection set minus the number of outlier stars. Outlier stars are those for which e exceeded 2.5 and these can be clearly seen in the figures as the upward tick at the right hand edge in all the plots. Because the number

	$\langle E \rangle \pm$ Stand. Dev.		$\sqrt{\langle E^2 \rangle}$		χ^2 Statistics	
THEORY	0.798 ± 0.603		1		49 ± 10	
Catalog	RA cos δ (″)	DEC (″)	RA cos δ (″)	DEC (″)	RA	DEC
FK5	0.807	0.606	0.651	0.519	1.009	0.832	248	90
FK5 EXT	0.647	0.517	0.540	0.451	0.828	0.703	82	234
AGK3RN	0.744	0.619	0.846	0.640	0.968	1.061	280	315
AGK3	0.490	0.407	0.473	0.389	0.637	0.613	2739	3030
ACRS	0.612	0.491	0.519	0.426	0.785	0.671	1141	2837
PPM	0.859	0.628	0.737	0.565	1.064	0.928	605	142
AGK3U	0.743	0.576	0.730	0.568	0.940	0.925	141	142

TABLE 2. Catalog comparison results.

of outliers can be different for the two coordinates, the normalization of the theoretical curves is sometimes slightly different.

REFERENCES

Boss, B., Albrecht, S., Jenkins, H., Raymond, H., Roy, A. J., Varnum, W. B., and Wilson, R. E. (1937). General Catalogue of 33342 Stars for the Epoch 1950, Carnegie Inst. of Wash. DC, Publ. No. 486.

Bucciarelli, B., Daou, D., Lattanzi, M. G., and Taff, L. G. (1992). *AJ*, **103**, 1689.

Bucciarelli, B., Lattanzi, M. G., and Taff, L. G. (1993). submitted to *ApJ*.

Bucciarelli, B., Taff and L. G. (1993). submitted to *ApJ*.

Bucciarelli, B., Taff, L. G., and Lattanzi, M. G. (1993). in press in *J. Stat. Comp. Sim.*

Corbin, T. E. 1978, in IAU Colloq. No. 48, *Modern Astrometry*, eds. F. V. Prochazka and R. H. Tucker (Univ. Obs., Vienna) pg. 505.

Corbin, T. E. and Urban, S. E. (1990). in IAU Symp. No. 141, *Inertial Coordinate System on the Sky*, eds. J. Lieske and V. Abalakin (Kluwer, Dordrecht) pg. 433.

Fricke, W. and Kopff, A. (1963). Veroff. Astron. Rechen-Institut Heidelberg, No. 10.

Fricke, W., Schwan, H., and Corbin, T. (1991). Veroff. Astron. Rechen-Institut Heidelberg, No. 33.

Fricke, W., Schwan, H., and Lederle, T. (1988). Veroff. Astron. Rechen-Institut Heidelberg, No. 32.

Heckmann, O., Dieckvoss, W., Kox, H., Gunther, A., and Brosterhus, E. 1975, AGK3. Star Catalogue of Positions and Proper Motions North of −2.5 deg Declination, Hamburg-Bergedorf.

Kendall, M. G. and Stuart, A. 1973, *The Advanced Theory of Statistics* Vol. 2, (Hafner Publ. Co., NY) Chap. 19.

Kopff, A. (1937). Veroff. Astron. Rechen-Instituts zu Berlin-Dahlem, No. 54.

Kopff, A. (1938). Abh. d. Preuss. Ak. Wiss., Phys.-math., No. 3.

Kopff, A. (1954). *MNRAS*, **114**, 478.

Morgan, H. R. (1952). Catalog of 5,268 Standard Stars, 1950.0 Based on the Normal System N30, *Astron. Papers. Amer. Eph. Naut. Almanac*, Vol. XIII, Pt. III.

Röser, S. and Bastian, V. (1989). PPM. Positions and Proper Motions of 181731 Stars North of −2.5 Degrees Declination for Equinox and Epoch J2000.0, Astron. Rechen-Institut, Heidelberg.

Taff, L. G. (1993). in press in *BAAS*, **25**.

Taff, L. G., Bucciarelli, B., and Lattanzi, M. G. (1990). *ApJ*, **361**, 667

Lick NPM program: NPM1 Catalog and its applications

By A. R. KLEMOLA, R. B. HANSON AND B. F. JONES

UCO, Lick Observatory, University of California, Santa Cruz, California 95064, USA

The Northern Proper Motion (NPM) program for Part I based on 899 fields for the sky outside the Milky Way is essentially complete. Computer–readable version will be available later in 1993. The resulting *Lick NPM Catalog*, denoted NPM1, contains nearly 149 000 stars spanning the blue magnitude range 7 to 18. In addition to the many astrophysically interesting stars taken from the Lick *Input Catalog of Special Stars*, NPM1 also contains a large number of brighter ($10 < m_b < 13$) and faint ($14 < m_b < 17$) randomly chosen stars suitable for galactic studies. The *rms* errors in proper motion for a single faint star are $0\rlap{.}''5$/cy and $0\rlap{.}''2$/cy for the absolute zero– point for an average epoch difference of 27 years (range 19 to 39 years). Immediate applications of the *NPM1* catalog include expanded studies in galactic motions, correction to precession, and luminosity calibration of the RR Lyrae–type stars. Preliminary work has started for Part II of the NPM program with 347 fields for the Milky Way.

1. Introduction

The goal of the Lick Northern Proper Motion (NPM) Program is the construction of a catalog of absolute proper motions, positions, and photometry for some 300 000 stars north of declination −23°. This huge project was started by W.H. Wright in the early 1930's, and has been carried on by several generations of Lick astronomers. The NPM program falls into two parts: Part I covers the 72% of the sky lying outside the Milky Way. Here, faint galaxies define the inertial reference frame for absolute proper motions. Part II covers the remaining 28% of the sky – the Milky Way; its measurement remains for the near future. A full description of the NPM program is given by Klemola, Jones, and Hanson (1987).

This paper reports the completion of Part I of the NPM program. The resulting catalog (which we denote NPM1) contains nearly 150 000 stars from measures in 899 of the 1246 NPM fields. Each NPM field was photographed with the Lick 51-cm double astrograph at two epochs between 1947 and 1988. The mean first and second epochs are 1950 and 1977; the average epoch difference is 27 years. The first-epoch plates were taken in the blue only; both blue and yellow plates were taken at the second epoch. The 6° × 6° fields lie in 5° declination zones from +90° to −20°. Each field overlaps its neighbors in all directions by at least one degree (except the −20° zone). (See Fig. 16.2 of Argue 1989 for the distribution of NPM fields on the sky). These plates have two exposures – one of 1.0 or 1.5 min. and the other of 2 hr. These exposure times, together with the use of a 4-mag. objective grating, permit measurement of stars over the blue magnitude range $8 < m_b < 18$.

2. Content of NPM Program

Several classes of objects are measured in the NPM program.

The NPM1 catalog contains some 94 000 stars chosen anonymously for the astrometric reductions and for statistical studies of stellar motions. These anonymous stars lie chiefly in two distinct magnitude ranges: $11 < m_b < 13$ (25 000 stars) and $14 < m_b < 17$

(69 000 stars), and comprise by far the largest subset in the NPM1 catalog. Additional anonymous stars were chosen in Kapteyn Selected Areas and Pulkovo Plan fields. The faint anonymous stars play an essential role in the proper motion reductions, where they are used to compute relative proper motion plate constants for each field. The bright anonymous stars are used in the reductions to "bridge" the two exposure systems on each NPM plate.

The NPM1 catalog contains some 28 000 positional reference stars which were used to derive equatorial coordinates from the measurements. For the northern sky we used the AGK3. For the southern sky we selected stars from the SAO catalog, using ACRS data for these stars for better accuracy. These catalog stars will also serve for later statistical studies.

The remaining 27 000 stars in NPM1 are chosen from the Lick *Input Catalog of Special Stars* (ICSS), which includes large numbers of stars of all classes gathered by Klemola from the astronomical literature. The ICSS contains over 116 000 entries of which many are duplicate, so that the total number of different stars is under 100 000 for the sky north of declination $-33°$. Specific classes of stars included in the ICSS are listed in Table II (Klemola *et al.* 1987).

Finally, there are over 50 000 faint galaxies, which define the inertial frame for the absolute proper motions. These are not included in the NPM catalog, but are retained in a separate file.

3. Description of the NPM 1 Catalog

The new NPM1 catalog contains 149 000 stars spanning the photographic magnitude range $8 < m_b < 18$. The magnitude distribution is bimodal; nearly half the stars are "bright" ($8 < m_b < 13$), measured on the short exposure and/or long-exposure grating images. Nearly half are "faint" ($14 < m_b < 18$), measured on the long exposure central images only. There is a distinct gap near $m_b = 13$. These stars were avoided in the NPM survey because they cannot be measured well in either exposure system.

The catalog is arranged in one-degree zones of declination, from $+89°$ to $-23°$. As in the AGK3, each star is assigned a running number within its zone, in right ascension order. Multiple measures from overlapping fields have been averaged to give one entry per star.

Positions are given for equinox B1950 and computed epoch 1950. Each star's entry includes the absolute proper motion ($''/$cy) and blue magnitude (m_b). For 97% of the stars the color ($m_b - m_v$) is also given. Other data given for each star are: the original mean epoch, a stellar class code, the number of NPM fields measured, and discrepancy flags for position, proper motion, and photometry. Finally, as an additional identification, the AGK3 (north) or SAO (south) number (if any) is given.

This is the content of the NPM1 catalog version now being prepared for delivery to the Astronomical Data Center. Detailed documentation and an updated description of the NPM reductions and catalog preparation will be published separately (Klemola, Hanson, and Jones 1993).

Collation of the full set of cross-identifications for the NPM stars is a large project now being completed by Klemola. The complete cross-identifications will be sent to the ADC as a separate file. The list of 50 000 NPM reference galaxies will be made available to interested users.

No final decision has been made about whether to publish a printed version of the NPM1 catalog. One possibility under consideration is to print only a "core catalog" of roughly 15 000 stars of well-defined astrophysical classes.

4. Error analysis of the NPM1 data

Compilation of the NPM1 catalog gave a large amount of data for assessing the random errors of the NPM proper motions, positions, and photometry. Error estimates were derived from the *rms* dispersions of multiple measures of some 60 000 stars measured on overlapping sections of NPM fields. The proper motion errors can also be compared with earlier estimates (Klemola et al. 1987, Hanson 1987).

The *rms* errors for the NPM absolute proper motions were computed from the field overlaps for each of the 114 declination zones from $-23°$ to $+89°$. The error in each component is about $0\rlap{.}''5$/cy, with little or no variation with declination. This agrees exactly with the earlier estimate (Klemola et al. 1987) of $0\rlap{.}''5$/cy for the NPM faint anonymous stars. The high quality of the NPM proper motions is maintained to the southern limit of the program. This important result is due in part to the longer epoch spans (approaching 40 yr) for the southernmost NPM zones.

The *rms* errors in right ascension and declination were also computed from the overlaps. Here the situation is more complex, for two reasons: First, the position errors at the catalog epoch 1950 are larger than at the original mean epochs ($\sim$ 1968), due to accumulated proper motion error. Second, there is substantial variation of the position errors with declination. For declinations $0°$ to $+70°$, the *rms* position error in each coordinate is about $0\rlap{.}''12$ at the original epochs, and about $0\rlap{.}''15$ at 1950. For the north polar cap the right ascension errors increase to $0\rlap{.}''20$. South of the equator, the position errors are more irregular, but again the right ascension errors increase above $0\rlap{.}''20$.

The *rms* errors for the NPM photometry, computed from the field overlaps, average about 0.2 mag. for the blue magnitude m_b, and 0.15 mag. for the color $m_b - m_v$.

The systematic errors of the NPM1 data are the subject of continuing study. Hanson (1987) derived preliminary estimates of these errors as a by-product of his solar motion and galactic rotation study. The external error of the proper motion absolute zero point in an individual NPM field was estimated at $0\rlap{.}''2$/cy. The overall systematic zero-point error was found to be $0\rlap{.}''06$/cy. These estimates used the NPM data for the faint anonymous stars in 617 NPM fields from $0°$ to $+65°$. New solutions, using all 899 fields from $-23°$ to $+90°$, give very similar results. Also under investigation are such questions as whether there is any systematic difference between the NPM bright star and faint star proper motions.

5. Distribution of reference galaxies

A question often asked about the NPM program is how the precision and accuracy of measurement of the faint reference galaxies compares to that of the stars.

The magnitude range of the 50 000 NPM galaxies extends over $14 < m_b < 18$, quite closely matching that of the faint anonymous stars. In the mean, the galaxies are only 0.5 mag. fainter. This minimizes the possibility of magnitude-dependent systematic errors in the proper motion zero-point corrections.

The *rms* errors of galaxy measurements can be found directly by examining the absolute proper motion distribution of the NPM galaxies. This is essentially just an error distribution, since the true proper motions of the galaxies are negligibly small. The galaxy proper motion errors are surprisingly small, amounting only to about $0\rlap{.}''9$/cy in each coordinate. This is less than a factor of two worse than the stars.

6. Applications of the NPM1 Catalog

Several important applications of the NPM data have been planned for many years. With the completion of NPM1, these studies are now underway at Lick.

6.1. *Solar motion and Galactic rotation*

A preliminary study of solar motion and galactic rotation was made by Hanson (1987), using the faint anonymous stars from 0° to +65°. Hanson discovered a remarkable variation of the solar apex location as a function of galactic latitude, tending (at higher latitudes) increasingly toward the direction of galactic rotation. Nearly all the change in the solar motion came from the Y component, with little or no change in the X or Z components. This study also yielded precise new values for the Oort constants of galactic rotation.

Hanson (1989) used a kinematic distance calibration for the NPM faint stars to convert this result into the variation of the V component of solar motion with distance from the galactic plane. This showed a large progressive lag behind circular galactic rotation with increasing distance from the galactic plane, caused by the changing mix of stellar populations.

These preliminary studies used only the NPM faint stars, with only $\frac{2}{3}$ of the NPM fields available, and no photometry. With the completion of NPM1, a more definitive study will be carried out, using many more stars over a much wider magnitude range, binning the stars by magnitude and color.

6.2. *Correction to precession*

One of the original motivations for establishing the NPM program by Lick astronomer W.H. Wright in the 1930's was to determine the correction to the precession constants using absolute proper motions measured with respect to galaxies. This objective continues with the new NPM1 catalog, which contains subsets of stars from the AGK3 (19 000 stars) and SAO (9000 stars) catalogs. Klemola and Hanson will derive the correction to precession by comparing the NPM1 proper motions for these stars with data from modern astrometric catalogs, such as the ACRS and PPM.

6.3. *Comparison with Hipparcos program*

Other important comparisons include that between NPM1 and Hipparcos catalogs. The NPM1 catalog contains 13 000 stars in common with the *Hipparcos Input Catalog* (HIC) in the magnitude range $8 < m_b < 12$. This comparison will have to be done very carefully, since many stars in the NPM1 catalog brighter than $m_b \simeq 9$ may prove to have been too bright for reliable measurement on the blue astrograph plates.

7. Other classes of stars in NPM1

A brief description is given here for some of the other stellar classes in the NPM1 catalog as a guide to future users.

The Lick ICSS contains about 20 000 variable stars of all types, of which nearly 3000 appear in the NPM1 catalog. Since NPM1 covers only the non-Milky Way sky, large numbers of stars showing strong concentration to the galactic plane are missing but will be measured later in Part II of the NPM program.

The field RR Lyr-type stars represent one of the important tracers of the galactic halo. They can be seen to remote distances. They possess information about the early history of the galaxy. They are essential in distance-scale calibrations. A total of 930 RR Lyr stars appear in the NPM1 catalog and are available for various luminosity and kinematic studies. However there remain several serious observational deficiencies before the proper motions can be fully exploited. Mis-classification of RR Lyr stars may prevail for a small fraction of these variables (Schmidt 1991). Accurate photometry and other parameters, including metallicity and radial velocity, are needed for many stars.

The designation FBO/FBS comprises a mixture of types, such as hot subdwarfs, white dwarfs, horizontal-branch (HB) stars, RR Lyr stars, QSO's, and galaxies. The NPM1 catalog includes 4800 FBO/FBS from various published surveys. This is an enormous astrometric resource waiting for proper exploitation. The great need now is for continued improvement on the astrophysical side, i.e., sorting this mixture into its proper components, and measurement of radial velocities and improved photometry where needed.

One component of the FBO's is the blue horizontal-branch stars (BHB) stars. The study of BHB stars is emerging as a valuable means of studying the galactic halo and the question of chemical gradients across the galaxy. Such stars are fairly easily recognizable from multicolor photometry to great distances and serve as valuable probes. The BHB candidates are now being discovered in large numbers from various multicolor photometric surveys. The selection of HB stars in the Lick program, initially including only about 50 HB stars from the Philip and other surveys of the 1960's, now includes many BHB candidates from the Beers, Preston, Shectman (1988) survey. It contains 4400 stars north of declination $-33°$. Unfortunately, because of the recentness of this publication, only the southern zones $-7°$ to $-23°$ contain these BHB candidates. The NPM1 catalog contains absolute proper motions and photometry for about 975 HB and HB-candidates.

The Lowell Survey provides the largest source of white dwarf stars and candidates (GD=1-1712) in the Lick ICSS catalog . Thousands of additional candidates are listed in Luyten's surveys but few are included in NPM1 because of difficulties in identification on the astrograph plates, as well as the fact that most are too faint (mag. 18+). The NPM1 catalog contains 1400 WD and WD-candidates.

Many red giant and candidate stars are included in the NPM program. A major source is the Dearborn objective-prism survey for the sky north of $-4°\!.5$. The Dearborn survey lists 44 076 K5-M4 stars and 3072 M5+ stars, which are mainly red giants. The ICSS includes a 5% sample of K5-M4 stars, or 2200 stars, as well as all 3072 M5+ stars, for a total of nearly 5300 Dearborn stars. The NPM1 catalog contains absolute proper motions for about 1800 Dearborn stars with $m_b = 9 - 12$. Such evolved stars may be useful for stellar motion studies towards the galactic halo.

The Lick ICSS also includes over 3000 carbon/CH, S, and R-type stars. Most lie in the Milky Way with only a small number found towards higher latitudes. Carbon and CH stars are important for the study of structure and kinematics of the outer halo in our Galaxy and in other galaxies. They are readily identifiable and can be seen to great distances. The NPM1 catalog contains 165 carbon stars which should be useful for halo studies. Imbedded in carbon star catalogs are the apparently rare dwarf carbon stars. Some may be recognized by their large proper motions. About 5-8 dwarf carbon stars are now known, with at least two appearing in the NPM1 catalog. We plan detailed examination of the carbon stars in NPM1 for other possible dwarf candidates.

Acknowledgements

We thank the National Science Foundation for its continued support of the Lick NPM program. Current work is supported by grant AST 92-18084.

REFERENCES

Argue, A.N., 1989, *The Hipparcos Mission. Pre-Launch Status.* Vol. II. p. 199 ESA Publications Division, c/o ESTEC, Noordwijk, The Netherlands

Beers, Preston, & Schectman, 1988, ApJS, 67, 464

Hanson, R.B., 1987, AJ, 94, 409

Hanson, R.B., 1989, BAmAS, 21, 1107

Klemola, A.R., Jones, B.F. & Hanson, R.B., 1987, AJ, 94, 501

Klemola, A.R., Hanson, R.B. & Jones, B.F., 1993, in preparation

Schmidt, E. G., 1991, AJ, 102, 1766

Southern Proper Motion Program: progress, scope, and accuracy

By WILLIAM F. VAN ALTENA[1], IMANTS PLATAIS[1], TERRENCE M. GIRARD[1] AND CARLOS E. LÓPEZ[2]

[1]Yale University Observatory, P.O. Box 6666, New Haven, CT 06511, U.S.A.

[2]Felix Aguilar Observatory, San Juan, Argentina

The Yale/San Juan Southern Proper Motion program (SPM) is the southern hemisphere extension of the Lick Observatory Northern Proper Motion program with respect to faint galaxies. Absolute proper motions and positions will be determined for approximately one million objects south of $\delta = -17°$. Recently, the purpose of the SPM has been broadened to include the creation of a secondary reference system of stars within the range 15 - 18 visual magnitude. To create the list of program stars, extensive use is made of existing databases and catalogs. Currently, we are finishing the measurements in a $\sim$1000 square degree region around the South Galactic Pole which will provide needed kinematical data for our Galaxy structure studies. Projected positional accuracy is expected to be around $0\rlap{.}''1$ at the mean epoch of 1980. The internal error of the absolute proper motions is 4 - 5 mas/yr. However, we still need to study extensively all possible sources of systematical errors before the results are finalized. For this purpose the Hipparcos positional catalog should contribute significantly and, conversely, results from the SPM and NPM programs will hopefully provide a link of the Hipparcos proper motion system to the extragalactic reference frame.

Introduction

The Southern Proper Motion program (SPM) was initiated in the early 1960's by D. Brouwer and J. Schilt as the complement to the Lick Northern Proper Motion program (NPM). The history of the SPM has been discussed elsewhere, (Wesselink 1974, van Altena et al., 1991), but here we wish to note the significant role played by Arnold Klemola in establishing the SPM program and in successfully bringing the NPM program to its current fruitful state, as described elsewhere in these proceedings. The SPM continues to benefit from the cumulative experience of the NPM program as we enter into the measurement phase of our program.

The SPM will produce positions, absolute proper motions, and V and B magnitudes for approximately one million stars. Objects to be measured in the SPM program include galaxies, astrometric reference stars, anonymous stars required in the reduction process, specific stars of special interest, subsets of special catalogs, and a large number of anonymous stars selected for the purpose of galactic structure study. The stars of special interest include variable and suspected variable stars, known high proper motion stars, and other individual stars with peculiar photometric or spectroscopic properties.

There are several problems in the area of galactic structure study which the SPM is intended to address. The most obvious application is in the determination of Oort's constants of galactic rotation, as has been recently done by Hanson (1987) using NPM data. Another area of study centers on the question of the existence and characterization of the thick disk component of the Galaxy. It is intended that the SPM will provide the necessary kinematic and photometric information to allow an accurate description of each of the major components of the Galaxy, in terms of their spatial distributions (luminosity function and scale heights), and their kinematic distributions (velocity ellipsoids and net rotation).

The SPM data will also yield a refined measure of the solar motion, corrections to the precession constants, and absolute proper motions for a number of star clusters. Plans are also underway to use the SPM and NPM data to link the Hipparcos proper motion frame to the extragalactic frame.

Finally, the scope of the SPM project has been expanded to include the creation of a faint, secondary astrometric reference catalog. This catalog will have a density of 15 to 20 stars per square degree at $V = 15 - 18$, with a positional accuracy of $\sim 0\rlap{.}''1$ at mean epoch 1980. It is felt that there is increasing need for such a catalog as can be seen, for instance, in the severe pointing requirements associated with the new multi-fiber-optic spectrographs used with some of the larger telescopes.

Plate material

The SPM plate material consists of blue (103 a-O emulsion) and yellow passband (103 a-G + GG515) plate pairs taken at two epochs using the 51 cm double astrograph at El Leoncito, Argentina. The first epoch plates were taken between 1965 and 1974. The second epoch observations were begun in 1987 and continue now, with ~ 200 of the 717 fields completed. The 17 x 17 inch plates with $55\rlap{.}''1$/mm plate scale cover an area of 6.3×6.3 degrees with approximately 1 degree overlap in α and δ between adjacent fields. Each plate contains a 2 hour exposure and a 2 minute exposure offset from the long exposure by about 2 mm. An objective grating is also used to produce diffraction images that are reduced ~ 4 magnitudes relative to the zero-order images. The limiting magnitudes are approximately 19 and 18 for the blue and yellow plates, respectively.

The multiple grating/exposure images allow us to connect the bright reference catalog stars to the faint galaxies which define the absolute proper motion frame. As with the NPM, this is accomplished in an indirect fashion. First, the system of the short exposure images is transformed to that of the long exposure using a set of "bridge" stars which are measurable in both systems. Relative proper motions are then determined by modelling the first-epoch plate into the second-epoch plate using faint anonymous stars to determine the differential plate model. Finally, the absolute proper motion zero-point correction is derived from the relative proper motions of the measurable galaxies.

Input catalog

Program stars to be included in the SPM are selected from a number of sources. The bulk of the stars are anonymous stars randomly selected for the purposes of our galactic structure studies and for the faint secondary reference catalog. An effort has been made to predict the number of stars, as a function of magnitude, that need be included in order to ensure that each component of the Galaxy (disk, thick disk, and spheroid) will be well sampled, in a statistical sense. This is done using a three-component Galaxy model code developed by R. Méndez, (see Méndez et al., 1992), which combines the current best estimates of galactic structure parameters with the known proper motion measuring accuracy of the SPM. For example, Table 1 shows the number and magnitude distribution of galactic structure stars to be measured in an extended region around the South Galactic Pole, (the region selected for the first SPM measurements and reductions).

A subset of these anonymous, galactic structure stars are chosen to also perform other tasks in the SPM reduction procedure. As mentioned earlier, "bridge" stars are used to link the systems of the long and short exposures, and well-distributed, faint anonymous

$V \pm 0.5$	Target Population*	Stars	Stars per field	Stars per deg^2
8	D	500	10	0.5
9	D	550	15	0.6
10	D	600	15	0.6
11	D	400	10	0.4
12	D,TD	7,000	175	7.0
13	D,TD	4,600	115	4.6
14	D,TD	3,500	90	3.5
15	D,TD,S	6,900	170	6.9
16	D,TD,S	3,900	100	3.9
17	D,TD,S	2,300	60	2.3
18	D,TD,S	800	20	0.8
Total		31,500	780	31.1

* D = Disk, TD = Thick Disk, S = Spheroid

TABLE 1. SPM program stars at the SGP

stars are used to perform the plate solutions, (see Platais et al., 1992 for more details). Added to this list of anonymous stars is all of the measurable galaxies, to provide the correction to absolute proper motion. Other program stars are selected from a variety of source catalogs. The IRS, ACRS, and PPM catalogs will provide astrometric reference stars. Also, an attempt is made to include all of the SIMBAD database objects above our plate limit as this is the most comprehensive compilation catalog available. Among other special interest stars are variables and suspected variables extracted from the GCVS and NSV catalogs, as well as Luyten's high proper motion stars from the NLTT. A special effort is made to identify and improve the positions for these groups of stars using the SPM plate material and the Yale Survey/Blink Machine. This invaluable work is enthusiastically performed by members of the Astronomical Society of New Haven. Finally, a variety of star catalogs with specific emphasis (e.g., photometry, metallicity, radial velocities) are included in our input list in whole or in part. The apparent magnitude distribution of all available objects in the SGP region as well as the distribution of the SPM program objects are shown in Figure 1.

The plates are measured with the Yale PDS microdensitometer in object-by-object mode. This mode provides the most accurate image positions but is relatively slow and requires an object input list with preliminary positions good to several arcseconds or better. The scanning speed of the PDS limits the number of objects per field to a few thousand. To obtain the preliminary positions, we make use of two existing catalogs, the COSMOS/UKST Object Catalog of the Southern Sky (which we shall refer to as the ROE Catalog), and the HST Guide Star Catalog (GSC). The ROE Catalog contains positions, magnitudes, and stellar/non-stellar classification down to $B_j = 21$. The GSC contains similar information but is less deep, ($B_j \approx 15.6$ at the South Galactic Pole).

The ROE and GSC catalogs are combined to form a master list of stars and galaxies, virtually complete to $B_j = 19$, the SPM plate limit. The ROE positions are tied into the GSC system using stars common to both catalogs. For many bright stars, more accurate positions (and proper motions) are adopted from catalogs such as the IRS, ACRS, PPM,

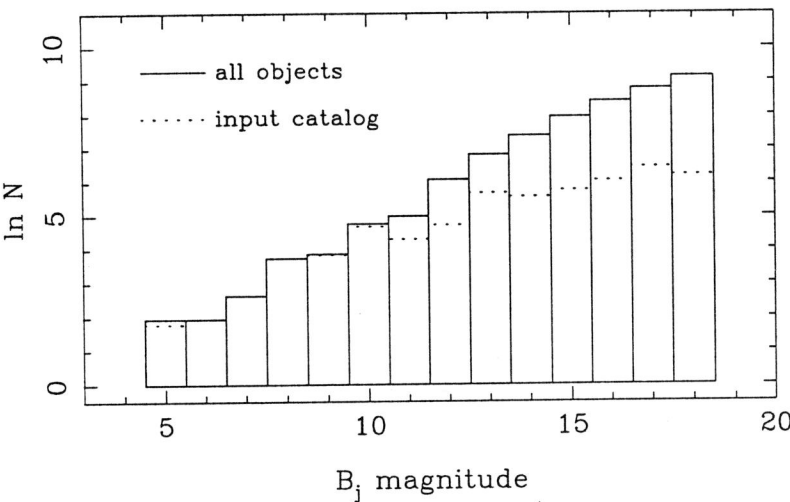

FIGURE 1. Magnitude distribution of all objects (solid line) and the SPM input catalog objects (dotted line) from a sample SPM field near the South Galactic Pole. Note the logarithmic scale.

etc. This ROE/GSC master catalog, it should be noted, is intended to support certain requirements in the pre-scanning stages of the SPM program and is not intended as a catalog of astrometric quality. It is invaluable in serving numerous functions in the preparation of the PDS Input Catalog:
• it is used as a base from which to randomly select the necessary anonymous stars described above,
• it serves as a master index for cross-identification with special catalogs or objects of interest extracted from SIMBAD,
• it provides a list of galaxies, necessary for the link to absolute proper motions,
• it allows us to predict potential image crowding/confusion problems due to the multiple exposures and grating images on the SPM plates.
With regard to this last point, a model of all images of a potential target star and its surrounding field is made to predict possible blends. Randomly selected (anonymous) stars are actually selected from those stars which do not have potential blend problems. It might also be noted that the ROE/GSC master catalog is quite useful for identifying objects of interest for which only a finding chart is available.

South Galactic Pole region

An extended region around the South Galactic Pole has been chosen within which to begin in earnest the SPM measurements and reductions. This region consists of 40 contiguous SPM fields covering ~1000 square degrees. The first several fields have helped serve as testbeds for developing the procedures by which the input catalog is compiled and processed through the measurement phase. As of this writing, 50 of the 160 plates have been measured. These plate measures are currently being reduced.

The formal standard error of the proper motions derived from several of the first fields around the SGP is 4 - 5 mas/yr for stars with $V < 16$. The uncertainty in the absolute proper-motion zero-point is 0.5 - 1.0 mas/yr depending on the number of measurable galaxies in the field. The internal precision of the equatorial coordinates is 0.10 arcsec (s.e.) per coordinate and per plate, at the plate epoch. The photometric accuracy of V and B is $\sim$0.1 mag.

We are currently exploring several methods for detecting and correcting for possible systematic errors in both the astrometric and photometric results. (Magnitude equation is of primary concern.) Due to the importance of this particular region for galactic structure study, our results for the extended South Galactic Pole region will be published separately from the remainder of the SPM, as soon as they are finalized.

Acknowledgments

The authors are pleased to thank Dr. Harvey MacGillivray for furnishing the relevant portion of the COSMOS/UKST Object Catalog of the Southern Sky. We are indebted to the members of the Astronomical Society of New Haven - C. Predom, R. DeMartino, M. Dzubaty, D. Kocyla and E. Wetherbee for their invaluable contribution to this project, improving the celestial coordinates of variable and NLTT stars. We also wish to thank R. Méndez for numerous helpful discussions and for providing and running the Galaxy model code used to establish criteria concerning the number of program stars. The SPM project is supported by grants from the National Science Foundation.

REFERENCES

Hanson, R. B. (1987) AJ, **94**, 409.

Platais I., Girard T. M., Méndez R. A., van Altena W. F., López C. E., MacGillivray H. T., Yentis D. J., Lattanzi M. G., Lasker B. M. (1992) In Astronomy from Large Databases II, eds. A.Heck, F.Murtagh, 455.

Méndez R. A., Platais I., Girard T. M., van Altena W. F. (1992) VII IAU Latin American Regional Meeting, Vina del Mar, Chile. Revista Mexicana Astron. y Astrofis. (in press)

van Altena W. F., Girard T. M., López C. E. (1991), Ap&SS, **177**, 59.

Wesselink A. J. (1974) In New Problems in Astrometry, eds. W. Gliese, C. A. Murray and R. H. Tucker, 201.

Proper motions from fourfold astrographic coverage of the northern hemisphere: first results

By G. A. IVANOV, V. S. KISLYUK, L. K. PAKULYAK,
T. P. SERGEEVA AND A. I. YATSENKO

Main Astronomical Observatory, Ukrainian Academy of Sciences, 252127 Kiev, Ukraine

The FON project is carried out using six wide-angle astrographs. About 95% of the plates have been obtained, and measurements of the Kiev part of the project has been carried out using the PARSEC automatic measuring machine. More than 250 plates have been measured and proper motions of about 400 000 stars in different areas of the northern hemisphere have been obtained. The accuracy of the results is also discussed.

1. Introduction

The project of fourfold coverage of the northern hemisphere of the sky known as FON (fotograficheskij obzor neba - photographic survey of the sky) was proposed by astronomers of the Main Astronomical Observatory (Kiev Observatory) of the Ukrainian Academy of Sciences (Kolchinsky and Onegina 1977). Observations are carried out by means of the same type C.Zeiss (Jena) astrographs with the same aperture (D=400 mm) and different focal distances, namely F=2000 mm (Kiev, Zelenchuk, Dushanbe, Zvenigorod) and F=3000 mm (Abastumani, Kitab).

In order to ensure the fourfold overlapping of the sky, six astrographs were combined in the four groups, as shown in Table 1. Each group photographs once the whole northern hemisphere. Between the groups plate centres are displaced for 2° in R.A. and/or Dec. Within each group the plate centres are displaced in R.A. in accordance with convergence of the meridians: i.e. 16 min. in declination zones from −2° to +60°; 32 min. from +60° to +75°; and 64 min. from +75° to +90°.

To consider magnitude equation there are two exposures of different duration (18 min. and 40 s) on each plate, shifted in R.A. and Dec. by about half a millimeter. The astrometric limiting magnitudes for two exposures are about $B = 15$–16 and $B = 12$–13, respectively.

Systematic observations for the FON project were started in 1982. The current status of the observations is shown in Table 1. Unfortunately, the disintegration of the USSR has had an influence on the continuation of observations. As can see from Table 1, Kitab and Dushanbe are most behind in the observations. Now Kiev is finishing observations of both the Kitab and Kiev zones.

2. Measurements and plate adjustment

We intend to apply the FON photographic collection first of all to the determination of positions and proper motions of all stars in the Astrographic Catalogue (AC). The AC is used in two ways: (i) as the first epoch of observations for determination of proper motions of stars, and (ii) as the input catalogue for the automatic measuring machine, PARSEC (Programming Automatic Radial-Scanning Coordinatometer). PARSEC, which has recently been constructed (Sergeev et al. 1991), measures the (x, y)

Overlap	Station	Dec from °	Dec to °	Plates required	Plates obtained
1	Kitab (Uzbekistan)	−4	+32	900	∼ 500
	Kiev (Ukraine)	+32	+90	952	1670∗
2	Zelenchuk (Russia)	−6	+90	1783	∼ 1700†
3	Abastumani (Georgia)	−6	+30	900	900
	Zvenigorod (Russia)	+30	+90	973	973
4	Dushanbe (Tajikistan)	−4	+90	1762	∼ 1370

Notes: ∗ Kiev photographs the whole northern hemisphere;
† Observational station of Kazan' Observatory (Tatarstan).

TABLE 1. Some characteristics of the FON project

coordinates of stellar images with accuracy better than 1 μm, as well as stellar magnitudes with an accuracy of 0.06 –0.14 mag. The speed of measurement varies from 400 to 900 stars per hour depending on the accuracy of the preliminary coordinates of the stars.

At the end of 1991 we started measuring the FON plates obtained in Kiev, owing to the availability of the automatic measuring machine there. First of all, the characteristics of the measuring machine and the real accuracy of the measurements were estimated. The accuracy of measurement is checked with a special test-plate on which there are some objects of different diameters with very sharp edges, regular forms and uniform density. Using this test-plate, accuracy of measurement with PARSEC is estimated to be about 0.7 μm.

The real accuracy of measurement depends on the dimensions of the images, their irregularities, and the errors of the measuring machine. Estimation of the real accuracy was carried out by comparison of measurements made of both the first and the second exposures on FON plates. The mean square differences of measurement on two exposures are 4.2 and 4.1 μm for (x, y). The mean square error of measurement of one star on a FON plate is estimated to be 2.9 μm (2.7 and 3.2 μm for the first and the second exposures, respectively).

As mentioned above, PARSEC permits the measurement of the diameters of stellar images with an accuracy better than 0.1 mag. However, it is only an instrumental possibility of PARSEC. Because of absence of sufficient numbers of photometric standards, the PPM catalogue (Röser and Bastian 1989) has been used for the calibration of measured stellar diameters. Mean square errors of stellar magnitudes obtained using two exposures are estimated to be 0.13–0.25 mag.

The procedure for plate adjustment is described briefly here. The PPM catalogue is used as the reference system. Images of stars in two exposures are measured for all PPM stars. We apply incomplete cubic polynomials as reduction models for each coordinate; i.e non-orthogonal plate models are applied for their adjustment. Plate parameters are determining from combined solutions of the connection equations for both exposures. Solution of the system of conditional equations is carried out by means of the Gramm-Schmidt method of orthogonalization for estimation of the number of significant terms in the reduction models. Mean errors of unit weight are 3.3 and 3.2 μm (or 0$\overset{''}{.}$33 and 0$\overset{''}{.}$32) for (x, y).

At present more than 250 plates of Kiev FON collection have been measured and

	FON–PPM	FON–CMC	Two exposures
	"	"	"
$\sigma_\alpha \cos\delta$	0.35	0.35	0.36
σ_δ	0.33	0.34	0.36
	"/yr	"/yr	"/yr
$\sigma_{\mu_\alpha} \cos\delta$	0.0040	0.0060	0.0055
σ_{μ_δ}	0.0040	0.0055	0.0052

TABLE 2. Comparison of FON data with PPM and CMC

proper motions of about 400 000 AC stars have been obtained. Because these plates are within declination zone 0° to 30°, the Astrographic Catalogue in the version by Fresneau (1983) has been used for the first epoch. Stellar proper motions obtained from eight randomly selected FON plates were compared with Carlsberg Meridian Catalogue No.4 (1989) and with PPM data. Table 2 shows the mean square differences in position and proper motion for these comparisons, as well as those obtained for the two exposure solutions.

Finally, the accuracy of FON positions and proper motions of stars from only one image is estimated to be 0".35 and 0".006/yr, respectively. If we use two images of stars on each plate, these estimations are 0".25 and 0".004/yr, respectively.

3. Conclusion

These results are preliminary. They cannot be used immediately for the study of kinematics and structure of the Galaxy. We intend later on to finish all observations and to continue measurements of FON plates. For processing of observations in declination zone higher than +30° we will use the machine-readable copies of the Astrographic Catalogue, carrying out this work in cooperation with the Astronomisches Rechen-Institut (Heidelberg) and the Sternberg Astronomical Institute (Moscow). It would also be desirable to use HIPPARCOS and TYCHO data for the plate adjustment.

REFERENCES

CMC, 1988, Carlsberg Meridian Catalogue La Palma Number 4. Copenhagen University Observatory, Royal Greenwich Observatory and Real Instituto y Observatorio de la Armada, San Fernando

Fresneau, A., 1983, AJ, 88, 1378

Kolchinsky, I.G. & Onegina, A.B., 1977, Plan of celestial photography with wide-angle astrographs, Astrometrija i Astrofizika (Kiev), 33, 11

Röser, S. & Bastian, U., 1989, PPM – Positions and Proper Motions of 181 731 stars north of −2.5 degrees declination, Astronomisches Rechen-Institut, Heidelberg

Sergeev, A.V., Sergeeva, T.P. & Riabokon, A.V., 1991, Ap&SS, 177, 329.

Improving proper motions in the southern hemisphere using CPC2

By N. ZACHARIAS†

University Space Research Assoc. (USRA), Washington DC, U.S.A.

The Second Cape Photographic Catalog (CPC2) consists of 276 131 star positions with mean epoch around 1968 and with an internal accuracy of about 60 mas. The current status and the use of this catalog for obtaining proper motions on the southern hemisphere is investigated.

1. Project overview

The Second Cape Photographic Catalog (CPC2) with mean epoch around 1968 has recently become available to the astronomical community. Paper I (de Vegt et al., 1993) gives full details of the project and paper II (Zacharias et al., 1992) describes the reduction and catalog construction details.

The CPC2 covers the entire southern hemisphere in a 4–fold overlap pattern on 5820 plates of 4°.1 field size, using a 530 – 640 nm spectral bandpass with limiting magnitude around V = 10.5.

To date only the conventional single plate adjustment (CPA) has been performed using the SRS reference star catalog. The average random catalog accuracy of 90 % of the stars is 60 mas per coordinate.

Because of the observational history, a significant epoch difference of overlapping plates (up to 8 years) is present around $\delta = -52°$, which has to be taken care of in a block adjustment (BA).

Table 1 gives a timetable for the entire CPC2 project.

2. Proper motions

The CPC2 is the most accurate major position catalog on the southern hemisphere and will improve the proper motions of about 276 000 stars when combined with other epoch data.

Table 2 gives a summary of major position catalogs (more than 100 000 stars) in the southern hemisphere.

An asterisk in the last column of Table 2 denotes the use of CPC2 for proper motions. The following procedure is suggested for optimal proper motion improvements.

(i) Use HIPPARCOS for the final CPC2 version. Because of the high accuracy and large number of reference stars, the internal x,y error of the CPC2 data will become the limit for the catalog positions, resulting in about $\sigma_{cat} = 50$ mas at epochs from 1962 to 1972.

(ii) Compile a new version of the ACRS for stars in the magnitude range of $V = 7...11^m$. Because of the epoch distribution and the accuracy of all catalogs available around 1997, this will be basically the TYCHO catalog (epoch = 1992, $\sigma_{cat} = 30$ mas) as second epoch and the CPC2 as first epoch. Thus a random error in these proper motions of $\sigma_\mu = 2.5$ mas/yr can be expected.

† Current address: U.S. Naval Observatory, 34th and Massachusetts Avenue NW, Washington DC 20392, U.S.A.

1954			first plans for a CPC2
1959			camera delivered
1962	-	1963	observations of the Cape Zone
1966	-	1968	observations of the zone $-30°$ to $-40°$
1967	-	1972	observations of all other zones
1978	-	1984	measuring of all plates
1984			publication of first results (Cape Zone)
	-	1985	all data from RGO to Hamburg
1987			block adjustment of the Cape Zone
1990			the entire CPC2[1] is available
1991			preparation for publication of results
1992			block adjustment simulations of entire CPC2 data
1993			BA solution with real data, analysing, optimising
1994	?		publication of CPC2[2], BA solution
1996	?		final CPC2[3], using the HIPPARCOS catalog

TABLE 1. Timetable of the CPC2 project

catalog	n* million	limit mag	epoch	accuracy mas
AC	4.0	13.0	1905	220
CPC2	0.25	10.5	1968	60
FOCAT-S	0.2	10.0	1982	123 *)
HIPP.	0.1	(10)	1992	2
TYCHO	1.0	12.0	1992	30
xxx	10.0	14.0	1995 ?	50
compilation catalogs:				
INCA	0.1	(10)		*)
ACRS	0.2	10.5		*)
PPM	0.2	10.5		*)

TABLE 2. Major position catalogs in the southern hemisphere

(iii) Reduce the AC with the new ACRS. The result will be a high precision first epoch for stars in the range $B = 11...13^m$. A random error of $\sigma_{cat} = 200$ mas can be expected (Urban & Corbin, 1993). Some zones are worth remeasuring and the potential accuracy gained by using modern measuring machines is currently being investigated (Winter, 1994).

(iv) Only a new observation catalog (1995?) will give the required second epoch for the stars in the range 11...13 mag and also a first epoch for at least double as many fainter stars. A catalog accuracy of $\sigma_{cat} \leq 50$ mas can be expected today. It is shameful to see that the only high precision position catalog which goes down to about $B = 13$ is the AC, observed around 1900!

(v) Rigorously evaluate the Schmidt surveys. This can be done only when high precision positions of enough stars down to at least $V = 14^m$ become available.

3. Status of block adjustment of CPC2

Simulations with the full CPC2 data have revealed the high potential precision which can be obtained by applying the rigorous block adjustment (BA) method (Zacharias, 1992).

With imperfections present in real data there is a great potential for bad or misleading BA solutions. Some notes of practical experience are worth mentioning.

- A large overlapping area of adjacent plates is highly desirable.
- A small epoch difference of all overlapping plates is required, or some preliminary proper motions must be applied prior to the BA. Including proper motions as parameters to be solved for is feasible in theory, but at least three well distributed epochs of roughly equal weight are required – far from current reality for global applications.
- The use of the appropriate physical model is mandatory.
- Refraction and apparent place routines have to be applied somehow prior to the BA; otherwise the overlapping plates x,y coordinates will not "match".
- A rigorous clean up of the input data is required, as well as various checks in the BA software to detect problems not discovered previously.
- If possible, a global pattern (hemisphere at least) should be used, which will provide a much "stronger" solution than a zonal pattern or quasi–rectangular area in the sky (Zacharias, 1992).

Thus, also with the BA, the dictum holds "think globally, act locally".

Only about 2 hours CPU time is required for a BA run of the entire CPC2 data with 877 483 images of 205 792 stars and 17 266 reference stars on 5687 plates. Using an 8-parameter plate model, 45 496 parameters have to be solved for simultaneously. About 15 MB of memory and less than 1 GB disk space were sufficient for this job, running with the recently finished installation of the HBAPP (Hamburg Block Adjustment Program Package) (Zacharias, 1987) on an HP 9000/750 system at USNO with some other users present on the system at the same time. The I/O speed was found to be the limiting factor for overall processing speed.

The first few runs of a CPC2 BA solution have been performed on the new system already. The analysis is in progress and final results will be published hopefully in 1994.

Acknowledgements

We are grateful to all contributors to the CPC2 project. Especially we would like to thank C.A. Murray, W. Nicholson and M.J. Penston from RGO, Cambridge, as well as Chr. de Vegt from Hamburg Observatory.

REFERENCES

de Vegt, C. Murray, C.A., Zacharias, N., Nicholson, W., Penston, M.J. & Clube, S.V.M. (1993). A&AS, **97**, 985

Urban, S.E & Corbin, T.E. (1993) The U.S. Naval Observatory AC reduction project. (These proceedings)

Winter, L. (1994). PhD. thesis, Univ. of Hamburg

Zacharias, N. (1987). PhD. thesis, Univ. of Hamburg

Zacharias, N. (1992). A&A, **264**, 296

Zacharias, N. de Vegt,C. Nicholson, W. & Penston, M.J. (1992). A&A, **254**, 397

Improving proper motions with the Carlsberg Automatic Meridian Circle

By C. FABRICIUS[1], L. V. MORRISON[2]
AND L. HELMER[1]

[1]Copenhagen University Observatory, Brorfeldevej 23, Tølløse DK-4340, Denmark

[2]Royal Greenwich Observatory, Madingley Road, Cambridge CB3 0EZ, UK

During its first eight years of operation the Carlsberg Automatic Meridian Circle on La Palma has observed accurate positions of more than 100 000 stars. For more than 85 000 of these stars, proper motions have been derived using the Astrographic Catalogue, the PPM and other catalogues. The typical mean error is $0\rlap{.}''003$/yr. A presentation is given of the programmes observed until now and of future directions.

1. Introduction

The Carlsberg Automatic Meridian Circle (CAMC) has been operated by the University of Copenhagen, the Royal Greenwich Observatory and the Real Instituto y Observatorio de la Armada en San Fernando at the Roque de los Muchachos observatory on La Palma since May 1984. Each year 100 000 observations are carried out resulting in some 12 000 positions and magnitudes of programme stars and between 1000 and 2000 observations of solar system objects.

The stars are observed one at a time with a moving-slit photoelectric micrometer (Helmer et al. 1991). The observations are compiled into yearly catalogues – the Carlsberg Meridian Catalogues – of which six volumes covering 1984-1990 have already been published. The seventh catalogue (CMC7) covering January 1991 to August 1992 will be published in 1993. The slit micrometer was replaced in 1988 by a new slit micrometer with higher precision and a better limiting magnitude. The catalogues, therefore, fall into two groups: CMC1-4 observed with the first micrometer and CMC5- onwards observed with the new micrometer.

The limiting magnitude for CMC1-4 is about V=13. It has improved to about V=15 for CMC5-7. The standard deviation, night-to-night, for the observation of one transit in the zenith has improved from $0\rlap{.}''184$ in CMC1-4 to $0\rlap{.}''145$ in CMC5-6 and $0\rlap{.}''123$ in CMC7. These improvements are due to the introduction of the new micrometer and to improved reduction methods introduced in CMC7.

The published catalogue positions are mean values of typically six transits and the final errors are consequently much better. In CMC7 a 13 mag. star 30° from the zenith has a mean error of $0\rlap{.}''08$ in right ascension and $0\rlap{.}''09$ in declination (CMC7 1993). For fainter or lower stars the accuracy deteriorates. Programme stars are not observed south of −45° of declination. This limit has now been revised to −40°.

Observing programmes with the CAMC are undertaken for one of two main reasons: either to improve and extend the stellar reference system or to improve proper motions for a group of stars by adding accurate modern epoch observations. Of course, many stars are observed for both reasons and all stars observed serve both purposes.

2. The CAMC Reference System

Between 60 and 100 FK5 stars are observed each night, and all observations are reduced to the FK5 system. This differential reduction process only involves fitting a few parameters, and as a result the Carlsberg positions are not strictly on the FK5 system, but rather on a smoothed version of the FK5 where the short periodic errors in the FK5 are removed. The CAMC reference system includes the IRS stars, and it played an important role in the construction of the system for the PPM (North) catalogue (Röser 1990). Also a dense net of faint reference stars is observed which is well suited to the reduction of photographic plates. Figure 1 shows the distribution of the 34 035 stars in CMC1-7 which are fainter than V=10.5. More than half of these stars are in fact fainter than V=11.5. Though the average density is 1 star per square degree some areas are rather sparse. In the coming years the CAMC will provide a uniform net of 11th magnitude stars at a density of 1 star per square degree, filling in the remaining gaps.

3. Constructing CAMC proper motions

The Carlsberg catalogues have a considerable epoch difference to almost all major astrometric catalogues and it was obvious right from the beginning of the CAMC project, that a significant contribution could be made towards improving existing proper motions or providing new proper motions. Table 1 shows the catalogues that are searched for older observations of the CAMC stars and the percentage of the CAMC stars that were found in each particular catalogue. The first four CAMC catalogues have been combined into one column. It should be noted that the CPC2 was not available for CMC1 and only available south of $-40°$ for CMC2-5. This explains the steep rise in its usage. The heading "Meridian Catalogues" includes the Perth70, the catalogues made by the Carlsberg instrument at Brorfelde 1964-1983 and also catalogues from Bordeaux, San Fernando and Bucarest. Overlap between these catalogues have not been accounted for in the table, so the numbers given are only approximate.

Beginning with CMC5, PPM(North) has superseded AGK2/3. The construction of the PPM also involved a re-reduction of the northern zones of the Astrographic Catalogue (Röser and Bastian 1988). Through a cooperation with Dr Siegfried Röser from Astronomisches Rechen-Institüt in Heidelberg, it has been possible to take advange of this re-reduction for northern stars not already in the PPM. For plate centres from $-2°$ to $+31°$ all the x,y coordinates exist in keypunched form, but north of that the stars have to be looked up in the printed catalogues. The plate constants of Günther and Kox (1970, 1972) have been keypunched and are used for the identification. All x,y coordinates are then treated by Dr Röser in the same way as PPM stars. This has meant that nearly all (98%) of the northern stars in the Carlsberg catalogues now get a proper motion as compared to 82% before. For the southern sky, however, the fraction of stars getting a proper motion has steadily decreased and is now at 66%. The reason for this drop is the fact that the IRS programme and other programmes of bright stars constituted a large fraction of the early catalogues, but now most of the stars observed are quite faint.

In CMC7 the situation will improve through the inclusion of PPM(South) and the southern zones of the Astrographic Catalogue which have recently been keypunched at the Sternberg Institute in Moscow. For CMC1-4 neither PPM nor the northern zones of the Astrographic Catalogue were available. However, 25 656 positions from CMC1-4 were included in the construction of the northern PPM. So for stars in CMC1-4 labelled as AGK3 stars, improved proper motions may be found in the PPM. In the southern

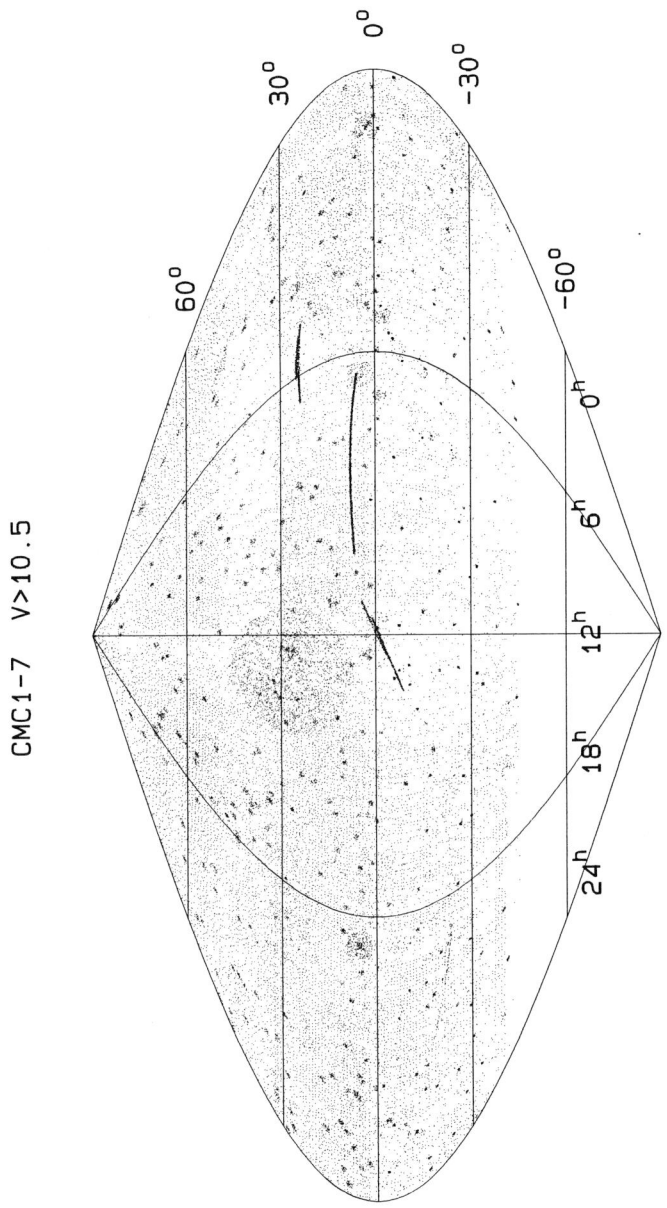

FIGURE 1. The distribution in the sky of the 34 035 stars in CMC1-7 fainter than V=10.5.

Catalogue	CMC1-4 %	CMC5 %	CMC6 %
Yale Zone Catalogues	52	23	17
AGK2/3	52	-	-
PPM North	-	44	33
Astrograph. Cat. North, non PPM	-	21	38
Astrograph. Cat. Cape zone	1	5	3
First Cape Photographic Catalogue	8	5	4
Second Cape Photographic Catalogue	2	6	18
General Catalogue	18	6	4
Meridian Catalogues (Perth, Brf, Bord)	29	11	9
Stars with proper motions in CMC	81	90	91
Northern CMC stars with proper motions	82	98	98
Southern CMC stars with proper motions	80	74	66

TABLE 1. The percentage of CMC stars found in each of the listed catalogues and the resulting percentage of CMC stars used to determine proper motions.

PPM about 20 000 positions from CMC1-5 were included and, again, improved proper motions can be found in the PPM for these stars.

The CMC proper motions were computed by fitting a least-squares line through the available positions, having assigned appropriate weights to each position and having transformed all positions to a common system. The details of this process are given in the printed introduction to each catalogue.

The mean error of a proper motion may be determined either from the residuals in the least-squares fit or from the weights given to each position. In CMC1-4 the residuals were used for all stars, whereas in CMC5 the residuals were used for non-PPM stars and weights for the PPM stars. In CMC6, and following volumes, only the weights are used. Using weights has the advantage of providing a mean error if only two positions are available, and the advantage of not occasionally giving spurious results. Figure 2 shows the distribution of proper motion mean errors in CMC6. The vast majority are better than $0''.003$/yr.

Figures 3 and 4 show the distribution of the 61 162 stars in CMC1-6 with respect to magnitude and spectral type having proper motions better than $0''.005$/yr. Starting with CMC7 many faint stars will be added. The distribution on the sky is fairly smooth, but with a higher density in the summer due to the meteorological conditions on La Palma.

Users of these proper motions should be aware that the Carlsberg catalogues are published in J2000.0 coordinates in accordance with the IAU (1976) System of Astronomical Constants. Because of the revision of the constant of precession and the elimination of equinox motion, there is a systematic difference between proper motions referred to the B1950.0 system (e.g. AGK3) and proper motions referred to the J2000.0 system (e.g. PPM or CMC) of up to $0''.005$/yr.

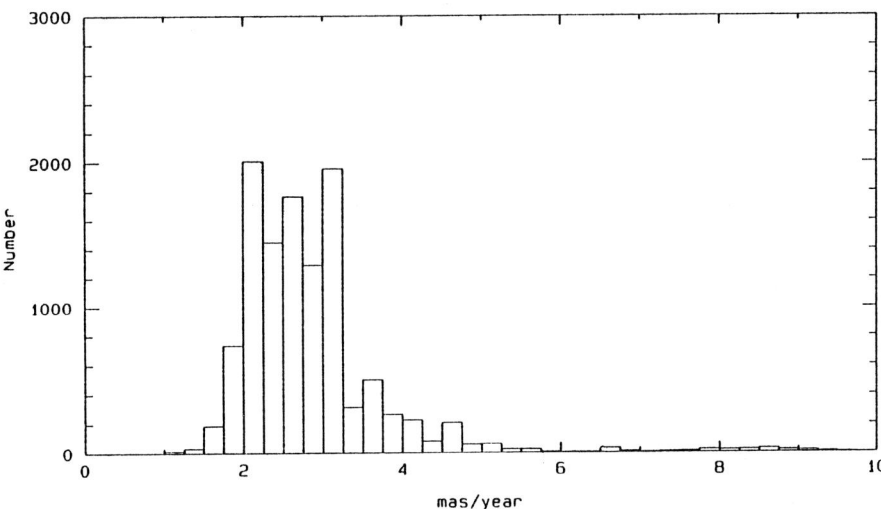

FIGURE 2. Distribution of the mean errors in proper motion for CMC6.

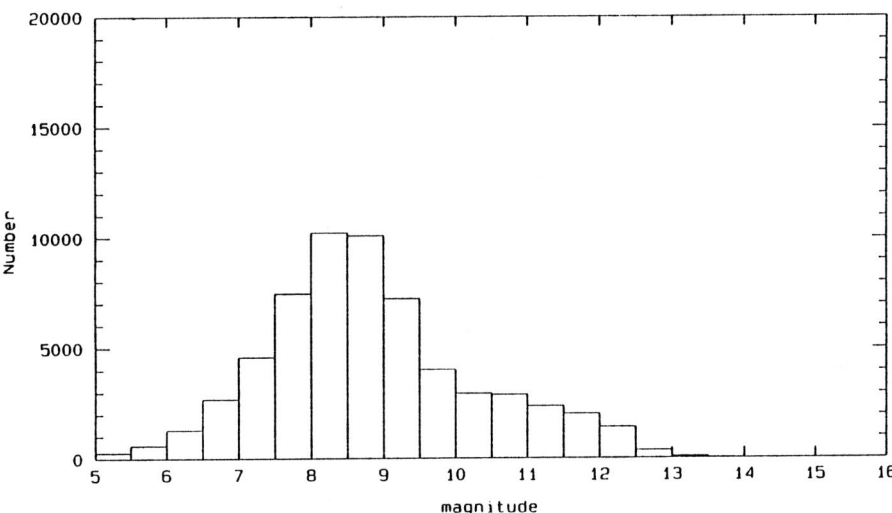

FIGURE 3. Distribution of stars by magnitude in CMC1-6. Only the 61 162 stars with proper motions better than 5 mas/yr are shown.

4. Observing programmes for Galactic kinematics

The present programmes on Galactic kinematics cover a wide range of projects. Table 2 lists the major programmes, their size and their present observational status.

A programme by J. Knude of 3000 A5-G0 stars in the North Galactic Pole area is 99% complete. A programme by D. Jones of 2000 red dwarfs from Stephenson's sample is now fully observed. These programmes are discussed separately at this meeting.

Two very large programmes of 10 000 F stars within 100 pc and of 9000 G stars (including the dwarfs within 30 pc) by E.H. Olsen are now fully observed. Photometry (uvbyβ) is nearly complete and radial velocities are available for the non-ZAMS and metal-weak F stars and for the G dwarfs (E.H. Olsen 1993). A subset of these stars has

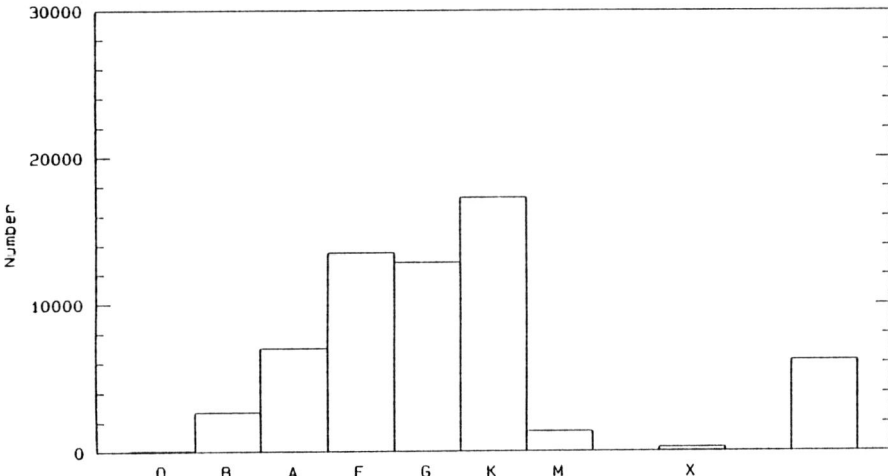

FIGURE 4. Distribution of stars by spectral type in CMC1-6. Only the 61 162 stars with proper motions better than 5 mas/yr are shown. "X" means other types than O...M. The last column represents unknown type.

No.	%	Programme
5000	86	Nearby O-B associations.
3000	99	A5-G0 North Galactic Pole.
2000	100	K&M dwarfs.
1200	100	High-velocity, metal-weak.
1500	100	Pop. II.
10000	100	F stars $r < 100$ pc.
9000	100	G dwarfs $r < 30$ pc., G giants.
400	89	Hot subdwarfs. White dwarfs.
3000	72	Gliese nearby stars.
900	99	Halo and disk dwarfs.
600	98	Galactic pole binaries.
260	27	RR Lyr

TABLE 2. The major CAMC observing programmes in Galactic kinematics. The table shows the number of stars and the fraction of each programme that was fully observed by June 1993.

recently been used in a detailed discussion of the chemical evolution of the Galactic disk (Edvardsson et al. 1993).

Also worth mentioning is a programme of 400 hot subdwarfs and white dwarfs (P. Thejll), a list of 1500 population II stars (A. Ardeberg), several open clusters, Gliese's nearby stars and a programme by Fresneau of 900 selected dwarfs. Except for the Gliese stars all these programmes are very near completion.

5. Future programmes and development

In the coming years we will see the completion of the HIPPARCOS and TYCHO catalogues. For a couple of decades HIPPARCOS will be an ideal standard catalogue for meridian work, replacing the FK5. The meridian circles observe with accuracies only slightly inferior to those of TYCHO, but for one or two decades observing large programmes brighter than the TYCHO limit, $V \sim 11$, will make little sense. The development of CCD micrometers for meridian circles (e.g. Stone 1993), however, opens up new possibilities. It is realistic to make a pole-to-pole coverage of the sky to a limiting magnitude of 17 using meridian circles. If the Palomar sky survey could be reduced to the HIPPARCOS system with an accuracy of $0\rlap{.}''25$ at its ~ 1950 epoch, proper motions could be obtained north of declination $-30°$ with an accuracy of $0\rlap{.}''006$/yr.

REFERENCES

Carlsberg Meridian Catalogue, La Palma, No. 1, 1985, Copenhagen University Observatory, Royal Greenwich Observatory, Instituto y Observatorio de Marina

Carlsberg Meridian Catalogue, La Palma, No. 2, 1986

Carlsberg Meridian Catalogue, La Palma, No. 3, 1987

Carlsberg Meridian Catalogue, La Palma, No. 4, 1989

Carlsberg Meridian Catalogue, La Palma, No. 5, 1991

Carlsberg Meridian Catalogue, La Palma, No. 6, 1992

Carlsberg Meridian Catalogue, La Palma, No. 7, 1993 (in press)

Edvardsson B., Andersen J., Gustafsson B., Lambert D.L., Nissen P.E. & Tomkin J. 1993, AA (in press)

Günther A. & Kox H. 1970, A&AS, 3, 85

Günther A. & Kox H. 1972, A&AS, 6, 201

Helmer L., Fabricius C. & Morrison L.V. 1991, Experimental Astronomy, 2, 85

Olsen E.H. 1993, private communication

Röser S. 1990, in: IAU Symp 141, Inertial coordinate system on the sky, eds. J.H. Lieske & V.K. Abalakin, Kluwer, Dordrecht, p.469

Röser S., Bastian U. 1988, A&AS, 74, 449

Stone R.C. 1993, IAU Symp 156, Developments in astrometry and their impact on astrophysics and geodynamics, eds. I.I. Mueller & B. Kolaczek, Kluwer, Dordrecht, 65.

Ground-based parallaxes for nearby stars

By H. JAHREISS

Astronomisches Rechen-Institut, Mönchhofstraße 12-14, D-6900 Heidelberg 1, Germany

Trigonometric parallax measurements constitute the basic data for nearby stars. The distance limits of the various catalogs of nearby stars published in the past were chosen in such a way that the trigonometric parallaxes could provide significant values. Whilst in the first half of our century mostly bright stars were selected for observing programmes, during recent decades trigonometric parallax work concentrated more and more on presumably nearby candidates as originating, for example, from proper motion surveys or spectroscopic surveys. Simultaneously, the accuracy of the measurements could be considerably improved. So, that at present a large amount of high precision parallax data for nearby white and red dwarf stars is available.

Nevertheless, the output of ground-based trigonometric parallaxes cannot compete with the output of other data. Photometric parallaxes become of growing importance. Their determination is far less time-consuming, and in the present edition of the *Catalogue of Nearby Stars* (CNS3) almost 50 per cent of the distances are based on photometric measurements.

HIPPARCOS parallaxes will become available very soon. Nevertheless, there will still be a need for ground-based trigonometric parallaxes during the next decades – at least for fainter stars.

1. Introduction

Having a look in the astronomical literature for publications on nearby stars, one can recognize that the term *nearby* may be applied by different authors with quite different meanings. Today *nearby* may even be applied to stars out to several hundred parsec. Already in the early 1920s three papers on nearby stars were published with quite different distance limits: E. Hertzsprung (1922) with $r_{lim} = 5$ pc, W. Luyten (1923) with $r_{lim} = 10$ pc, and J. Haas (1923) with $r_{lim} = 15$ pc. Later on it was after all Peter van de Kamp (1953) who updated very regularly the list of stars within 5 pc.

Then in early 1950, Wilhelm Gliese, stimulated by van de Kamp, started his work on nearby stars. In view of the fact that the average parallax error of the then recently published *General Catalog of Trigonometric Stellar Parallaxes* (Jenkins, 1952) was estimated by E. Hertzsprung (1952) to be $\pm 0\overset{''}{.}016$, he came to the conclusion that a distance of 20 pc would still be appropriate to get fairly realistic parallaxes.

The first catalogue CNS1 (Gliese, 1957) contained 915 systems with altogether 1094 components within 20 pc. Table 1 lists also the content of several following compilations of nearby stars until the most recent CNS3 (preliminary version). The increasing number of objects reflects the considerable progress achieved since 1957. Within 20 pc the number of known stars has practically doubled during the past 35 years. Table 1 also shows that this increase of newly detected nearby stars is even steeper for the last decade. Nevertheless, we still know only 50 per cent. of the stars predicted by a commonly accepted stellar luminosity function for the solar neighbourhood.

The second edition CNS2 (Gliese, 1969) was extended to 22 pc, and the recently released preliminary version of the CNS3 (Gliese & Jahreiß, 1991) intends to list all known stars within 25 pc. It contains 3264 systems with altogether 3803 components. Yet, we estimate that only 3120 of these stars are really nearer than the catalogue limit. Some properties of the CNS3 were already described in Jahreiß & Gliese (1993).

reference	$\pi \geq 0\rlap{.}''050$	$\pi \geq 0\rlap{.}''045$	$\pi \geq 0\rlap{.}''040$
Gliese (1957)	1094		
Gliese (1969)	1277	1585	
Woolley et al. (1970)			2150
Gliese & Jahreiß (1979)	1476	1848	
Gliese & Jahreiß (1991)	2031	2570	3120

TABLE 1. Number of objects in different catalogues of nearby stars.

source	N	$\pi_{91} - \pi_{old}$	$\Delta\pi_{old}$	$\Delta\pi_{91}$
CNS1 (1957)	816	+3.0	11.5	6.3
CNS2 (1969)	1325	−0.3	8.7	7.3
GJ-stars (1979)	180	−0.4	4.3	3.9
new (r < 25 pc)	264			5.2

TABLE 2. Average error of trigonometric parallaxes in mas.

2. Trigonometric parallaxes

The basic distance parameter is the trigonometric parallax. Most of the 806 CNS1 parallaxes came from the GCTSP (Jenkins, 1952) containing 8 853 trigonometric parallax measurements for 5 822 stars. But due to Schlesinger's programme, that covered all stars brighter than 5.5 mag, except stars of spectral class earlier than A and late-type giants, the GCTSP contains too many stars with too small parallaxes: $\pi_{med} = +0\rlap{.}''018$ compared with an average error of $\pm 0\rlap{.}''016$. All parallaxes were transformed to the system of the Allegheny Observatory. A procedure which was later on critized by several authors (see Strand, 1963) since it seemed to be not justified and required negative systematic corrections for practically all other parallax series.

With respect to these findings the observatory corrections were removed in the second edition CNS2 (Gliese, 1969), and the new parallaxes were calculated according to the precepts proposed by Strand (1963).

The differences in the parallax system between CNS1 and CNS2 can be seen in Figure 1a. The original 806 parallaxes used by Gliese in 1957 are no longer available in digitized form. Therefore, GCTSP parallaxes from the 1961 version were taken. This increased the number of parallax stars to 816, but, certainly this does not affect the results presented here.

One can recognized that there are several distinct offsets from zero up to $+0\rlap{.}''006$, almost independent of M_V. This reflects very nicely the systematic corrections applied to the GCTSP parallaxes. The first line in Table 2 summarizes these corrections to a systematic difference of $+0\rlap{.}''003$ between the GCTSP (1952, 1961) and the YPC (1991). Apart from these systematic effects, Figure 1a shows almost no scatter. In other words, only very few new trigonometric parallaxes were determined for CNS1 stars during the Sixties.

In the early Seventies the first results of the USNO astrometric reflector – installed at Flagstaff – became available and also other observatories concentrated their programmes on nearby candidates. So that within only ten years 180 new trigonometric parallaxes larger than $0\rlap{.}''045$ were published (Gliese and Jahreiß, 1979). These new parallaxes were determined with considerably increased accuracy. The average mean error for the CNS1

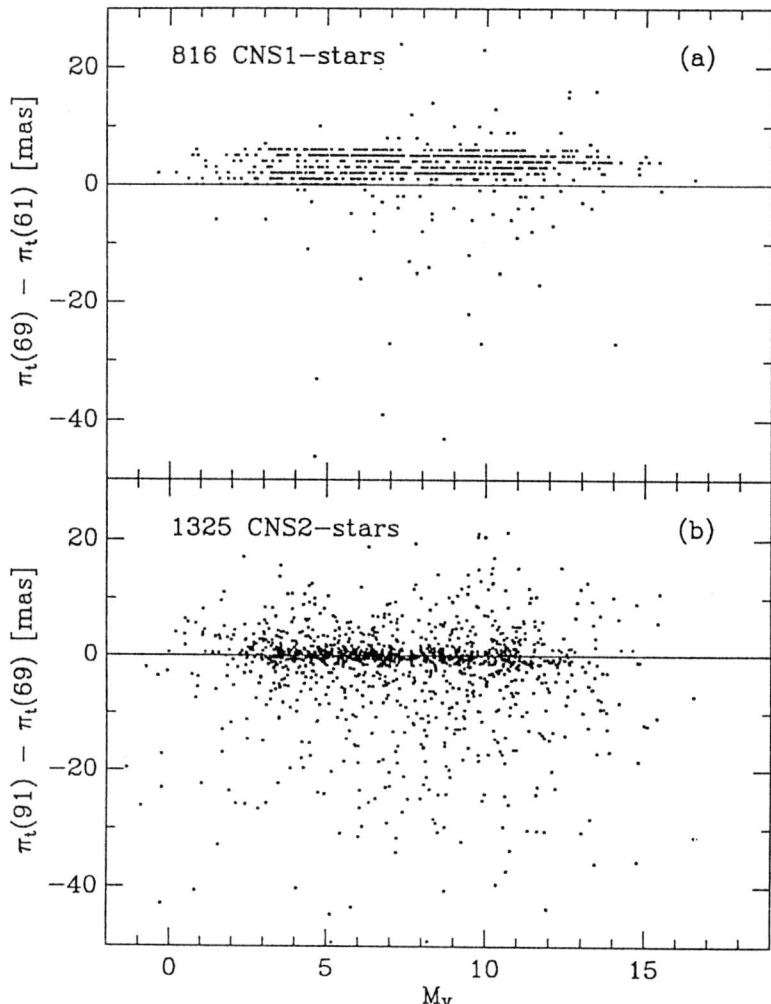

FIGURE 1. Differences of new and old trigonometric parallaxes as a function of absolute visual magnitude M_V (CNS3). (a) CNS2 - CNS1 parallaxes for 816 CNS1 stars; $\pi_t(61)$ from GCTSP and its supplement (1961). (b) CNS3 - CNS2 for 1325 CNS2 stars; $\pi_t(69)$ as listed in CNS2 and $\pi_t(91)$ from YPC (1991).

stars is 11.5 mas, and for the CNS2 stars 8.7 mas – already indicating an improvement in the later measurements. It has dropped to only 4.3 mas now (see column 4 in Table 2.) Weighted means were calculated throughout Table 2 to reduce the influence of outliers on the results.

The new *General Catalog of Trigonometric Stellar Parallaxes* YPC (van Altena et al. 1992) – containing 13 936 measurements for 7 874 stars – constitutes the principal source of CNS3 parallaxes. Figure 1b shows the difference between CNS2 and CNS3 for 1325 stars. There are no systematic differences. (Table 2 lists an average difference of −0.3 mas). In other words, the YPC introduces no new system. Convincing arguments for an omission of systematic corrections are presented by Professor van Altena in the following paper.

On the other hand, a significant scatter is present in Figure 1b with a slight tendency to negative values. This is very probably due to the fact that in a volume-limited sample, stars with positive parallax errors are over-represented, i.e. on the average the parallaxes are too large. This effect is more pronounced for less accurate measurements, so, that new parallaxes which are of higher accuracy should correct for it. The last column of Table 2 shows an average error of 5.2 mas for the 264 newly added stars within 25 pc. But, it is even more important that 50 per cent. of these stars are fainter than $m_V = 12\overset{m}{.}45$, which is about the magnitude limit of the HIPPARCOS satellite.

3. Photometric and spectroscopic parallaxes

Spectroscopic parallaxes were already taken into account in 1957 for the CNS1 in order *to complete the material.* Thus 109 stars without trigonometric parallaxes were added, and for another 161 stars trigonometric and spectroscopic parallaxes were combined in a resulting parallax with the intention to improve their distances.

In the course of the compilation of the CNS2 in 1969 also broad-band photometry UBV RI was also added where it was available. Colour-luminosity calibrations carried out by Gliese (1971) then allowed the determination of additional photometric parallaxes. Thus again 121 stars with unknown π_t could be included, and another 328 trigonometric parallaxes were *improved* with the help of spectroscopic and/or photometric parallaxes.

During recent years a lot of photometric observations have been carried out on high proper motion surveys or spectroscopic surveys of red dwarf stars. This resulted in an immense increase in *non-trigonometric* nearby stars. In the present version of CNS3 1165, stars were included with photometric or spectroscopic distance estimates, and for an additional 734 stars the photometric parallax was found to be superior to the trigonometric one.

The contribution of spectroscopic or photometric parallaxes to the compilations of nearby stars rose from 30 per cent. in 1957 to 40 per cent. in 1969, and to 50 per cent. in 1991. The importance for the completeness of the CNS3 is demonstrated in Figure 2. If only trigonometric parallaxes had been considered (dotted histogram), the number of bright stars $M_V < 5^m$ would be over-estimated by 25 per cent. This is due to the selection effect at the catalogue limit where the positive parallax errors dominate. This effect is strenghtened by the poor parallaxes of the bright nearby stars. Figure 2 shows that the contribution of the photometric parallaxes reaches its maximum at $M_V = 12^m$ with almost 60 per cent. Whereas, the distances of the very faint stars $M_V > 15^m$ rely almost exclusively on trigonometric parallaxes.

There may be two arguments against the inclusion of photometric or spectroscopic parallaxes. First, the distance modulus may be corrupted by the presence of undetected companions. This can be circumvented only by carrying out systematic searches for unseen companions with radial velocity work or other means. Second, the broad-band colours allow only a calibration of the *mean* main sequence for red dwarf stars. In so doing, the cosmic dispersion which is of the order of $0\overset{m}{.}5$ is suppressed. Nonetheless, additional spectroscopic or photometric means are required to study the fine structure of the lower main sequence in more detail. Therefore, certain sub-samples of the CNS3 can at least be regarded as pre-selection of promising candidates for further observation.

4. Conclusions

The average error of the parallax of a nearby star decreased from 14 mas in 1957 to 11 mas in 1969, and to 8 mas in 1991. Very soon HIPPARCOS will supply us with trigonometric parallaxes having errors in the order of 2 mas. Not in the HIPPARCOS *Input*

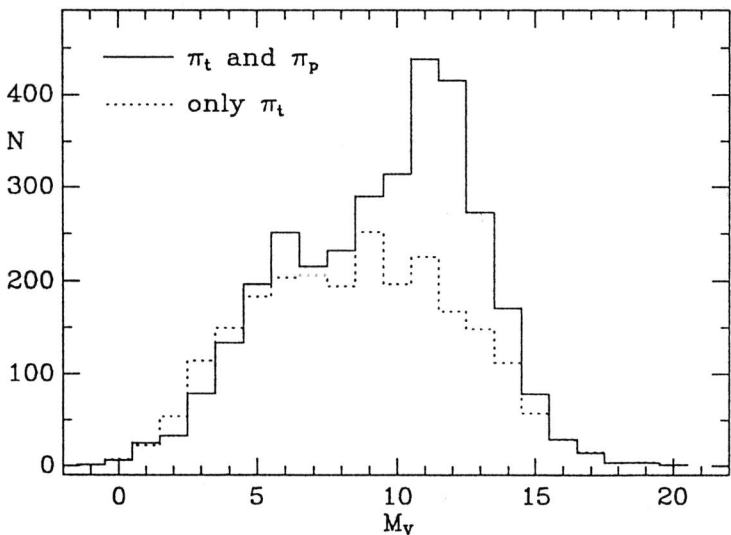

FIGURE 2. Number of stars within 25 pc as a function of absolute visual magnitude M_V. Full line: trigonometric and photometric or spectroscopic parallaxes are included. Dotted line: only trigonometric parallaxes are included.

Catalogue are only 15 per cent of the stars in our working file accessible to HIPPARCOS; i.e. with visual magnitudes brighter than about $12^{\mathrm{m}}\!.5$.

Table 3 summarizes the amount of trigonometric parallaxes published by the principal observatories since the very beginning of photographic parallax work early this century. In view of HIPPARCOS it presents the almost final result of a long chapter of astronomical observations. Indeed, it can be seen, that many important parallax series were terminated in the past and some others will very probably terminate in the near future.

There is a necessary trend towards CCD parallaxes which allows one to observe more fainter stars. But, there also is a trend to very small parallax series making it very difficult, if not impossible to study systematic effects. And, there still exists a north-south discrepancy. Already in 1923, Schnauder (1923) stated that *87 per cent of all parallaxes relate to stars in the northern hemisphere*. Hopefully, things will improve soon, and we must not wait until ROEMER (Høg, these proceedings) provides us in the next century with the predicted high accuracy astrometric and photometric measurements of all stars down to $m_V = 17^{\mathrm{m}}\!.5$.

REFERENCES

Gliese, W. (1957) Katalog der Sterne näher als 20 Parsek für 1950.0, Mitteilungen des Astronomischen Rechen-Instituts Serie A Nr. 8

Gliese, W. (1969) Catalogue of Nearby Stars. Edition 1969, Veröffentlichungen des Astronomischen Rechen-Instituts Heidelberg Nr. 22

Gliese, W. (1971) Relations between spectral types, colours, and absolute magnitudes of nearby main-sequence stars. Veröffentlichungen des Astronomischen Rechen-Instituts Heidelberg Nr. 24

Gliese, W. and Jahreiβ, H. (1979) Nearby Star Data Published 1969–1978, A&AS **38**, p. 423

Haas, J. (1923) Die nächsten Fixsterne, Veröffentlichungen der Universitätssternwarte Berlin-Babelsberg, Band **3**, Heft 3

Observatory	1950	1962	1968	1978	1993
Allegheny	1642	242	130	53	20:
Cape	1404	428		4	83
Greenwich	759	1		51	28
Lick				109	16
McCormick	1669	230	60	79	91
McC. South.					28
Mt. Wilson	470	1	7		
Sproul	343	50	155	151	737
US Naval Obs.			1	497	411
US Naval CCD					72
Van Vleck	187	54	13	269	122
Yale	1734	295	56	46	13
Yerkes	364	24		76	34
other	295	37	4	30	30:
sum	8853	1362	426	1365	1685

TABLE 3. Sources of trigonometric parallaxes.

Hertzsprung, E. (1922) List of stars nearer than 5 parsecs, BAN **1**, p. 21

Hertzsprung, E. (1952) Accuracy of Yale parallaxes, Obs. **72**, p. 242

Jahreiß, H. and Gliese, W. (1993) The Third Catalogue of Nearby Stars: Completeness and Stellar Kinematics. in Proceedings *Workshop on Databases for Galactic Structure*, Swarthmore, Penn. eds. A. G. D. Philip, A. R. Upgren, L. Davis Press, in press

Jenkins, L. F. (1952) General Catalogue of Trigonometric Stellar Parallaxes, Yale Univ. Obs., New Haven, Conn.

Luyten, W. (1923) A study of the near-by stars. Annals of the Harvard College Observatory Vol. **85**, No. 5. p. 73

Schnauder, G. (1923) Ergebnisse und Stand der Parallaxenforschung, in Ergebn. der exakten Naturwissenschaften **2**, p. 19

Strand, K. Aa. (1963) Trigonometric stellar parallaxes, in Stars and Stellar Systems III. Basic Astronomical Data. ed K. Aa. Strand, Univ. of Chicago Press p. 55

van Altena, W. F., Lee, J. T., and Hoffleit, D. (1992) General Catalogue of Trigonometric Stellar Parallaxes. Yale Univ. Obs., in preparation

van de Kamp, P. (1953) PASP, **65**, p. 73

Woolley R., Epps, E. A., Penston, M. J. and Pocock, S. B. (1970) Catalogue of Stars within twenty-five parsec of the Sun. Royal Obs. Annals No. 5.

The new edition of the Yale Parallax Catalogue:
A look at systematic errors

By W. F. VAN ALTENA, J. T. LEE AND E. D. HOFFLEIT

Yale University Observatory, P.O.Box 6666, New Haven, CT 06511, U.S.A

The new edition of the General Catalogue of Trigonometric Parallaxes contains 15 430 determinations of parallaxes for 7888 stars. Auxiliary data including cross-identifications, positions, photometry, spectral types and luminosity classes, and data on the duplicity and variability of the stars are given for each star. In addition, a reference to the publication for each parallax is given so that users can examine the original paper to extract further information. In this paper we describe the compilation of the data and the comparison of parallaxes with independent determinations so that we could investigate the possible existence of systematic errors in the YPC.

1. Introduction

In earlier publications (van Altena, Lee and Hoffleit, 1988, 1991a) we described our work leading up to the publication of the CD-ROM "The General Catalogue of Trigonometric Parallaxes, Preliminary Version, 1991" (van Altena, Lee and Hoffleit 1991b). In this paper we will briefly review that work and examine the possible existence of systematic errors in the General Catalogue of Trigonometric Parallaxes, Fourth Edition (van Altena, Lee and Hoffleit, 1993b). The weighting system for the parallaxes determined at various observatories is discussed in van Altena, Lee and Hoffleit (1993a).

In the previous edition of the YPC by Jenkins (1952, 1963), systematic corrections were derived between the various observatories and those corrections were applied before calculating the weighted mean parallaxes. However, both Schilt (1954) and Strand (1963) were of the opinion that the corrections were poorly established and not due to true systematic errors; they therefore recommended against their use. Much later, Lutz, et al. (1981) came to the same conclusions using more extensive analyses and similarly recommended against the use of the corrections where the physical cause of the differences could not be identified. In this paper we will discuss our analyses, and those of others, to see what evidence exists for systematic errors in the parallaxes.

2. Data Bank

The data bank for the new edition of the YPC was generated by three independent literature searches for publications listing trigonometric parallaxes resulting in a bibliography of approximately 500 citations. From these publications we extracted the relative parallax and proper motion, right ascension and declination, mean magnitude of the reference stars, reduced magnitude of the parallax star, cross-identifications, data on the star about its duplicity, variability and other interesting characteristics. Using the positions and cross-identifications, we then searched about 50 catalogues for improved positions and proper motions, UBV photoelectric photometry, and MK Spectral Types and Luminosity Classes. Finally, additional catalogues were searched to improve the completeness of our cross-identifications. Quality assurance has been enforced by proofreading, computer searches for obvious blunders and making pre-publication versions of the YPC available to the astronomical community for its use, and help in detecting errors.

3. Evaluating systematic differences in the YPC

3.1. *Correction from Relative to Absolute Parallax*

Some of the stars in the YPC have more than one measurement; it was therefore necessary to consider the possibility of systematic differences occurring between the measurements before combining the data into a most probable value for the parallax. However, since the mean magnitude of the reference stars varies between observatories and also evolves with time at each observatory as the individual program goals change, it was first necessary to correct each relative parallax to absolute. For that purpose, advantage was taken of the renewed interest in galactic modeling and new corrections to absolute were calculated for each star using a three component model consisting of a disk, thick disk and a spheroid that represented the best estimates at that time for the distribution of the constituents of our galaxy. The corrections to absolute parallax computed with that model are discussed by van Altena, Lee, Hanson and Lutz (1988) and will be published in the near future.

3.2. *Caveats for the interpretation of the systematic differences*

In the following section we will discuss the mean differences between the parallaxes determined at different observatories, or with respect to Spectro-photometric parallaxes, and refer to them as systematic differences, or systematic errors. In most cases, the means are not statistically significant due to limited sample sizes but in others the mean differences are found to be significant. In the later cases, the main issue is whether on not the mean difference is due to some physical cause or is due only to selection effects or biases in the data samples. These issues have been discussed in detail by Hanson and Lutz (1983) so we will only briefly summarize some of the points here.

The objectives and limits of the various major parallax surveys have been extracted from the literature and are summarized by Hanson (1980) and Lutz, et al. (1981). The original observational programs were modified at times during the programs and it is possible to document some of the changes, but others are not well defined. A few representative problems will be adequate to illustrate potential problems in the interpretation of the derived differences. Allegheny and Yale were both directed initially by Schlesinger and the programs defined by him to be surveys of the stars brighter than 5.5, with some spectral type limitations. However the stars common to both programs are of necessity confined to the equatorial band due to the geographic location of Allegheny in the USA and Yale in South Africa; those stars will have been observed at moderately large zenith distances in opposite directions. Cape and Yale were both located in South Africa, but Cape concentrated on fainter high proper motion stars. The number of stars common to the latter two programs was later expanded by the addition of some fainter stars to Yale and brighter stars to Cape, but it is possible that the stars added were selected with knowledge of the other observatory's previously determined parallax. That procedure could easily introduce a systematic difference that would be present only in the stars common to the two programs but not in the balance of the stars. A third example for the future might be investigations relative to the highly precise HIPPARCOS Astrometric Satellite parallaxes. The stars in common to the two catalogues will be limited primarily to the stars brighter than 8th magnitude, with few fainter stars because of the small aperture of HIPPARCOS. The assumption that differences determined from the brighter stars are also applicable to the fainter stars in the YPC has questionable validity. Finally, due to the heterogeneous nature of the data in the YPC, any global analysis attempting to characterize the overall systematic difference of the YPC with respect to some other standard should be looked at with great caution.

3.3. *The Observatory Pairs Method*

Since there are no absolute standards with which to compare the parallaxes, we must examine different methods of evaluating systematic differences and look for consensus among the methods. For this new edition, we have used two independent methods to evaluate them: the Observatory Pairs method (Vasilevskis 1966); and the Spectro-photometric parallax method. We have already discussed the first method in previous publications (see for example: van Altena 1986) and refer here to Figure 1 in that publication. Hanson (1980) had previously identified a magnitude effect in the Allegheny parallaxes that is apparent in that figure for stars brighter than about magnitude 6.5. Based on corroborative data from K-line parallaxes, Spectro-photometric parallaxes, cluster parallaxes and the Observatory Pair data, a mean magnitude dependent correction was defined and applied to the Allegheny parallaxes. A similar, but more poorly defined correction has also been applied to the Dearborn Observatory parallaxes. Analyses repeated with more data for this paper show that the mean differences for many observatories are functions of the magnitude and the number of plates used in the parallax determination.

3.4. *The Spectro-photometric Parallax Method*

For the method of Spectro-photometric parallaxes, we rely on the fact that for distant stars the value of the derived parallax is much more precisely known than the trigonometric parallax for the same star. The difference between the two parallaxes for the same star is dominated by the error in the trigonometric parallax and it is possible to compare, 1) the derived external error with the published internal error, and 2) the derived systematic difference with respect to the presumed correct zero-point of the Spectro-photometric parallaxes. Since the Spectro-photometric parallaxes were calibrated using trigonometric parallaxes, it is necessary to consider the possibility of circularity in our comparisons. However, since the Spectro-photometric parallaxes were calibrated using only large parallaxes and we are restricting ourselves to small parallaxes (see the next paragraph) there is little danger of circularity in our analyses. Using the YPC data bank, with the assistance of G. K. Torrington, we derived systematic difference estimates for all observatories with sufficient numbers of stars with UBV photometry and MK Spectral Types and Luminosity Classes. Solutions were analyzed using Luminosity Class III, III with various subclasses such as IIIa and IIIb, and V. An insufficient number of stars with classes I, II, and IV were availabe to obtain reliable results. The individual solutions yielded similar results, therefore all class III and subtypes, and class V stars were analyzed together in one solution.

Each star was first corrected for interstellar reddening using the intrinsic color and luminosity calibrations (UBV: Schmidt-Kaler 1982; VRI: Weis 1986, 1987, 1988, and Upgren, private communication) for the MK Spectral Type and Luminosity Class and the UBVRI photometry. The trigonometric and photometric parallaxes were intercompared and Probability Plots used to estimate the dispersion of the differences and the systematic difference for each observatory or observatory subset. By restricting the photometric parallaxes to some maximum value we are able to limit the contribution of the error in the photometric parallax to a negligible amount consistent with maintaining a reasonable number of stars in the sample. For most cases that limit was set at $0''.015$, which for a cosmic dispersion of 0.6 mag. in the luminosities at a given spectral type results in an error in the derived photometric parallax of $0''.004$ which is small in the context of the average trigonometric parallax error of $0''.015$. The derived dispersion of the differences was then corrected for the inferred error of the photometric parallaxes to yield an estimate for the external error of the parallax for the sample. (For the cases of the newer parallaxes with internal errors comparable to $0''.004$, there are generally too few MK Spectral Types and Luminosity Classes, the existing calibrations are too poorly defined, or the calibration stars dominate the sample and hence it is not possible to derive reliable external errors or systematic differences by this approach.) The results of that analysis of the errors are reported by van Altena, Lee and Hoffleit (1993a).

An examination of the Zero-points in Table 6 of Hanson and Lutz (1983) for stars with K-line spectroscopic parallaxes shows that the Zero-points change with the adopted maximum spectroscopic parallax limit. We have found a similar result for our investigations using Spectro-photometric parallaxes. Hanson and Lutz (1983) suggest that the bias is due to selection effects in the stars used in the analysis, but whatever the cause it indicates that considerable care must be taken in interpreting the results from investigations of this type. To say that they are due to systematic errors in the system of an observatory's parallaxes would be a gross simplification of the situation.

4. Summary and Recommendations

There appear to be well documented cases for systematic differences as a function of magnitude for the Allegheny and Dearborn parallaxes and corrections for those effects have been applied in the new edition of the YPC. Aside from that, no other systematic differences have been applied to the parallaxes. We believe that the other differences found in our investigations, and those of others, are due to a variety of selection effects that are probably variable, too poorly defined, and therefore not amenable to correction. Possible candidates for physical causes for systematic differences might be: special techniques used in the observations and reductions; differential atmospheric refraction; and thermal effects on the telescope lens between the evening and morning observations, which might introduce magnitude or chromatic aberration changes. We do find that the systematic differences are in some cases a function of magnitude, the number of plates used in the parallax series, the technique used to derive the systematic differences, and the sample of stars used in the analysis.

We recommend that since the systematic differences are a function of many quantities, including the specific sample, that they should not be applied to stars outside the sample. The analysis of a sample of weighted mean parallaxes from the YPC is unlikely to yield meaningful results due to the heterogeneous nature of the data within the YPC. An analysis of bright stars contained in the YPC is unlikely to shed any light on the nature, distribution, errors, or systematic differences of faint stars in the YPC.

Finally, if you are concerned about the potential of a zero-point error in the parallaxes used in your investigation, then we recommend that you use: 1) modern high quality parallaxes, which do not seem to show any discernible systematic differences; or 2) wait for the HIPPARCOS parallaxes!

Acknowledgements

We would like to thank Mr G.K. Torrington for his dedicated assistance in carrying out most of the calculations related to the determination of the systematic differences derived from the Spectro-photometric parallaxes. This research has been supported in part by grants from the National Science Foundation.

REFERENCES

Hanson, R. B. 1980. MNRAS, **192**, 347.

Hanson, R. B. & Lutz, T. E. 1983. MNRAS, **202**, 201.

Hertzsprung, E. 1952. Observatory, **72**, 242.

Jenkins, L. F. 1952. General Catalogue of Trigonometric Stellar Parallaxes, Yale University Observatory, New Haven.

Jenkins, L. F. 1963. Supplement to the General Catalogue of Trigonometric Stellar Parallaxes, Yale University Observatory, New Haven.

Lutz, T. E., Hanson, R. B., Marcus, A. H. & Nicholson, W. L. 1981. MNRAS, **197**, 393.

Schilt, J. 1954. AJ, **59**, 55.

Schmidt-Kaler, Th. 1982. In Landolt-Börnstein New Series Group VI: Astronomy, Astrophysics and Space Research, **2b**, ed. K. Schaifers & H.H. Voigt, (Spring-Verlag, Berlin, Heidelberg, New York).

Strand, K. Aa. 1963. In Basic Astronomical Data, ed. K. Aa. Strand (Univ. Chicago Press, Chicago) 55.

van Altena, W. F. 1986. In IAU Symp. 109, Astrometric Techniques, ed. H. K. Eichhorn & R. J. Leacock (Reidel, Dordrecht) 183.

van Altena, W. F. & Lee, J. T. 1988. In IAU Symp. 133, Mapping the Sky, ed. S. Debarbat et al., (Reidel, Dordrecht) 269.

van Altena, W. F., Lee, J. T., Hanson, R. B. & Lutz, T. E. 1988. In Calibration of Stellar Ages, ed. A. G. D. Philip (L. Davis Press, Schenectady) 175.

van Altena, W. F., Lee, J. T. & Hoffleit, E. D. 1991a. abstract of a talk given at the XXIst General Assembly of the IAU, in *Transactions of the IAU XXIB, Buenos Aires 1991*, ed. J. Bergeron (Dordrecht, Kluwer) 235.

van Altena, W. F., Lee, J. T. & Hoffleit, E. D. 1991b. The General Catalogue of Trigonometric Parallaxes, Preliminary Version, 1991 in *Astronomical Data Center CD-ROM Selected Astronomical Catalogs, Volume I*, ed. L. E. Brotzman, S. E., Gessner, J. M. Mead & M. E. Van Steenberg (Goddard Space Flight Center, Greenbelt).

van Altena, W. F., Lee, J. T. & Hoffleit, E. D. 1993a. In Workshop on Databases for Galactic Structure, ed. A. G. Davis Philip & B. Hauck (L. Davis Press, Schenectady) (in press).

van Altena, W. F., Lee, J. T.& Hoffleit, E. D. 1993b. The General Catalogue of Trigonometric Parallaxes, Fourth Edition, (Yale University Observatory, New Haven).

Vasilevskis, S. 1966. ARA&A, **4**, 57.

Weis, E. W. 1986. AJ, **91**, 626.

Weis, E. W. 1987. AJ, **93**, 451.

Weis, E. W. 1988. AJ, **96**, 1710.

USNO parallax program: directions, results and applications

By C. C. DAHN

U.S. Naval Observatory, PO Box 1149, Flagstaff, Arizona 86002, U.S.A.

The USNO parallax program has measured trigonometric parallaxes for over 1100 stars, primarily using photography. The transition to CCD detectors began with a TI800 chip in the mid-1980s and now continues with a Tek2048 detector. Further progress toward delineating the dwarf, subdwarf and degenerate sequences is illustrated here, with particular emphasis on new results for the sdF–G stars. The relationship between these stars and the extreme late-type subdwarfs is slowly being clarified by combining recent astrometric work with new spectrophotometry interpreted through new model atmosphere calculations.

The USNO 61–inch Astrometric Reflector began operation at the Flagstaff Station in March 1964. For the first 20 years, the parallax program used only photography and generally targeted stars in the somewhat limited brightness range $10 < V < 16$. Program extensions were initiated in the mid-1980s to permit observations of (1) bright stars ($V < 10$) by using neutral-density, magnitude-compensating filters (Dahn *et al*, 1988), and (2) very faint stars (generally $16 < V < 20$) by using a TI800 CCD detector (Monet *et al*, 1992).

The quality of the USNO photographic parallax results has improved significantly over the years due to several factors. First, in 1970 the program began using optically flat filters mounted in the plateholders only a few millimeters in front of the emulsion. Second, since the early 1980s greater attention has been given to obtaining the fully-exposed images (i.e., central densities > 2.5) needed for best results with the SAMM measuring machine. And third, a gradual transition has been made to finer-grain emulsions (i.e., from 103a-type to IIa-type, and finally to hypersensitized IIIa-type) as they became available. As a result of these changes, typical parallax precisions for completed series have improved from $\sim \pm 4$ mas to $\sim \pm 2$ mas (m.e.). In the most recent list (Harrington *et al*, 1993) there are 34 stars with formal relative parallax uncertainties between ± 1.2 and ± 2.0 mas.

The formal precisions of the TI800 CCD parallax determinations continue to be typically $\sim \pm 1.0$ mas, with the "best" results yielding formal uncertainties between ± 0.5 and ± 0.7 mas. In comparing the qualities of CCD and photographic results it is important to note, however, that all of the CCD determinations have been corrected for differential color refraction (see Monet *et al*, 1992) whereas the photographic solutions include no such treatment. Since the majority of the photographic fields were exposed in a blue bandpass (IIIa-J + GG385) — where the change in refraction as a function of wavelength is significantly steeper than across the broad red bandpass used for the CCD observations — additional improvement to the photographic results might be expected from a similar treatment.

Observations using a Tek2048 CCD were initiated early in 1992. This chip provides a field 11.0 arcminutes on a side compared with the 2.7 arcminute field of the TI800, resulting in better reference star configurations. It also has deeper full-well capacities and, hence, permits observations covering a larger range in brightness. Although the limited epoch ranges of the observations to date do not permit full solutions for parallax, preliminary reductions based on mapping reference frame stars together indicate that precisions comparable to the TI800 results will be readily obtainable.

The success being achieved by HIPPARCOS, combined with demands on USNO telescope time and various pressures to abandon photography altogether, have resulted in a cutback in bright star work. Efforts continue to devise a CCD-based detector with magnitude-compensating capabilities. However, the first attempt – employing an array of 6 CCDs (Monet, 1990) – proved unsuccessful due to the poor performance of the Fairchild CCD222 in this application. Hence, limited photographic capabilities are being maintained until a magnitude-compensating CCD system is operational.

1. USNO parallax results

In total, 980 trigonometric parallax solutions have been published to date. An additional 66 stars on the photographic program have been completed and approximately 20 more are nearing completion. Hence, a tenth list of photographically determined parallaxes will be ready for publication sometime in mid-1994. Roughly 100 more stars have now been completed using the TI800 CCD and a second list of these results is anticipated for early 1994.

Figure 1 shows the M_V vs V–I color-magnitude diagram for a selection of 510 stars with published USNO parallaxes along with 170 of the completed or nearly completed stars. The stars with published parallaxes in the figure were selected for meeting the following criteria: formal mean errors in M_V of a) $\leq \pm 0.20$ for the dwarfs, b) $\leq \pm 0.40$ for the subdwarfs, and c) $\leq \pm 0.15$ for the degenerates. (Giants and subgiants are shown, irrespective of the parallax quality, due to the small numbers of such stars.) All points are plotted with $\pm$ 1 m.e. error bars but, due to the scale of the diagram, they often do not show.

Progress continues toward delineating the various sequences — degenerate, dwarf, and subdwarf. Regarding the degenerates, we still find no convincing evidence that this sequence extends significantly fainter than $M_V \sim 16$. However, progress is being made in extending it into the region brighter than $M_V \sim 11$. The 5 brightest members shown (i.e., those with $M_V < 8.4$) are the central stars in the planetary nebulae S216, PW1, A21, NGC6720 (Ring), and NGC6853.

An additional 14 faint dwarfs have been established with $M_V > 15.0$. However, only one new star with $M_V > 18.0$ (LP412−31; $M_V = 18.41$) has been identified. Our lack of success to date in identifying additional very low luminosity dwarfs — such as LHS2924 ($M_V = 19.58$) and LHS2065 ($M_V = 19.15$) — from candidates selected from the Palomar Proper Motion Survey (Luyten 1979) supports the use of infrared photometry as a much more efficient tool for selecting such targets (Tinney, 1993).

New members of the extreme subdwarf sequence discussed in Monet *et al* (1992) have been identified and the sequence remains clearly delineated from the dwarf and mild subdwarf stars. Although stars undoubtedly exist between the two, they are clearly fewer in number — a feature that can not be explained solely by the selection of the candidate parallax stars from proper motion surveys. Two stars — LHS2100 ($M_V = 15.86$; V−I = 3.45) and LP671−3 ($M_V = 15.38$; V−I=3.29) — have been established as very late subluminous objects, although neither exhibits the very large tangential velocity typical of the extreme subdwarfs.

2. New parallaxes for brighter subdwarfs

The field sdF–G stars provide the potential calibrators for the distances to galactic globular clusters. They are also important for studies of light element nucleosynthesis (cf., Magain, 1989; and Gilmore *et al*, 1991). Table 1 presents completed or nearly

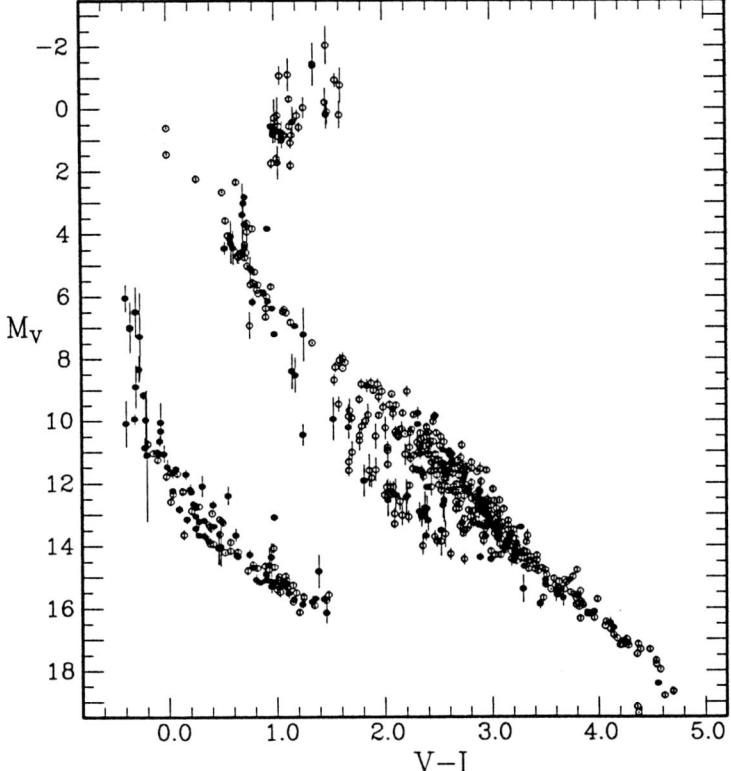

FIGURE 1. The M_V vs V−I color-magnitude diagram for a selection of 680 stars with (a) published USNO parallaxes (open circles; see text) or (b) unpublished but completed or nearly completed USNO parallaxes (filled circles).

completed USNO photographic parallax results for bright field sdF–G stars, along with the weighted mean values from previous parallax determinations. In order to emphasize differences between the USNO values and the previous determinations, the latter values were extracted from an earlier preliminary version of the new Yale Catalogue since the NASA ADC version (van Altena et al, 1991) already includes USNO data (completed and preliminary) in those means. The tabulated M_V and B−V values are based on USNO results; the quoted [Fe/H] values are primarily from Cayrel de Strobel et al (1992).

The tabulated USNO parallaxes are, in general, smaller than previously reported values. Figure 2 shows the more luminous (M_V < 10.6) non-degenerate stars from Figure 1 in the M_V versus B−V color diagram in order to compare the sdF–G subdwarfs from Table 1 with the theoretical isochrones from Bergbusch & VandenBerg (1992). Based on the USNO parallaxes, two of these stars — HD140283 and HD198300 — are not subdwarfs at all but have evolved into subgiants. It is interesting that the mean parallax for HD140283 quoted in the Yale Catalogue was derived from five independent determinations (generally in good agreement) and places the star solidly in the subdwarf region at M_V = 5.24. However, the value of the star's surface gravity derived from detailed spectroscopic abundance analyses (Magain, 1989) supports the USNO result (i.e., that the star is a subgiant) and is *not* consistent with a subdwarf interpretation. Within the formal uncertainties of present parallax determinations, there is generally accept-

	π_{abs} (milliarcseconds)				
Star	USNO	Yale Cat.	M_v	B–V	[Fe/H]
HD6582	133.9 ± 2.3	134.5 ± 2.4	5.78	0.70	−0.7
HD25329	53.7 ± 1.8	54.8 ± 4.7	7.20	0.85	−1.8
HD64090	38.0 ± 2.4	45.6 ± 3.1	6.17	0.62	−1.8
HD74000	8.6 ± 2.4	21.1 ± 6.1	4.30	0.41	−1.8
HD84937	14.8 ± 1.4	29.4 ± 6.4	4.15	0.38	−2.1
HD103095	111.6 ± 2.2	116.0 ± 3.3	6.66	0.76	−1.4
HD140283	14.5 ± 1.7	40.4 ± 5.4	3.02	0.49	−2.5
HD149414	22.6 ± 3.5	50.9 ± 9.6	6.38	0.74	−1.3
HD188510	22.8 ± 1.3	–	5.61	0.60	−1.8
HD194598	18.6 ± 2.0	19.3 ± 2.1	4.69	0.48	−1.6
HD198300	12.1 ± 2.2	–	3.94	0.58	−0.9
HD201891	29.9 ± 2.3	38.2 ± 5.0	4.75	0.51	−1.4
HD250792	14.1 ± 2.3	37.4 ± 6.4	5.08	0.60	–
BD+14°2513[†]	25.6 ± 5.0	–	6.92	0.66	–
BD+34°2476	6.4 ± 1.5	22.4 ± 13.8	4.09	0.38	−2.3
G183−11	8.2 ± 2.0	–	4.44	0.40	−2.1

[†] – USNO photographic parallax for cpm companion, LP495−171

TABLE 1. Data for brighter F–G field subdwarfs

able agreement between the observations and the theoretical isochrones although there definitely is a tendency for the data points to fall slightly above the appropriate loci. Parallaxes good to ≤ ±1 mas are needed for an unambiguous comparison between most of the sdF–G stars in Table 1 and model interior calculations.

3. The sdF–G/extreme late-type subdwarf connection

Astrometry alone can not fully define the relationship between the brighter field sdF–G stars and the extreme late-type subdwarfs. A collaborative effort involving J. Liebert (Steward Obs.), D. Kirkpatrick (McDonald Obs.), F. Allard (Univ. Montreal), H. Harris (USNO) and Dahn is attempting to derive metal abundances for a selection of the extreme subdwarfs by fitting model atmosphere calculations to spectrophotometric observations. Observations covering $\lambda\lambda$ 4700–8700 Å at a resolution of $\sim$18 Å have been obtained for 19 extreme subdwarf stars using Multiple Mirror Telescope. Attempts to fit the models of Allard (1990) to these observations did not yield satisfactory agreement between the major hydride (MgH, CaH, FeH) and oxide (TiO, VO) band features. However, improvements to the models are in progress (cf., Allard, 1993). To date available models have only included abundances with [m/H] $\geq$ −2.5. Based on our most recent fit attempts, models as metal poor as [m/H] = −3.0 or even −3.5 might be required for stars as extreme as LHS169, LHS1970 and LHS205a. While globular clusters with [m/H] = −2.5 are very rare (cf. Carney et al, 1990) numerous field subdwarfs with metallicities covering the range −3.5 ≤ [m/H] ≤ −2.5 have been identified in recent years (Laird et al, 1988; Ryan & Norris, 1991). If the extreme late-type subdwarfs indeed turn out to have [m/H] ≤ −2.5 this might support the steeper luminosity function for metal-poor field stars suggested by Richer & Fahlman (1992). The analysis of this material continues.

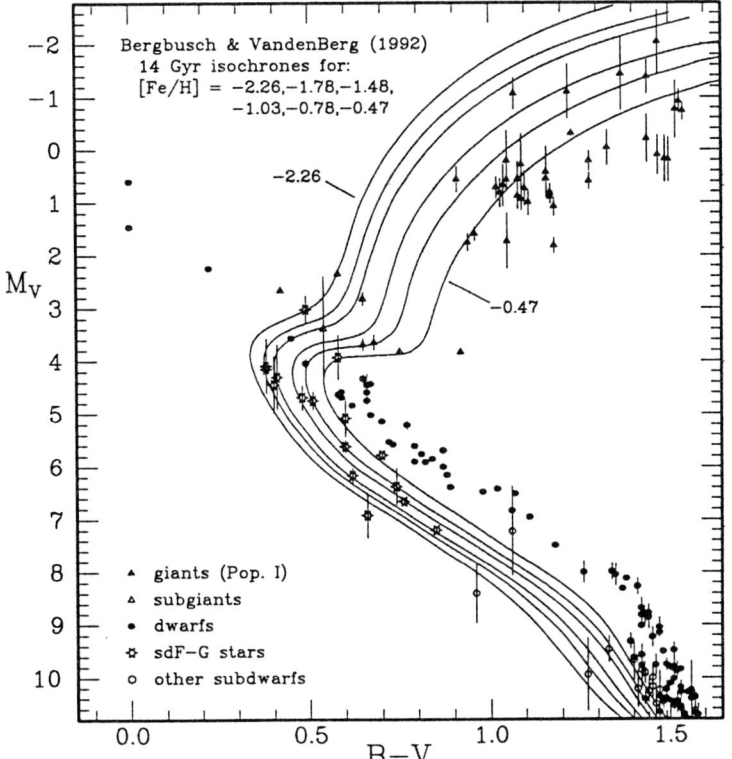

FIGURE 2. The brighter sdF-G subdwarfs compared with the Bergbusch and VandenBerg (1992) 14 Gyr isochrones in the M_V versus B-V color-magnitude diagram.

4. Future needs for ground-based parallax determinations

The enormous success being realized by HIPPARCOS in determining parallaxes for 118,000 stars at the $\sim \pm 1.8$ mas precision level does not eliminate the need for precise ground-based parallax determinations. While the HIPPARCOS results will undoubtedly satisfy the majority of the needs for parallactic corrections to catalogued star positions, there remain numerous problems for which parallaxes with sub-mas precisions are required. The sdF-G stars discussed above are but one example. Although the HST Astrometry Team has demonstrated the capability to measure parallaxes at the ± 0.5 mas level using the Fine Guidance Sensors, it is unclear whether significant numbers of stars will be measured by this technique.

Finally, investigators employing V-I for photometric distance estimators must be aware of the very significant spread in luminosity over the dwarf/subdwarf regions in the range $1.3 < V-I < 3.0$. Arguments based on star-type frequency and tangential velocity constraints should be employed to reduce uncertainties in photometrically derived luminosities.

Acknowledgements

The USNO parallax program is a team effort and the author thanks the other members — Harold Ables, Harry Guetter, Bob Harrington (d. 23 Jan. 1993), Hugh Harris, Varkey

Kallarakal, Chris Luginbuhl, Alice Monet, Dave Monet, Jeff Pier, Betty Riepe, Fred Vrba, and Dick Walker — for their contributions to the material presented here.

REFERENCES

Allard, F. 1990. Model atmospheres for M-dwarfs. Ph.D thesis, Ruprecht Karls Universität, Heidelberg.

Allard, F. 1993. Molecular opacities in M dwarf atmospheres. In *Molecular opacities in the stellar environment, IAU Colloquium No. 146*, ed. U. Jorgenson, in press. New York: Springer-Verlag.

Bergbusch, P. A. & VandenBerg, D. A. 1992. Oxygen-enhanced models for globular clusters stars. II. Isochrones and luminosity functions. *The Astrophysical Journal Supplement Series*, **81**, 163–220.

Carney, B. W., Latham, D. W. & Laird, J. B. 1990. A survey of proper-motion stars. X. The early evolution of the galaxy's halo. *The Astronomical Journal*, **99**, 572–89.

Cayrel de Strobel, G., Hauck, B., Francois, P., Thevenin, F., Friel, E., Mermilliod, M. & Borde, S. 1992. A catalogue of [Fe/H] determinations, 1991 ed. *Astronomy and Astrophysics Supplement Series*, **95**, 273–336.

Dahn, C. C., Harrington, R. S., Kallarakal, V. V., Guetter, H. H., Luginbuhl, C. B., Riepe, B. Y., Walker, R. L., Pier, J. R., Vrba, F. J., Monet, D. G. & Ables, H. D. 1988. U.S. Naval Observatory parallaxes of faint stars. List VIII. *The Astronomical Journal*, **95**, 237–46.

Gilmore, G., Edvardsson, B. & Nissen, P. E. 1991. First detection of beryllium in a very metal poor star: A test of the standard big bang model. *The Astrophysical Journal*, **378**, 17–21.

Harrington, R. S., Dahn, C. C., Kallarakal, V. V., Guetter, H. H., Riepe, B. Y., Walker, R. L., Pier, J. R., Vrba, F. J., Luginbuhl, C. B., Harris, H. C. & Ables, H. D. 1993. U.S. Naval Observatory parallaxes of faint stars. List IX. *The Astronomical Journal*, **105**, 1571–80.

Laird, J. B., Carney, B. W. & Latham, D. W. 1988. A survey of proper-motion stars. III. Reddenings, distances, and metallicities. *The Astronomical Journal*, **95**, 1843–75.

Luyten, W. J. 1979. NLTT Catalogue. Volume I. Minneapolis: University of Minnesota Press.

Magain, P. 1989. The chemical composition of the extreme halo stars I. Blue spectra of 20 dwarfs. *Astronomy and Astrophysics*, **209**, 211–25.

Monet, D. G. 1990. The USNOFS array sensor. In *CCDs in Astronomy*, Astronomical Society of the Pacific Conference Series, **8**, ed. G. H. Jacoby, pp. 11–7. San Francisco: Astronomical Society of the Pacific.

Monet, D. G., Dahn, C. C., Vrba, F. J., Harris, H. C., Pier, J. R., Luginbuhl, C. B. & Ables, H. D. 1992. U.S. Naval Observatory CCD parallaxes of faint stars. I. Program description and first results. *The Astronomical Journal*, **103**, 638–64.

Richer, H. B. & Fahlman, G. G. 1992. Low-mass stars in the spheroid of our galaxy. *Nature*, **358**, 383–6.

Ryan, S. G. & Norris, J. E. 1991. Subdwarf studies. II. Abundances and kinematics from medium resolution spectra. *The Astronomical Journal*, **101**, 1835–64.

Tinney, C. G. 1993. The faintest stars: Trigonometric CCD parallaxes. *The Astronomical Journal*, **105**, 1169–78.

van Altena, W. F., Lee, J. T. & Hoffleit, D. 1991. General Catalogue of Trigonometric Stellar Parallaxes, Preliminary Version. New Haven: Yale University Observatory.

An analysis of the accuracy of the parallaxes in the 1952 Yale General Catalogue

By K. AA. STRAND

3200 Rowland Place NW, Washington DC 20008, U.S.A

The Yale General Catalogue of Stellar Parallaxes(1952) contains, with few exceptions, the results obtained following the Schlesinger method. The data from three of the major contributors to the Catalogue has been compared with the USNO parallaxes of bright stars. While the material is limited, the comparison confirms the results previously found by statistical means.

1. Accuracy of parallaxes

The purpose of this paper is to call to attention the large external errors of the trigonometric stellar parallaxes in the Yale General Catalogue (1952) as to those obtained by the U.S. Naval Observatory (USNO) and by the Hipparcos space satellite.

The ninth list of USNO parallaxes obtained with the 155cm astrometric reflector contains bright stars photographed through filters with nichrome attenuated spots (Harrington et al. 1993). The standard errors of these parallaxes range from 1.2 to 4.1 mas with an RMS of 2.42mas. Lindegren (private communication, Nov. 1992) made a comparison of 55 stars with the USNO parallaxes in common with Hipparcos parallaxes, and found an RMS difference of 3.10 mas with the observed differences consistent with the quoted standard errors, which for Hipparcos was 1.74 mas. There was no systematic difference between the two sets of parallaxes.

On the basis of this, a comparison of the USNO parallaxes with those in the General Catalogue would determine their external errors or accuracy. In view of the limited number of the USNO bright star parallaxes the comparison was limited to three of the four main contributors to the Yale Catalogue: Allegheny, McCormick, and Yale Observatories. No attempt was made to deal with systematic differences as this would require a much larger material.

Comparisons were made of the relative parallaxes of the USNO parallaxes and the Yale parallaxes, using the values of the parallaxes on the right-hand pages of the Yale Catalogue. Adjustment was made for any systemactic difference between the two series.

Table 1 lists the results of this comparison, showing the external errors with their standard errors for each observatory. The last line shows the comparison with the Yale Catalogue. In this case absolute parallaxes were compared. In contrast to the relative parallaxes, there was no systematic difference between the two series.

The results arrived at show a general agreement with those obtained by Hertzsprung (1952) of +16 mas from a simple statistical method, and with those of Upgren (1978) from a comparison of the trigonometric parallaxes with their spectroscopic ones, based upon MK spectra and V photometric magnitudes.

2. Acknowledgement

I am indebted to the late Dr. R.S. Harrington for providing the data on the USNO parallaxes in advance of publication.

Allegheny	±11 mas	±1.3 mas	62 par.
McCormick	±16	±1.9	70
Yale	±14	±2.2	40
1952 Cat.	±11	±1.3	68

TABLE 1. External errors of parallaxes

REFERENCES

Harrington, R.S. et al., 1993, A.J., 105, 1571
Hertzsprung, E., 1952, The Observatory, 72, 242
Jenkins, L.F., 1952, General Catalogue of Trigonometric Stellar Parallaxes, (New Haven, CN)
Upgren, A.R., 1978, IAU Colloquium No.48, 69.

Parallaxes for the non-astrometrist

By CHRIS TINNEY[1], NEILL REID[2] AND JEREMY MOULD[2]

[1] European Southern Observatory, Karl-Schwarzschild-Straße 2, W-8046,
Garching-bei-München, Germany

[2] California Institute of Technology, 105-24, CA 91125, U.S.A.

The outstanding success of the CCD parallax programme at the USNO (Monet et al. 1992) has prompted the initiation of several small scale parallax programmes in recent years – e.g. the programmes of Ruiz et al. at CTIO, Ianna et al. at Mt Stromlo and our programme at Palomar. Rather than trying to measure parallaxes for large numbers of stars of all types, these programmes are targetted at smaller samples of stars of particular scientific interest to the observers carrying out the programmes. We discuss the first results from the CCD parallax programme being carried out on the 60″ telescope at Palomar. We find that parallaxes with accuracies of $\leq 0\rlap{.}''004$ can be obtained in a few years using standard CCD equipment on a common-user telescope.

1. Introduction

CCDs make almost ideal devices for doing relative astrometry on intrinsically faint objects. They are much more sensitive than photographic plates – allowing the observation of objects ~ 3 magnitudes fainter for a given telescope and observing conditions. They also allow observations much further to the red (essential for minimising the effects of atmospheric differential colour refraction). They have no moving parts – whereas photographic plates must be digitised with a scanning machine (with all the uncertainties that introduces), CCDs directly produce a digital image suitable for subsequent processing. Lastly, whereas photographic plates have an irregular matrix of grains, CCDs have a matrix of pixels whose location and size is determined to high precision by the fabrication process of CCDs. Of course, no device is perfect. CCDs have a relatively limited dynamic range – about 3-4 magnitudes for well exposed reference and programme objects. And while information can be obtained from saturated photographic images, no system has yet been derived for obtaining high quality positions from saturated CCD images. Their other disadvantage is small size, which limits their field of view to $\sim 5' - 10'$. (Though evidence is beginning to indicate that the atmosphere may limit high precision relative astrometry to scales not much larger than this in any case (Monet 1993)).

These advantages combine to make CCDs the detector of choice for certain astrometric programmes – specifically those targetted at very faint objects. (The limited dynamic range and small field of view makes it almost impossible to obtain sufficiently distant reference stars for objects brighter than $\sim$ 10th magnitude). Moreover, these astrometric observations can be carried out with the common-user equipment now provided at most telescopes. This means that astrometric (and in particular, parallax) programmes can be targetted by observers at specific classes of *scientifically* interesting objects, rather than having to rely on the currently over-taxed USNO programme.

2. The Palomar parallax programme

We were particularly interested in the stars at the bottom the H-burning main sequence (Very Low Mass stars, or VLMs, with $M \lesssim 0.1 M_\odot$). These are possibly the most poorly

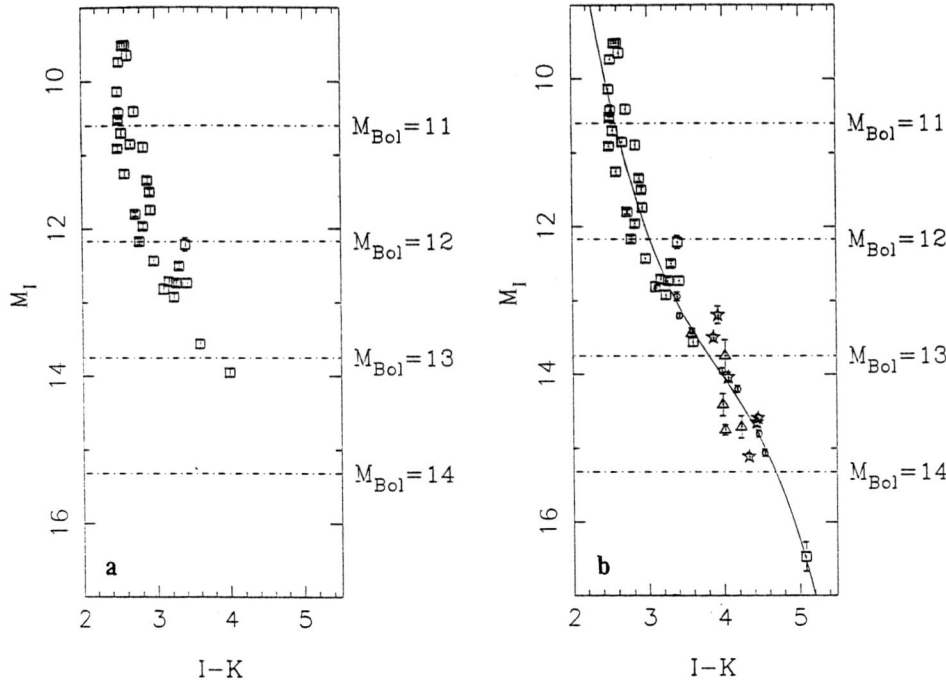

FIGURE 1. (a) The M_I/I–K diagram as known and based on parallaxes available *circa 1985*. (b) The current M_I/I–K diagram, including parallaxes from the Palomar (*stars*, Tinney 1993a), USNO (*squares and circles*, Monet et al. 1992) and Mt Stromlo (*triangles*, Ianna 1992) programmes.

understood of all stars. They have bolometric luminosities so small ($M_{Bol} > 14$) that even using the largest telescopes and the most sensitive detectors available, they can only be detected within ~ 40pc of the sun. As a result, only a small number of such stars are known, and the number well studied is even smaller. In particular, their colour-magnitude relations were very poorly defined. In Figure 1(a) we show the M_I/I–K colour-magnitude relation for the lowest mass stars as it existed *circa* 1988. As the figure clearly shows, not a lot was known about the relation bewteen colour and absolute magnitude for VLM stars, making the interpretation of the existing luminosity- and mass-function survey data (all based on luminosities extimated from colours) almost impossible. It was this situation which we sought to improve with our parallax programme at Palomar.

The details of our programme have been published elsewhere (Tinney 1993a) – the following is a short summary. Our programme contains about 25 stars with I $\lesssim 17$, most of which have been only recently discovered by photometric surveys (Tinney 1993b, Irwin et al. 1991, Irwin 1992). Our observations are carried out at five epochs per year on the Palomar 60″ telescope, using a 1024^2 unthinned Tektronix CCD. This CCD has 24μm square pixels, giving a scale of 0″.372/pixel. All observations were made through a Gunn *i* filter ($\lambda_{eff} \approx 7900$Å, $\Delta\lambda \approx 1000$Å) mounted ~ 15cm in front of the CCD.

Images were centroided using DAOPHOT. We found that each observation could be centroided to about 1/200th of an image size. That is, for a star of I=16.5, a single 300 s exposure in 1″.0 seeing gives a position good to ≈ 5 mas. Observations were not carried out in seeing worse than 2″.5. It was found to be not worth the time spent reducing the data. Over the three years of our programme we expect to obtain parallaxes with uncertainties of 1-2 mas. Over the two years for which data have been reduced to-date we obtain 2-3 mas uncertainites (Tinney 1993a). For the one object we have measured in common with the USNO (VB10) agreement was found within our estimated 1-σ uncertainties (Monet et al. 1992). The greatest difficulty in carrying out a programme for these stars is dealing with the effects of the differential colour refraction (DCR) introduced by the atmosphere. Put simply, the reference stars have a shorter effective wavelength through the Gunn i filter than the much redder programme stars. This results in the programme stars suffering less atmospheric refraction than the reference stars. However, so long as observations are made reasonably close to the meridian, this effect can be substantially corrected. We do this by observing each object as it rises and sets on a single night. The motion introduced by DCR can then be measured and calibrated out on subsequent nights. We found that our use of a Gunn i filter gave us a significant reduction (a factor of about 4) in the amount of DCR observed, compared with that seen at the USNO, where a much wider filter ($\lambda_{eff} \approx 6900$Å, $\Delta\lambda \approx 3000$Å) is in use. This means that while we need to take longer exposures, we have considerably more flexibility in scheduling, since we can observe ≈ 4 times further from the meridian for a given DCR effect. For a non-dedicated facility we regard this flexibility as being more important. (In any case, longer exposures allow differential seeing effects over the field of view to be averaged out and so should increase astrometric accuracy (Monet 1993).)

3. Results and conclusions

The first results of our work, together with those of the USNO and Mt Stromlo stars suffering less atmospheric programmes are shown in Figure 1(b). The improvement over Figure 1(a) is dramatic. The new data allows us to make firm statements about the absolute magnitudes of VLM stars. Figure 1(b) shows almost all the known VLM stars (including the faintest known object, GD165B). None of them are particularly convincing brown dwarf candidates, as they all lie in the range of luminosities consistent with being 'transition region' objects – i.e., objects intermediate between true main sequence stars, and true brown dwarfs which never burn H (Tinney 1993a).

It is clear that small, scientifically motivated, parallax programmes can be successfully (and relatively easily) carried out on common user telescopes using CCD detectors. We recommend the use of 'narrow' filters (i.e. $\Delta\lambda \lesssim 1000$Å) in programmes without a dedicated telescope, because of the extra scheduling flexibility this allows. Lastly, we hope that the success of our programme – despite our 'astrometric naïveté' – will encourage others to consider work in this most fundamental of fields.

REFERENCES

Ianna, P.A., 1992. Developments in Astrometry, IAU Symposium No. 156, ed. I.I. Mueller, Kluwer

Irwin, M., McMahon, R.G., Reid. N. 1991. MNRAS, 252, 61p

Irwin, M. 1992. *private communication*

Monet, D. G. 1993. Poster paper at this meeting

Monet, D. G. et al. 1992. AJ, 103, 638

Tinney, C.G. 1993a. AJ, 105, 1169

Tinney, C.G. 1993b. ApJ, 414, *in press*.

HIPPARCOS proper motions and parallaxes †

By L. LINDEGREN[1], J. KOVALEVSKY[2]
AND M.A.C. PERRYMAN[3]

[1] Lund Observatory, Box 43, S-22100 Lund, Sweden

[2] Observatoire de la Côte d'Azur, CERGA, Avenue Copernic, F-06130 Grasse, France

[3] Astrophysics Division, ESTEC, Keplerlaan 1, Postbus 299, 2200-AG Noordwijk, The Netherlands

Since its launch in August 1989 the European Space Agency's astrometry satellite HIPPARCOS has accumulated about two years' worth of high-quality astrometric and photometric information on its observing programme of 118 000 stars. We present an overview of the proper motion and parallax results obtained from preliminary analyses of the first 18 months of the observations. Extrapolation to a solution using all collected data yields a median external standard error of 2 mas/yr for the final proper motions and 1.8 mas for the parallaxes. A link to the extragalactic reference frame must be fixed during 1995 in order to allow the final HIPPARCOS catalogue to be published according to an agreed schedule.

1. Introduction

The basic astrometric parameters obtained with the ESA HIPPARCOS satellite are the trigonometric parallax and the components of the position and annual proper motion. According to the mission goals these should be determined for about 120 000 stars with an accuracy of 2 mas at 9th magnitude. Detailed information on the satellite, the observing programme and the data reductions is found in Perryman *et al.* (1989). A first assessment of the performance in orbit was given in Perryman *et al.* (1992) and subsequent papers in the May (I) 1992 issue of *Astronomy & Astrophysics*. The Tycho experiment, utilizing the auxiliary starmapper detectors to derive two-colour photometry and (less accurate) astrometry for many more stars, is separately reviewed by Høg *et al.* in the present volume.

The processing of HIPPARCOS data collected during one apogee passage (some 3 to 8 hours) allows to determine the one-dimensional relative positions (or 'abscissae') of stars along a fixed reference great circle. Geometric instrument parameters, in particular the 'basic angle' between the two viewing directions, are determined simultaneously with the abscissae, which are thus free from the effects of (slowly) changing instrument behaviour. The astrometric parameters of the programme stars are in the end determined by combination of the abscissae obtained on many different reference great circles. The 'great-circle reductions' and the 'sphere solution' process are internally quite rigid, so that the angular separation of any pair of programme stars becomes a well-determined quantity, even when the angle is large. This is the basis for the determination of a consistent system of coordinates over the whole sky, as well as of *absolute* parallaxes, using the variation the parallax factor across the sky. The measurements do however leave undefined the precise orientation and rotation of the resulting coordinate system. Since practically no extragalactic sources are observed, external information must be used to link this coordinate system to an extragalactic reference frame (§6).

† Based on observations made with the ESA HIPPARCOS satellite

2. Status of observations and reductions

Routine science data collection started on 26 November 1989, some 15 weeks after the launch. Except for a few short breaks the average data collection rate remained on a fairly good level (60–70%, see Fig. 1) until August 1992, when a series of hardware and operational problems made it necessary to suspend observation and put the satellite in a safe mode. Observations were resumed by the end of October 1992 after a complex but successful reprogramming permitting the satellite to be operated with only two gyros functioning. The average data collection efficiency was however lower than before, about 40%. Operations in this mode continued until another critical failure occurred in mid-March 1993. A contingency plan for zero-gyro operation had in the meantime been worked out, but attempts to initiate this mode have so far been thwarted by other problems. At the time of writing it is unlikely that more science data will be collected. Thus, a total of 2.0 years converged attitude time was accumulated over a period of 3.3 years. For comparison, the pre-launch accuracy assessment (Perryman et al. 1989) budgeted about 2.25 years of actual data for the nominal 2.5-yr mission.

The reduction consortia FAST and NDAC have successfully reduced the first half of the data, covering about 18 months up to mid-1991. The reductions include determinations of the positions, proper motions, trigonometric parallaxes and magnitudes for about 90 000 'single' stars (within the resolving capacity of the HIPPARCOS instrument) and preliminary double-star solutions for about 10 000 more objects. One-dimensional positions were also obtained for 47 minor planets observed at several epochs during the period. There remain some 18 000 stars for which no acceptable solutions were obtained thus far. Some of them have not been observed a sufficient number of times during the 18 months, others may be more complex objects (multiple stars, etc) for which special solutions will be made at a later stage. As more data are incorporated into subsequent reductions the accuracy will improve and good solutions should be found for nearly all programme stars.

The discussion below concerns only the 90 000 'single' stars for which the consortia give concordant results. Since the two solutions are very similar, but not identical, we consider a simple mean of the FAST and NDAC results.

3. Formal errors from current solutions

The formal standard errors provided by the reduction consortia are derived from the least-squares solutions and thus depend entirely on the internal characteristics of the data and estimation process. The distributions are shown in Fig. 2. The median standard errors are $\sigma_\pi = 1.83$ mas for the parallaxes, and $\sigma_{\mu\lambda} \cos\beta = 3.36$ mas yr^{-1}, $\sigma_{\mu\beta} = 2.85$ mas yr^{-1} for the proper motions in ecliptic longitude (λ) and latitude (β).

4. External errors

Several comparisons with independent (ground-based) data were made in order to check the accuracy of the parallaxes. (At present only the parallaxes are readily subjected to such checks: no optical positions of comparable accuracy exist to date, while comparisons with the best ground-based proper motions are confused by regional errors of the ground-based data and the short time baseline of the present solutions.) Among the more important tests are: (1) the parallaxes of about 45 stars believed to be situated in the Magellanic Clouds (true $\pi \sim 0.02$ mas), (2) some 60 bright stars in common with the latest parallax list of the U.S. Naval Observatory (Harrington et al. 1993), and (3)

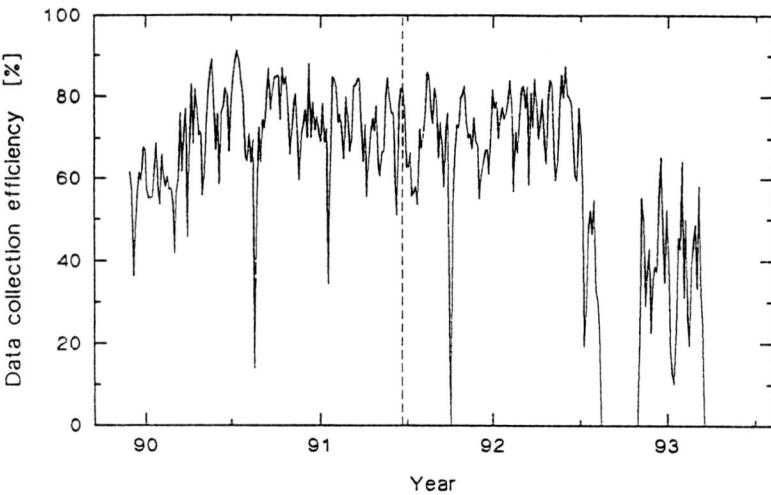

FIGURE 1. HIPPARCOS data collection efficiency. The curve shows the fraction of time (averaged in intervals of four days) in which the real-time attitude determination on-board the satellite remained in convergence – a necessary (but not sufficient) condition for the collection of valid science data. The dashed line marks the end of the data subset used for the 18-month solutions.

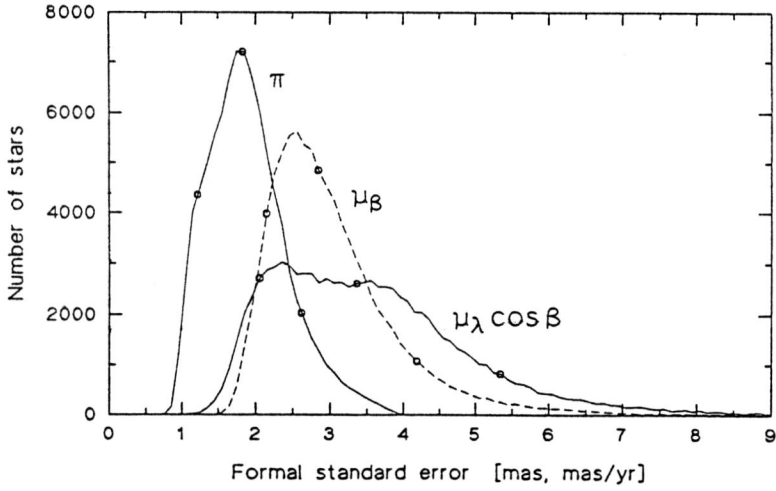

FIGURE 2. Distribution of formal errors (estimated standard deviations) in parallax and proper motion for the 18-month solutions. The small circles mark the 10th, 50th and 90th percentiles of the distributions.

a statistical study of the distributions of all 90 000 parallaxes divided according to their formal errors, in particular the tail of negative values. All of these tests indicate that the external errors are in the range 1.0 to 1.2 times the formal errors. Because of the way the astrometric parameters are computed in a single least-squares process we believe that this conclusion essentially holds also for the proper motions.

5. Predicted final accuracy

Using the collection efficiency data $f(t)$ shown in Fig. 1 it is a simple matter to extrapolate the accuracies obtained in the present solutions. If data up to $t = T$ are included the overall accuracy will depend on the moments $m_k = \int^T f(t)\exp(-t/\tau)t^k \mathrm{d}t$ for $k = 0, 1, 2$. The exponential factor with $\tau = 6$ yr (Lindegren & Kovalevsky 1990) accounts for instrument degradation due to particle radiation. The variances of the parallaxes (and of the positions at mean epoch m_1/m_0) improve as m_0^{-1}, while the proper motion variances improve as $m_0/(m_0 m_2 - m_1^2)$. The resulting expected improvement from 1991.5 to 1993.25 is a factor 0.82 for σ_π and 0.51 for σ_μ. Including a conservative factor 1.2 for the ratio of external to formal errors, this gives the following estimates for the final median standard errors:

$$\sigma_\pi = 1.8 \text{ mas}, \qquad \sigma_{\mu\lambda}\cos\beta = 2.0 \text{ mas yr}^{-1}, \qquad \sigma_{\mu\beta} = 1.7 \text{ mas yr}^{-1}.$$

6. Link to the extragalactic reference frame

The positions and proper motions of the HIPPARCOS output catalogue must be put on a reference frame as close as possible to the conventional extragalactic frame described in IAU (1991) Resolution A4 (Recommendation II). In principle this amounts to the determination of the components of the instantaneous orientation and of the rotation of a provisional HIPPARCOS frame relative to the extragalactic frame, and the subsequent application of the adopted orientation/rotation vectors to the provisional positions and proper motions. This determination is only possible through external observations linking individual HIPPARCOS stars to the extragalactic frame.

The schedule for the publication of the catalogue requires that the reference frame to be adopted for HIPPARCOS is fixed already in early 1995. A special working group is set up to bring together the necessary expertise and data from the 'outside' world of astrometry, and to act as a unique interface with the HIPPARCOS project for this specific problem. Because of the wide range of techniques, precisions and numbers of objects involved, the task of establishing this link has been divided into three subtasks:

(a) Radio observations of HIPPARCOS radio stars in the VLBI frame, and Hubble Space Telescope (FGS) observations of HIPPARCOS stars with respect to quasars.

(b) Photographic observations of optical counterparts to VLBI sources relative to HIPPARCOS stars.

(c) Photographic proper motion surveys including HIPPARCOS stars observed relative to external galaxies (Lick, Yale, Kiev).

Subtask (a) will determine both the orientation and rotation of the HIPPARCOS frame; (b) will mainly contribute to determining the orientation and (c) exclusively to the rotation of the frame. Each subtask will make its independent determination of the link components. These are then compared, evaluated and synthesized by the working group and applied to the HIPPARCOS catalogue formed as a combination of the final FAST and NDAC solutions.

7. Acknowledgements

The excellent data produced by the HIPPARCOS mission are the result of an enormous effort by many individuals and organizations throughout almost two decades. At this point it is a particular pleasure to thank the HIPPARCOS operations team at ESOC, Darmstadt (Germany), whose dedication, skill and enthusiasm have kept the satellite alive long enough to fulfil its scientific goals.

REFERENCES

Harrington, R.S., Dahn, C.C., Kallarakal, V.V., Guetter, H.H., Riepe, B.Y., Walker, R.L., Pier, J.R., Vrba, F.J., Luginbuhl, C.B., Harris, H.C. & Ables, H.D. 1993. U.S. Naval Observatory Photographic Parallaxes. List IX. *Astron. J.*, **105**, 1571–80.

IAU 1991. Proceedings of the Twenty-First General Assembly. *Transactions of the International Astronomical Union*, **XXIB**, 41–63.

Lindegren, L. & Kovalevsky, J. 1992. Accuracy predictions and final prospects for the HIPPARCOS mission. *Highlights of Astronomy*, **9**, 425–8.

Perryman, M.A.C., Hassan, H., Turon, C., Lindegren, L., Murray, C.A., Høg, E. & Kovalevsky, J. 1989. *The HIPPARCOS mission, Vol. I–III.* ESA Publications Division, ESA SP-1111.

Perryman, M.A.C., Høg, E., Kovalevsky, J., Lindegren, L., Turon, C., Bernacca, P.L., Crézé, M., Donati, F., Grenon, M., Grewing, M., van Leeuwen, F., van der Marel, H., Murray, C.A., Le Poole, R.S. & Schrijver, H. 1992. In-orbit performance of the HIPPARCOS astrometry satellite. *Astron. Astrophys.*, **258**, 1–6.

Tycho astrometry of one million stars †

By E. HØG, V. V. MAKAROV AND H. PEDERSEN

Copenhagen University Observatory, Ostervoldage 3, DK-1350, Copenhagen K, Denmark

The Hipparcos satellite's star mapper gives photon counts in two spectral channels simultaneously. The transit times and the signal amplitudes for each star across two groups of four slits are derived and used for astrometry and photometry, respectively, and this constitutes the Tycho project. Astrometric processing of the first four months of observations shows that the complete Tycho mission will result in positions, annual proper motions and parallaxes for 1 000 000 stars with an accuracy about $0''\!.03$ rms at 11 mag., improving to $0''\!.006$ for stars brighter 9th mag.

---

1. Introduction

The Tycho experiment of the Hipparcos satellite (Perryman *et al.* 1992) supplies data containing astrometric and photometric information.

The complete photon counts from the B and V channels of the star mapper, called the Tycho counts, are subject to a number of processing steps. The transit time of a star crossing the star mapper slits is the basic astrometric datum obtained. In the astrometry process, this observed transit time is compared with the predicted transit time obtained from the satellite attitude and the approximate position of the stars in the Tycho Input Catalogue Revision (TICR) which is based on the first 12 months of Tycho observations. The geometry of the star mapper slits is, however, calibrated by using the much more accurate positions in the provisional Hipparcos Output Catalogue (HOC) obtained from NDAC, based on the first 1.6 years of Hipparcos observations. This provides a close connection of the Tycho reference system to the Hipparcos system at a common epoch. The mathematical formulation of the astrometry task was outlined by Høg *et al.* (1992a).

The observational data are analyzed by the Tycho Consortium as described by Høg *et al.* (1992b). The Tycho counts for the first 2.5 years of the mission have been received from ESA and the first processing has been completed. The astrometry processing is performed at Copenhagen University Observatory (CUO) based on 'identified transits' received from Astronomisches Institut Tübingen, the HOC produced at Lund Observatory in 1992 and the TICR produced at Observatoire de Strasbourg. The transits are based on the satellite attitude reconstruction at the Royal Greenwich Observatory and CUO and connected to the TICR at the Astronomisches Rechen-Institut in Heidelberg.

TICR contains positions at the epoch 1990 of one million stars with an accuracy about 60 milli-arcsec (mas), as has been found by comparison with the much more accurate HOC. TICR is designed for use in the further Tycho data reduction, but not for publication because it contains a few percent of spurious 'stars'. They will be rejected in the astrometry processing by the use of all the years of Tycho observations. It is intended to publish a list of the stars in TICR with approximate positions which would be useful in preparing the measurement of photographic plates, for example. The plate reduction could then more quickly be repeated when the final Tycho Astrometric Catalogue becomes available about 1997.

It is planned to produce a Tycho Reference Catalogue (Röser and Høg 1993) containing proper motions with an accuracy of about 3 mas/yr derived from the Tycho results and a new reduction of the Astrographic Catalogue, based on the final Hipparcos Catalogue.

† Based on observations made with the ESA Hipparcos satellite

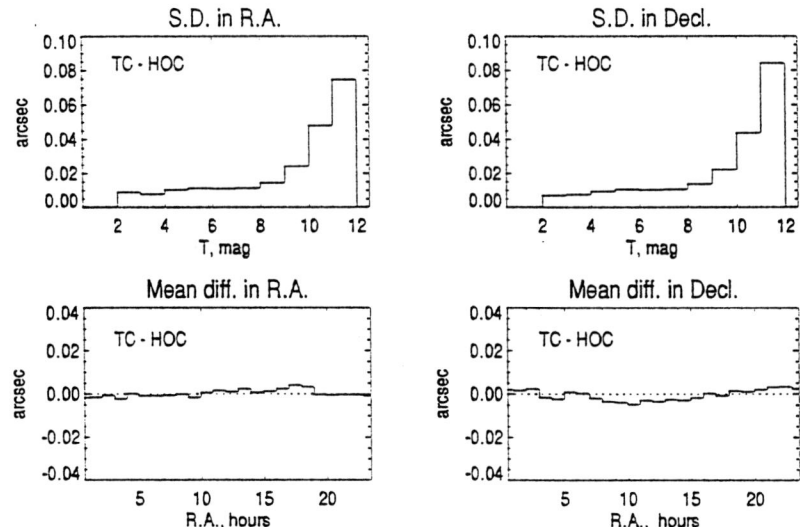

FIGURE 1. Accuracy of Tycho positions from four months of the mission. Accidental and systematic errors (upper and lower half, respectively) were derived from 28 312 stars having at least 20 observations and equations with condition number less than 3.0.

2. Analysis of four months of data

Tycho astrometry has so far only dealt with identified transits from the first four months of the mission, sufficient to evaluate the processing.

The accuracy of single transits was studied using observation sets or 'days' of about 24 hours, corresponding to one tape received from ESA. The uncertainty is due to the photon noise of the observation and to the satellite attitude, slit geometry and TICR position that were used in predicting the transit time. The uncertainty from slit geometry and TICR positions was eliminated by the use of (preliminary) calibration parameters and by the introduction of the HOC positions. The remaining uncertainty was formally modelled as due to photon noise and attitude errors, assuming other modelling errors to be negligible. The photon noise is represented as a function of the amplitude counts of the transit and the background counts by a formula derived before launch using the single slit response functions. Since the photon noise dominates at faint stars and becomes negligible at bright stars compared with attitude noise, the two contributions can be separated.

The resulting photon noise error was about 10 percent smaller than given by the original theory. The rms attitude error corresponds to 27 mas along scan and 50 mas perpendicular to scan. Both errors are quite satisfactory for Tycho astrometry, but the error along scan requires further study since it should be only a few mas from the great circle solution. The error model for single transits to be used in further processing was adjusted to accord with these results. It is used in the weighting of transits and in further analysis of accuracy. It is also used in rejection of outliers by a 3 sigma criterium in which the uncertainty of the TICR position is also taken into account.

Systematic errors of the positions depending on magnitude and color index can only arise if the single transits are affected by such errors and this is not expected from the theoretical properties of the instrument. A dedicated (provisional) study based on six days of observations shows that both kinds of systematic errors in terms of 3 sigma limits are less than 5 mas per mag. at the level of single transits. The resulting errors in positions would be much smaller due to the averaging by scans in different directions.

The calibration of slit geometry by means of HOC positions gave constant parameters within about 10 mas during the four months, so that, at present, only one set of parameters for the whole interval is used.

With this calibration, the positions were derived from the four months of observations. The resulting Tycho Catalogue (TC) was studied with respect to accidental and systematic errors by comparison with the HOC positions. Figure 1 shows the standard deviations of R.A. and Dec. as a function of magnitude. The Tycho magnitude (T) was derived in the TICR production from the $B + V$ Tycho counts and corresponds approximately to $(B + V)/2$. The errors are almost equal in R.A. and Dec., as was expected. A separate study has shown that the error decreases nearly with the square root of the number of observations n, at least up to $n = 50$. This leads to the following supposedly conservative prediction of the final Tycho astrometric accuracy. An accuracy at the mean epoch about $0\rlap{.}''03$ rms is expected for stars at $T = 11$ mag. The standard error of the positions increases by a factor of two per magnitude for stars fainter than $T = 9$ and the error will be about $0\rlap{.}''006$ for brighter stars. Parallaxes and annual proper motions will obtain nearly the same accuracy.

Systematic errors as a function of R.A. are shown in Figure 1. The sine dependence indicated in Dec. is equivalent to a relative rotation of the present coordinate systems of Hipparcos and Tycho by 5 mas around an axis in the equatorial plane. This is probably a result of the provisional nature of the attitude used in Tycho processing at this stage.

Acknowledgements

This work was supported by the Danish Space Board.

REFERENCES

Høg E., Bastian U., Hansen P.C., van Leeuwen F., Lindegren L., Pedersen H., Saust A.B., Schwekendiek P., Wagner K. 1992a, A&A, 258, 201.

Høg E., Bastian U., Egret D., Grewing M., Halbwachs J.L., Wicenec A., Bässgen G., Bernacca P.L., Donati F., Kovalevsky J., van Leeuwen F., Lindegren L., Pedersen H., Perryman M.A.C., Petersen C., Scales D., Snijders M.A.J., Wesselius P.R. 1992b, A&A, 258, 177.

Perryman M.A.C., Høg E., Kovalevsky J., Lindegren L., Turon C., Bernacca P.L., Crézé M., Donati F., Grenon M., Grewing M., van Leeuwen F., van der Marel H., Murray C.A., Le Poole R.S., Schrijver H. 1992, A&A, 258, 1.

Röser S., Høg E. 1993, Tycho Reference Catalogue (TRC): Positions and Proper Motions of One Million Stars, in: *Workshop on Databases for Galactic Structure* Swarthmore College May 17-19, 1993.

A test of preliminary HIPPARCOS parallaxes using photometric parallaxes of distant stars

By R. WIELEN, H.-H. BERNSTEIN, C. DETTBARN, R. HERING, R. JÄHRLING, H. LENHARDT, AND H. G. WALTER

Astronomisches Rechen-Institut, Mönchhofstraße 12–14, D-69120, Heidelberg, Germany

We describe a method of testing HIPPARCOS parallaxes by using very accurate photometric parallaxes of distant stars. We discuss the statistical and systematic accuracy with which a common zero-point error in the HIPPARCOS parallaxes can be determined from these photometric distances. We predict that a zero-point correction can be obtained with an uncertainty of less than ±0.1 mas. We have applied the method to preliminary HIPPARCOS parallaxes. The zero-point error of these preliminary HIPPARCOS parallaxes is very small, probably insignificant, and can be determined indeed with an accuracy of better than ±0.1 mas. The reliability of the mean errors quoted for the preliminary HIPPARCOS parallaxes is confirmed by the comparison with the photometric parallaxes.

1. Introduction

The test of HIPPARCOS parallaxes with very accurate photometric parallaxes of distant stars has two basic motivations: (1) the general validation of the results obtained with the ESA Astrometric Satellite HIPPARCOS, and (2) to determine, if necessary, a zero-point correction common to all trigonometric parallaxes measured by HIPPARCOS.

2. Validation of HIPPARCOS results

The data reduction procedure derives from the HIPPARCOS observations five astrometric parameters for each star: two positional coordinates, two proper motion components and one parallax. It is clearly desirable to have at least for a subset of all HIPPARCOS stars independent checks on the reliability of the HIPPARCOS results by using ground-based data of highest accuracy.

This task is most difficult for the HIPPARCOS positions: the typical accuracy of the final HIPPARCOS positions is expected to be better than ±2 mas. Most ground-based positions have much higher errors. In the near future, only VLBI positions of a few radio stars, contained also in the HIPPARCOS observing program, may allow to check the relative distances of these stars on the sphere with an accuracy of significantly better than 1 mas.

The situation is much better for HIPPARCOS proper motions, for which a final accuracy of about ±2 mas/yr or better is expected. While this accuracy is certainly a very important improvement over ground-based results for most of the 118 000 HIPPARCOS stars, for the subset of about 1500 stars contained in the Basic FK5, the FK5 proper motions, with an estimated accuracy of about ±1 mas/yr, are very probably more accurate than even the final HIPPARCOS proper motions. Hence these FK5 proper motions should allow a significant test of the HIPPARCOS results. The comparison is, however, complicated by systematic errors in the FK5 system and by the relative rotation between the FK5 system and the HIPPARCOS system. Furthermore, the FK5 results are probably not *much* better

	Group	Number of stars			
		in HIC	r_{phot} available	$r_{phot} \geq$ 0.5 kpc	$r_{phot} \geq$ 1.0 kpc
(1)	Magellanic Clouds	48	48	48	48
(2)	Cepheids	322	204	196	157
(3)	RR Lyrae stars	239	106	98	58
(4)	Open star clusters	1602	1573	694	382
(a)	probable members	279	278	130	74
(b)	possible members	480	479	361	206
(c)	uncertain members	228	224	135	78
(d)	non-members	31	21	17	13
(e)	unknown	584	571	51	11
(5)	OB stars	1051	1051	1051	1051
Total (without redundancies)					
(1) — (4a)		882	633	469	336
(1) — (4b) and (5)		2331	2081	1799	1511

TABLE 1. Distant stars selected

than the final HIPPARCOS results, and the quoted errors of the FK5 have been derived rather indirectly, leaving some room for uncertainties.

The most significant test of a HIPPARCOS astrometric parameter can be carried out for the HIPPARCOS parallax. For a well-chosen subset of more than thousand stars at great distances from the sun, we can derive photometric parallaxes which are more accurate than the expected final accuracy of HIPPARCOS parallaxes (±2 mas or better) by a factor of about 10, i.e. by a whole order of magnitude. Even if we were not interested in parallaxes as such, the chance to check this astrometric parameter with nearly 'error-free' ground-based results should be strongly used. The general outcome of such a comparison of HIPPARCOS parallaxes with these much better, independent data should be at least also fairly representative for HIPPARCOS proper motions, since both astrometric parameters are derived from 'local' variations in the stellar positions.

3. A possible zero-point error p_0 in all HIPPARCOS parallaxes

Due to a thermal perturbation of the HIPPARCOS satellite by the varying direction of the sunlight, there may be a slight zero-point error, p_0, common to all the HIPPARCOS trigonometric parallaxes. The reason for this is discussed e.g. by Lindegren et al. (1992). We shall give here a short indication only of why a zero-point error may occur in the HIPPARCOS parallaxes.

Table 1 lists for each group the number of stars in the HIPPARCOS catalogue. The basic quantity used to derive the astrometric parameters from HIPPARCOS observations is the observed abscissa v of a star along a reference great circle. The differential effect of the parallax of a star on the abscissa v is given by

$$\delta v_{par} \approx -p_{star} sin W sin(v_{star} - v_{sun}), \qquad (3.1)$$

where p_{star} is the parallax of the star, and v_{star} and v_{sun} are the abscissae of the star and the sun respectively at the time of observation. W is the angle between the pole of the reference great circle (RGC) and the sun. Since the pole of the RGC is always close to the direction of the spin axis of HIPPARCOS, which itself has a fixed angle of about 43° with respect to the sun, the angle W is always rather constant at about 43°.

Unfortunately, there is another effect which also produces a variation δv with $sin(v_{star} - v_{sun})$. The sun is heating the spinning, non-axisymmetric satellite unevenly. This thermal effect can produce a perturbation $\delta v_{thermal}$ in the abscissae v as a function of the aspect of the sun, measured by the angle $v_{star} - v_{sun}$.

$$\delta v_{thermal} = f(v_{star} - v_{sun}). \quad (3.2)$$

The unknown function f has to be determined empirically. f should be normally periodic in its argument, and hence can be fourier-analysed:

$$\begin{aligned} \delta v_{thermal} = & \quad K_0 \\ & + K_1 sin(v_{star} - v_{sun}) + K_2 cos(v_{star} - v_{sun}) \\ & + K_3 sin2(v_{star} - v_{sun}) + K_4 cos2(v_{star} - v_{sun}) \\ & + \cdots . \end{aligned} \quad (3.3)$$

The (constant) coefficients K_2, K_3, K_4, $\cdots$ can be determined empirically during the sphere solutions as global instrumental parameters. The coefficient K_1, however, cannot be distinguished from a constant zero-point error p_0 in all the HIPPARCOS parallaxes, since both p_0 and K_1 produce the same sinusoidal effect in v, varying proportional to $sin(v_{star}-v_{sun})$. Therefore, K_1 is not solved for in the HIPPARCOS data reduction. If K_1 is different from zero, this introduces a constant zero-point error p_0,

$$p_0 \approx -K_1/\sin 43° = -1.47 K_1, \quad (3.4)$$

into all the HIPPARCOS parallaxes. The resulting zero-point error p_0 has to be determined empirically from *external* data, after the HIPPARCOS data reduction. For this purpose, photometric parallaxes of well-chosen, distant stars are much better suited than ground-based trigonometric parallaxes, since both the random and the systematic errors of the former are much smaller than those of the latter.

4. Distant stars with accurate photometric parallaxes

The samples of stars for which we can obtain very accurate photometric parallaxes, are listed in Table 1. They are ordered in a sequence of increasing random and systematic errors in the photometric parallaxes p_{phot}: from stars in the Magellanic Clouds over Cepheids, RR Lyrae stars, members of open star clusters to OB stars. In order to minimize the absolute error of p_{phot}, we should use only stars with photometric distances r_{phot} larger than an appropriate limit, say 0.5 kpc or 1 kpc. This corresponds to an upper limit p_{lim} for p_{phot} of 2 mas or 1 mas. The typical random errors of r_{phot} or p_{phot} for the selected groups of stars are probably $\pm 20\%$ or better. This corresponds to an absolute random error of p_{phot} of ± 0.4 mas at the limit of 0.5 kpc and of ± 0.2 mas at 1 kpc. Since the mean distance is typically twice as large as the limit, the average mean error of p_{phot} is only half of the value at these limits, namely ± 0.2 mas or ± 0.1 mas for the limits 0.5 kpc or 1 kpc. Hence these random errors of p_{phot} are negligibly small in comparison to the errors ε_{hipp} of HIPPARCOS parallaxes. The systematic error in p_{phot} is probably half of the quoted random error of p_{phot}, at least for the groups (1) to (4): 10

ε_{hipp} [mas]	2.0	1.5	1.0
N[stars]	$\varepsilon_{hipp}/\sqrt{N}$ [mas]		
20	0.45	0.34	0.22
50	0.28	0.19	0.14
100	0.20	0.15	0.10
200	0.14	0.09	0.07
500	0.09	0.06	0.04
1000	0.06	0.04	0.03
2000	0.04	0.03	0.02

TABLE 2. Random error of p_0 or of a mean parallax obtained from N stars

%, or 0.2 mas at the limit 0.5 kpc, corresponding to 0.1 mas for $<r_{phot}> = 1$ kpc. The galactic OB stars may have slightly larger errors.

Input Catalogue (HIC), the number of stars in HIC for which we have accurate photometric distances available at present (perhaps more in the future), and the number of these stars with r_{phot} larger than 0.5 kpc and 1 kpc. Some of these stars are binaries and should be excluded because of their less precise parallaxes. Some stars may have to be excluded for other reasons. Nevertheless, we should expect to have sufficiently accurate photometric parallaxes available for more than 1000 stars. The distribution of these stars on the sky is, however, far from being uniform. Most of them belong to the galactic belt. Only RR Lyrae stars are scattered all over the sky.

From the photometric parallaxes p_{phot} and the corresponding HIPPARCOS parallaxes p_{hipp}, we can determined the zero-point correction p_0 as the weighted mean difference,

$$p_0 = <\Delta p> = <p_{hipp} - p_{phot}>, \quad (4.5)$$

using the individual values of $1/\varepsilon^2_{hipp}$ as the weights of p_{hipp} and of $\Delta p = p_{hipp} - p_{phot}$, since the error in p_{phot} is always negligible with respect to the mean error ε_{hipp} of p_{hipp}.

The mean random error Δp_0 of the zero-point correction p_0 is of the order of $\varepsilon_{hipp}/\sqrt{N}$. From Table 1 and Table 2, we can estimate that, for the final HIPPARCOS parallaxes, Δp_0 should be smaller than ± 0.1 mas, perhaps as small as ± 0.04 mas. Such a small random error of p_0 is only meaningful, if the *systematic* error of p_0 is also small. A systematic error of p_0 could be due to a systematic error in the photometric distances and parallaxes, e.g. caused by a systematic error ΔM in the calibration of the absolute magnitudes used for deriving r_{phot}. In Table 3, we give the systematic error Δp_0 of p_0 as a function of ΔM for various mean parallaxes or mean distances of the stars used in determining p_0. Table 3 shows that the *systematic* error can easily be the dominant source of uncertainty in p_0, if the random error in p_0 is smaller than ± 0.1 mas, as expected from Table 2. Hence it is important to find the best compromise between the demand of highest systematic quality of the photometric distances and the requirement of a sufficiently high number of stars, which is influenced both by the selection of appropriate stars and by the chosen upper limit p_{lim} for p_{phot}. Presently, a good choice seems to us to use $p_{lim} = 2$ mas ($<r_{phot}> \sim 1$ kpc) for the groups (1) to (4b) of Table 1, and $p_{lim} = 1$ mas ($<r_{phot}> \sim 2$ kpc) for the OB stars (group (5)).

ΔM		0.1	0.2	0.3	0.5	1.0
$\Delta p_{phot}/p_{phot}$ [%]		4.6	9	14	23	46
$<p_{phot}>$ [mas]	$<r_{phot}>$ [kpc]			Δp_0 [mas]		
2.0	0.5	0.09	0.18	0.28	0.46	0.92
1.0	1	0.05	0.09	0.14	0.23	0.46
0.5	2	0.02	0.05	0.07	0.12	0.23
0.2	5	0.01	0.02	0.03	0.05	0.09

TABLE 3. Systematic error Δp_0 in the derived zero-point correction p_0 due to a systematic error ΔM in the absolute magnitudes used for deriving p_{phot}

5. A method for eliminating systematic errors from the zero-point correction p_0

In order to study the most probable kind of a systematic error in p_0, let us assume that all the photometric distances of a group of stars used are wrong by the same factor f, e.g. due to a systematic error ΔM in the absolute magnitudes used (see Table 3:

$$r_{phot,used} = f \, r_{phot,true}. \tag{5.6}$$

It is easy to show that this leads to

$$\Delta p = p_{hipp} - p_{phot,used} = p_{0,f} + (f-1) \, p_{phot,used}, \tag{5.7}$$

where $p_{0,f}$ is the zero-point correction for $f \neq 1$. Hence $p_{0,f}$ and f can be derived from a linear fit of Δp versus $p_{phot,used}$, taking again individual weights $(1/\varepsilon_{hipp}^2)$ into account. The random error of $p_{0,f}$ will be in general larger than that for $f \equiv 1$ (discussed in Section 4), since $p_{0,f}$ is essentially determined by extrapolating the observed values of Δp to $p_{phot} = 0$, using an uncertain slope $(f-1)$. By contrast, if the zero-point correction is derived by assuming $f \equiv 1$ (as in Section 4), p_0 corresponds to a mean Δp at the mean parallax $<p_{phot}>$. According to our experience, the mean random error of p_0 from Eq. (5.7) is typically twice as large as that from Eq. (4.5). The systematic error of p_0 may, however, be smaller if we allow for $f \neq 1$. On the other hand, we should always verify whether or not a formal result for f, if significantly different from one, is real and in accordance with our general knowledge about the photometric calibration of this group of stars.

6. General results obtained by using preliminary HIPPARCOS parallaxes

We have applied the methods described above to preliminary HIPPARCOS parallaxes which have been derived by the FAST Consortium on the basis of non-iterated solutions for the five astrometric parameters from HIPPARCOS observations covering the first 18 months of the mission. The zero-point correction p_0, derived from the preliminary HIPPARCOS parallaxes, is always small and insignificant, if judged by the criterion $|p_0| < 3\Delta p_0$. Typically, $|p_0|$ is of the order of Δp_0. Allowing for a systematic error in the photometric distances as described in Section 5 by a factor f, we also derive very small values of p_0.

We conclude that (1) the preliminary HIPPARCOS parallaxes do agree within the quoted errors with very accurate photometric parallaxes of distant stars, and (2) that a common zero-point error p_0 of the preliminary HIPPARCOS parallaxes is extremely small, probably insignificant, and can be determined very accurately, with a precision of better than ±0.1 mas. Such an accuracy is actually desirable if one discusses the mean parallax of many stars (see Table 2, e.g. of nearby open clusters or of specific samples of astrophysically interesting stars. We hope that the results for the more accurate final HIPPARCOS parallaxes will be even better. In any case, already the tests of the preliminary HIPPARCOS parallaxes prove the superb quality of the HIPPARCOS results.

REFERENCES

Lindegren, L., van Leeuwen, F., Petersen, C., Perryman, M.A.C., & Söderhjelm, S. 1992. A&A, **258**, 131.

Status of Hubble Space Telescope Fine Guidance Sensor Astrometry

By P. D. HEMENWAY [1,2,3], W. H. JEFFERYS [4], AND G. F. BENEDICT [1] et al. †

[1] McDonald Observatory, University of Texas, Austin, Texas, USA

[2] Center for Space Research, University of Texas, Austin, Texas, USA

[3] Dept. of Physics and Astronomy, University of St. Andrews, Fife, Scotland

[4] Astronomy Department, University of Texas, Austin, Texas, USA

The Fine Guidance Sensors (FGS) of the Hubble Space Telescope (HST) are measuring relative positions with internal accuracies of 2 – 4 mas *rms*, per observation, and the separations of double stars down to separations as close as 20 mas. The method of observation, the interaction of the instrument with the spacecraft, the resulting precautions that must be exercised, the calibrations, and the current ongoing science program are described. We urge that more astrometry be done with the HST by the astronomical community.

1. Introduction

The Fine Guidance Sensors (FGS) of the Hubble Space Telescope (HST) are beginning to be used to make astrometric measurements. The details of the instrument and the projected accuracies have been given in numerous papers (*cf.* Bradley et al. (1991), Benedict et al. (1992), and papers cited therein.) This paper reviews the current state of astrometry with the FGSs including the internal accuracies achieved, the calibrations, and the science program of the HST Astrometry Science Team. We end with a plea for more astrometric use of the FGS by the astronomical community.

The FGSs are optical Koesters prism interferometers (Figure 1) fed by a 5″ field stop on the sky. The "transfer function" is the response of the interferometer as a function of positional offset of the target from the null position. Figure 2 shows two transfer function scans of a single star. The ordinate is the fine error signal (FES) derived from a combination of the outputs of the two photomultiplier tubes shown in Figure 2, of the form $(A - B)/(A + B)$. Before the light enters the interferometer, it passes through two "star selectors" which move the aperture around the sky on a curvilinear grid. The areas on the sky available to the FGSs are indicated in Figure 3. Ideally, for positional work two FGSs are used for guiding, while one (FGS3) is used as the astrometer. The areas are called 'pickles' because of their shapes and cover about 67 square arcminutes on the sky. The projected accuracy of the FGSs for measuring relative positions has always been about 2.7 mas per observation *rms* sum of two coordinates. We will only discuss the FGS used for astrometry (FGS3) unless otherwise stated.

2. Astrometric calibration

The following calibrations are required for various forms of astrometric measurements with the FGS. Not all calibrations are needed for all types of measurement:

† Other contributing authors: E.P. Bozyan, R.L. Duncombe, O.G. Franz, L.W. Fredrick, T. Girard, R. Kloepper, T. Mailoux, P. Shelus, W. van Altena, Q. Wang, L. Wasserman and A.L. Whipple

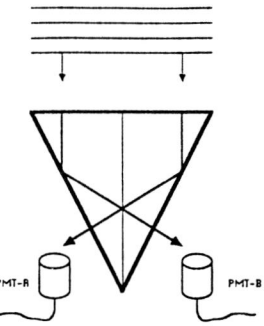

FIGURE 1. Schematic diagram of a Koesters Prism Interferometer.

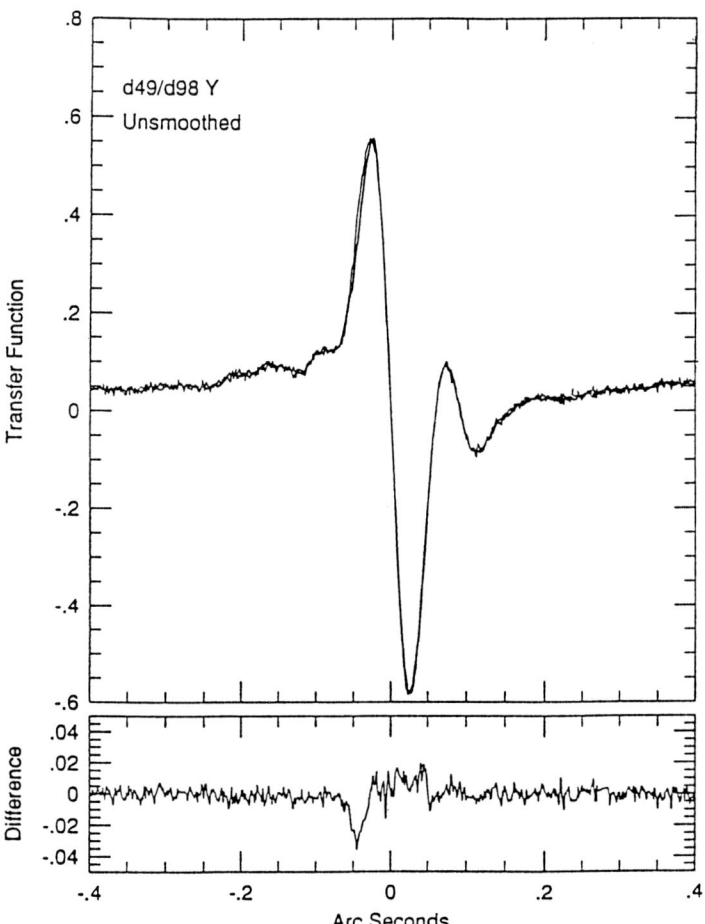

FIGURE 2. Superposition of two scans of a single star separated by 40 days.

(i) Optical Field Angle Distortion
(ii) "Plate Scale"
(iii) "Lateral Color" (chromatic aberration)
(iv) Filter wedge (cross filter calibration, for use with the Neutral Density Filter, for example)

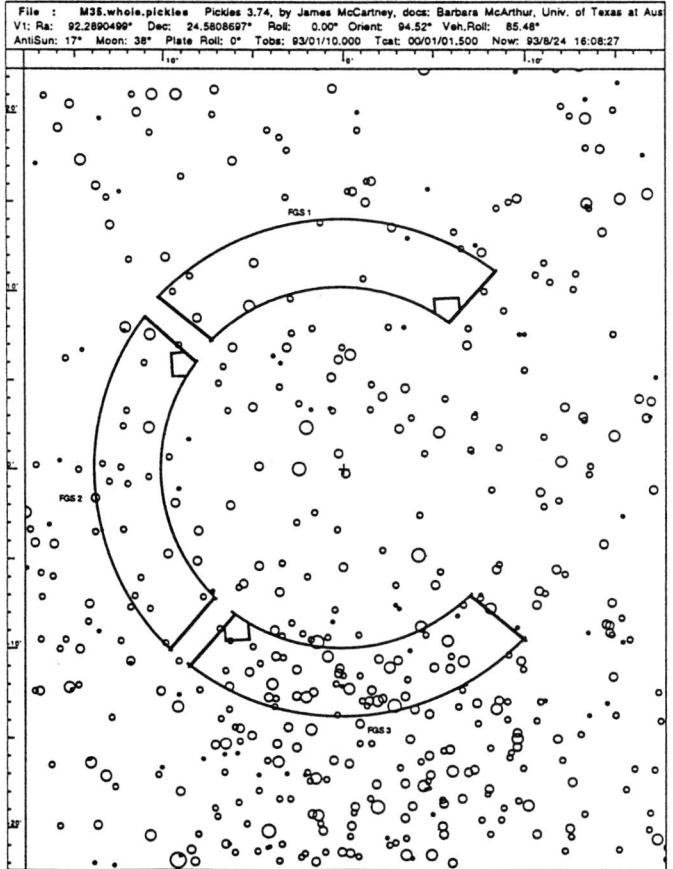

FIGURE 3. The HST field of view with the three FGS pickles shown.

(v) Single Star transfer function (for multiple star work)
(vi) Monitor changes in all of the above (Geometric Stability).

2.1. Optical Field Angle Distortion (OFAD)

The most difficult and time-consuming calibration is the OFAD. To reach the milliarcsecond level of accuracy we apply the technique of overlapping plates (*cf.* Jefferys, 1981). Approximately 30 stars in M35 (shown in Figure 4) are measured at 18 separate pointings during 20 orbits (two are repeats for check purposes). Typically, the pointings are separated by an arcminute on the sky. The data are reduced using the GAUSSFIT analysis program. In Figure 5, the pickle has been rectangularized, and the measurements of the three check stars are shown in each pointing, for the 20 pointings of the OFAD. Two orbits had large apparent drifts associated with them (orbits 16 and 20) and were not used in the final analysis. The scale of typical drift motion is 10 mas or less during an observation set (= one orbit = one "plate"). The connected dots indicate separate measured positions for the same check star, usually at the beginning, in the middle, and at the end of an observation set.

Figure 6 shows the design distortions relative to one point in the pickle. They show the level of distortion that must be removed to get to the milliarcsecond (mas) level of accuracy. Figure 7 shows the residuals from the OFAD measurements, solving for the

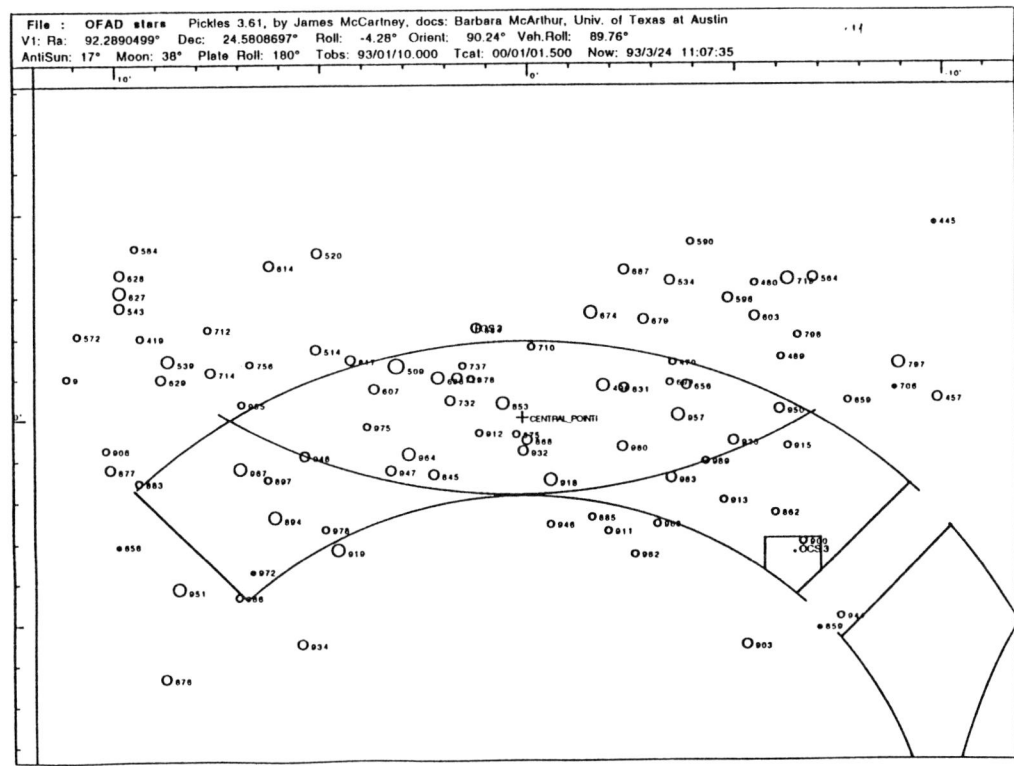

FIGURE 4. The field of FGS3, plus the stars of M35 used in the OFAD.

coefficients of the 'design' fifth order polynomial. We are now at the 3-4 mas level of accuracy for the relative error of a measurement (2 coordinates) in the FGS reference frame. One can use the covariance matrix of the parameters to estimate the contribution of the error of the solution to a measured position at a given location in the field. Figure 8 shows a contour plot of this contribution from the OFAD. The contour levels are at 1 mas intervals. We are below the 1 mas level over most of the pickle, in our knowledge of the distortions.

2.2. *Plate Scale*

Currently we use the design focal length of HST to determine the conversion from engineering units (star selector coordinates) to arcseconds on the sky. The number is equivalent to the classical "plate scale" of the telescope and is the coefficient of the linear term in the OFAD polynomial in each coordinate. An accurate determination is being performed by measuring the separation of two HIPPARCOS stars. The accuracy should be 2 mas/10 arcminutes, or better. Ultimately we hope to use a minor planet to determine the plate scale to 0.5 mas/15′.

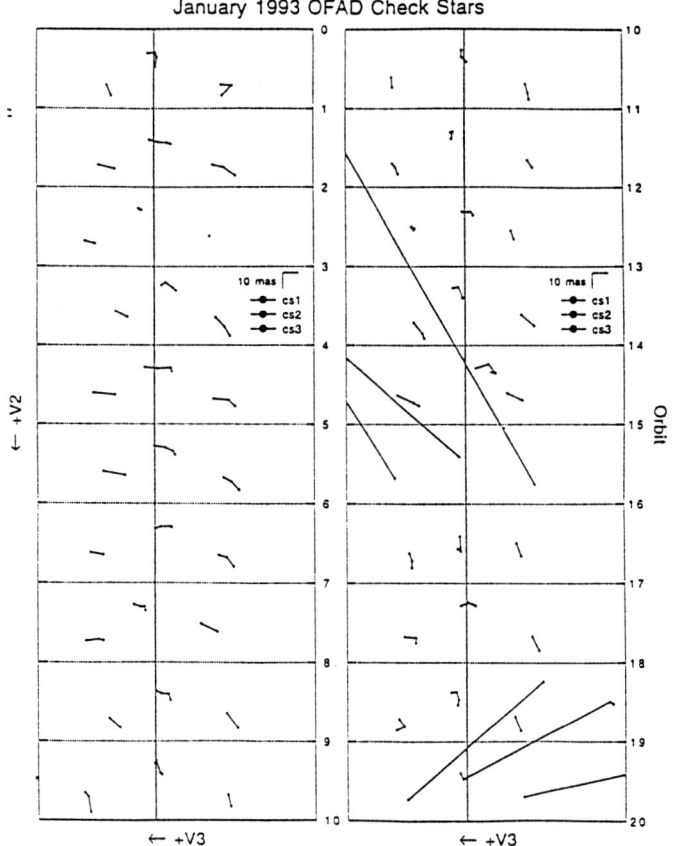

FIGURE 5. "Rectangularized" pickles for the 20 orbits of the OFAD.

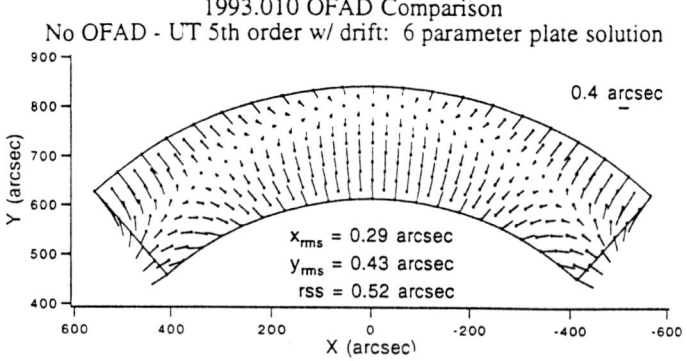

FIGURE 6. Full uncorrected differential distortion (design) in the pickle.

2.3. Lateral Color

The design of the FGS has zero chromatic aberration on the axis of the interferometer. However, because of slight misalignments of refractive elements in the FGSs and any mismatchs of the photomultiplier tubes in each interferometer, an offset of several mas between a red star and a blue star can exist. This 'lateral color' affect is being calibrated

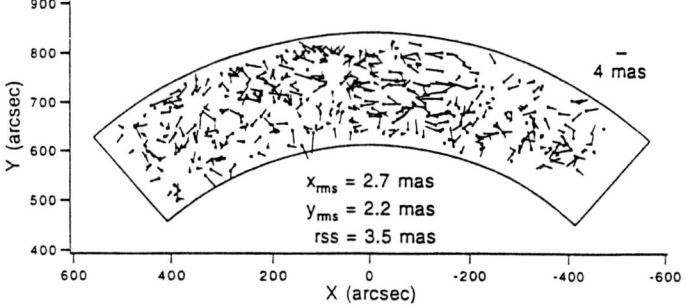

FIGURE 7. Vector plot of residuals to the fit for the OFAD measurements.

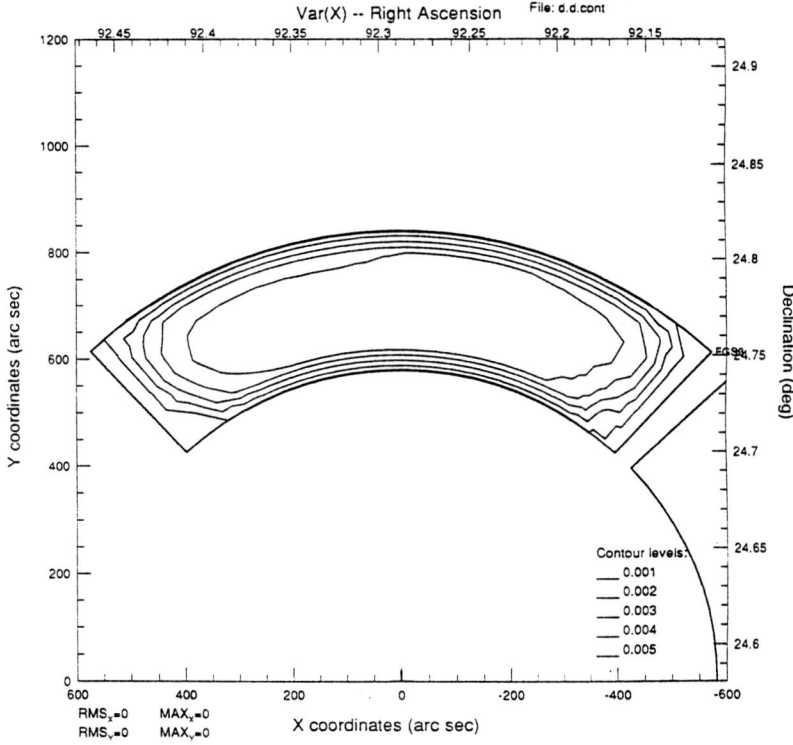

FIGURE 8. Contour plot of the error of the fit in x.

from measurements of a 4-star asterism containing one very red and one blue star. The asterism was measured in three orientations of HST separated by approximately 120 degrees in roll.

2.4. Filter Wedge

Five filter wheel positions exist in each FGS in a collimated beam. The astrometer has in the fifth position: a full aperture clear position = the "clear" filter, a $\frac{2}{3}$ aperture clear position = the "pupil" position, a yellow filter, an orange filter = the "red" filter, and a neutral density 2.0 filter = the "ND" position. Repeated measurements on the same 9th magnitude star indicate an 4 mas offset between these two filters. We are preparing a rigorous cross filter calibration which probably will be performed after the refurbishment mission.

2.5. Temporal Stability

2.5.1. Long term stability

Between December 1990 and December 1991, coarse track data showed temporal changes which amount to the order of $0''.1/\text{yr}$ over the length of the pickle. In order to monitor this scale-like change, a set of stars in M35 is being measured monthly or bi-monthly to monitor this change and determine if any other significant changes in the OFAD are taking place. So far, the distortions appear constant to within the errors of measurement.

2.5.2. Short Term Stability

During an orbit over which astrometric measurements are being made, the FGS may appear to move (usually as a solid body) with respect to the guiding FGS(s), with a typical size being 5-10 mas. Some of this apparent motion has been attributed to motion of the secondary mirror due to thermal/mechanical changes. To correct for this apparent motion, several stars are measured near the beginning, at the middle, and at the end of the observation set. The motion of the FGS is determined and its signature is determined from the observations in the observation set.

2.5.3. Transfer Function

Slight changes in the shape of the transfer function affect the analysis of multiple stars, particularly at close separations. Such changes are compensated by making scans of a single star before or after scans of a multiple system.

3. Problems

HST has had several hardware problems which have affected the astrometric capability. We have already mentioned the problems of temporal changes and how we compensate for them. Among the other problems are:

(a) FGS2 was removed from service as a guider because the current necessary to drive the star selectors became very large. One conjecture is that Teflon (registered trademark) flakes have accumulated in the races for the bearings. Normally guiding is accomplished with two FGSs in fine lock. The guiding algorithm was changed to use FGS1 with a guide star to stablize the pointing and the gyros were used to maintain roll.

(b) In order to prevent FGS3 from failing in the same way as FGS2, the star selectors will be moved more slowly than previously. Thus we will have fewer observations of fewer "reference stars" in a given observation set.

(c) HST is an open energy-momentum system. Energy from the sun is used to drive the reaction wheels to point the telescope. The buildup of excess momentum in the reaction wheels is 'dumped' into the earth's magnetic field using two magnitometers on the side of the spacecraft. One of these has failed. The result is that the pointing of HST is restricted so that HST never gets into a situation where it could not get to a safe position with respect to the sun in the event of another failure. Currently the restrictions are from 55° to 150° from the sun. Parallax observations are not in jeopardy at this time.

4. Astrometric science programs

Examples of the Astrometric programs being carried out with the FGS are:

(i) Multiple Stars: HST can measure separations with mas accuracy, position angles, and magnitude differences down to 13th magnitude; for brighter pairs, separations can be determined for stars closer than 20 mas.

(ii) Proxima Centauri: We began our "engineering tests" by looking at the field surrounding Proxima Centauri. Continuing astrometric measurement of this field has produced the following results:

- 36 observations spread over 1.2 years yield a parallax and proper motion with a current error of 0.6 mas, that match the ground based values determined from 130 observations over 45 years.
- A narrow peak in the periodogram of the residuals around 80 days is tantalizing. We are continuing observation and intensive study of this system.

(iii) We are measuring parallaxes of a few Hyades stars with the goal of obtaining an accuracy of 0.5 mas rms in the relative parallaxes of the individual stars.

(iv) We are using the FGSs to measure the motion of HIPPARCOS stars with respect to QSOs, BL Lac objects, and some AGNs, in order to "stop" the rotation of the HIPPARCOS instrumental reference frame with respect to an extragalactic quasi-inertial reference frame. So far we have four good observation sets. In one case with a VLBI position for a QSO and a HIPPARCOS Input Catalogue (HIC) position for the star, the measured and computed separations differ by only 0″.04. In two cases of positions with respect to AGNs, the separations differ by 0″.16 and 0″.02. (The expected error from the HIC is of the order of 0″.1-0″.2 rms.) Finally, in one case, for the three repeated measurements on an AGN, the separation difference is 0″.4, but two of the repeated measurements fall within 10 mas of each other while the third differs from those two by more than 0″.1. The raw data show that the FGS clearly "locked onto" something specific in all three cases. The best current conjecture is that the FGS "locked onto" different "hot spots" within the nucleus of the galaxy.

(v) We have a program to measure the parallax of some central stars in planetary nebulae.

(vi) We have a program to measure the relativistic gravitational bending effect using Jupiter as the gravity source.

(vii) A GO program is underway to measure the parallax amd motion of a few stars which may exceed the escape velocity of the Galaxy.

5. WE NEED YOU!

The amount of technical support in funding and people-power at the HST Science Institute for each instrument depends directly on how much that instrument is used for astronomical observation. During a recent cycle, the FGSs had 3% of the observing time as the prime science instrument, yet it is one of six prime science instruments. Therefore, theoretically, it requires 17% of the total support. While we believe that the science coming from FGS astrometry is unparalled in the history of our science, and that the results will make vital contributions to our understanding of many areas including the cosmic distance scale, absolute magnitudes, and the reference frame, we MUST have more people using the accurate and unique astrometric capabilities of the FGSs, or we will lose the support necessary to continue to the astrometry with HST. So, please, think of how you can use the FGSs with a capability to measure relative positions to 3 mas rms per observation, and separations closer than 20 mas, so we can continue with the science we have just begun. The FGSs are an astrometric tool for all of us.

Acknowledgements

This paper is based on NASA/ESA Hubble Space Telescope observations obtained at the Space Telescope Science Institute which is operated by the Association of Universities for Research in Astronomy Inc., under NASA Contract No. NAS5-26555.

We acknowledge essential support from NASA Contract NAG5-1603, NASA Grant NAGW 3111, and STScI Grant GO-3918.01-091A.

REFERENCES

Bradley, A., Abramowicz-Reed, L., Story, D, Benedict, G. & Jefferys, W., 1991, PASP, 103, 317

Benedict, G. F., Nelan, E., Story, D., McArthur, B, Whipple, A. L., Jefferys, W. H., van Altena, Wm. F., Hemenway, P. D., Bradley, A. & Duncombe, R. L., 1992, PASP, 104, 958

Jefferys, W. H., 1981, AJ, 86, 149.

Stellar kinematics: the Galactic context

By GERARD GILMORE[1] AND ROSEMARY F. G. WYSE[2]

[1]Institute of Astronomy, Madingley Road, Cambridge CB3 0HA, UK

[2]Dept. of Physics & Astron., John Hopkins Univ., Baltimore MD 21218, USA

Heaven has no colour of any kind, no centre and no sides – how can you hope to find its rules?
Liu Tsung-yuan (773-819)

Modern models of Galaxy formation make fairly specific predictions which are amenable to detailed test with galactic kinematic and chemical abundance data. For example, popular Cold Dark Matter models 'predict' growth of the Galaxy about a central core, which should contain the oldest stars. Later accretion of material forms the outer halo and the disks, while continuing accretion will continue to affect the kinematic structure of both the outer halo and the thin disk. A massive dark halo which is flattened and probably triaxial will arise naturally. This modern picture, which contains aspects of both the monolithic ('ELS') and the multi-fragment ('Searle-Zinn') pictures often discussed in chemical evolution models, makes some specific predictions which are amenable to test with astrometric data.

1. Galaxy formation: an overview

Many of the assumptions utilised in Galactic kinematic analyses are a disguised form of the physics of galaxy formation. To appreciate the plausibility and validity of those assumptions it is helpful to recall current models of galaxy formation. These models then identify those aspects of stellar kinematic analyses where the simplifying assumptions which are necessary to provide tractable problems are least likely to be soundly based, and thus where greatest caution need be taken. Conversely, testing the assumptions in Galactic kinematics tests current models of galaxy formation and evolution. An excellent recent review of models of galaxy formation and evolution is provided by Silk & Wyse (1993), where details and references may be found.

The most popular models of galaxy formation at present involve cold dark matter (CDM) as the dominant contribution to the gravitational potential. It is perhaps worth noting that CDM models are popular partially because they are more tractable that the Hot DM alternative and hence one can calculate more things and hence actually have some testable predictions. Many small 'condensations' of this CDM form early in the expansion of the Universe. Many of these, but perhaps not all, acquire a baryonic gas component as ordinary mass cools and falls into the local potential well. Large galaxies then form by the merger of a significant number of these sub-units.

Two basic paths then are available for the formation of what we see today as a large galaxy: a large number of small units may accrete onto a dominant 'core', or many comparable units may merge, with no single unit being identifiable as "the" proto-galaxy. In both these cases the merger process may be effectively complete at very early times, or it may be continuing at a significant rate at the present.

A picture of 'galaxy formation' at high redshifts then becomes the merger of CDM halos. In such a merger the (probably) hot diffuse gas in each mini-halo is assumed in the simplest case to be shock heated up to the virial temperature of the newly-formed composite halo. It will then assume a smooth $\rho \propto r^{-2}$ density profile, and eventually cool onto the density centre, which will be occupied by the largest pre-existing mass of cool condensed gas, if such existed. More complex situations, particularly involving secondary infall, can of course be envisaged, but are not fundamental to the discussion here. If a substantial amount of cool condensed gas were available in each mini-halo before merger then that gas mass may survive for some time as a definable condensation, and may form new stars and chemical elements in effective isolation from its surroundings. Eventually dynamical friction and/or cloud-cloud collisions will destroy such condensations, and the contents will disperse. Any gas will become part of the centrally condensed, well-mixed system, any stars will move along dispersion orbits in the larger potential well, as will the dark matter. The timescale on which such sub-units will be destroyed, and their contents mixed into the larger background, is comparable to a dynamical time, so significant substructure is not expected to be a long-lived phenomenon (Gilmore & Wyse 1993; Tremaine 1993).

This blend-and-stir process will have been common at high redshifts, when a rain of dwarf 'proto-galaxies' was normal weather for a budding giant. It continues today, at a rate which may still be significant for some galaxies. The term 'significant' here is worth some thought: in the central regions of galaxies masses are large, timescales are short and dynamical friction effective. Thus significant changes to a galaxy require mergers of components of comparable mass. It has been suggested that such mergers would destroy the thin disk of a galaxy like the Milky Way, as argued recently by Ostriker. If this argument were correct then normal late-type spirals must have completed the bulk of their merger events at very early times.

In the outer parts of galaxies mass densities are low, the fraction of the total luminous galaxy which is seen is very small, and timescales are comparable to a Hubble time. Thus one expects relatively little fossil kinematic structure to be visible in the central regions of normal galaxies, but it is probable that a large fraction of the outer parts of a large galaxy is a recent (on kinematic timescales) acquisition from afar. Fundamentally, the central regions of a galaxy need not be related in any obvious way to the outer parts of that same galaxy.

The dark matter which makes up the dominant contribution to the gravitational potential will have its spatial distribution modified by the rearrangement of the baryonic gas during this merger process. The gas dissipates its energy radiatively, and sinks towards the centre of the potential well. The resulting increase in central concentration of a significant part of the mass will alter the orbital distribution of the dark matter particles. Detailed simulations of this effect (eg Dubinski 1993) predict triaxial distributions with $b/a \geq 0.7; c/a \approx 0.5$.

A triaxial mass distribution in the halo predicts observable effects on stellar kinematics, and in turn can be tested from kinematic analyses. The most direct of such tests include ellipticity in the Galactic disk and non-stationarity of the kinematics of the solar neighbourhood (Kuijken & Tremaine 1993), both of which are probably seen, and which therefore support the case for a triaxial dark halo.

As with all sophisticated models, one should ask before proceeding if any robust evidence exists to support the fundamental assumptions and conclusions. In this case the evidence in favour of CDM models is marginal rather than conclusive (Ostriker 1993). However, not only do careful dynamical studies confirm the dominance of galactic potentials by an extended dark halo (Gilmore, Wyse, & Kuijken 1989; Fich & Tremaine 1991) but direct evidence for continuing mergers is provided by the impending merger between the Galaxy and the Magellanic Clouds. Considerations of the kinematic remnants of such a merger to a later observer may well be premature, but they are certainly relevant.

2. Is the Galactic Bulge the Galactic Seed?

A specific choice presented by CDM galaxy formation models is that the Galactic Bulge is either (part of) the central core about which the Galaxy grew, or it is a later arrival. Discussion of the chemical evolutionary implications of these choices has been provided by, among others, Wyse & Gilmore (1992,1993). One significant result of their analysis is the use of the distributions of specific angular momentum in the Galaxy's stellar populations to identify those which can in principle and those which cannot be related in an evolutionary sense.

2.1. Angular momentum

Low angular momentum is a defining characteristic of the Galactic stellar halo. Wyse & Gilmore (1992,1993) adopted the analysis of the rotation velocity of the stellar halo by Ryan & Norris (1991), who analysed their own data together with other extant samples, concluding that $<V> = 30 \pm 10$ km/s, independent of metallicity. Following Frenk & White (1980), they then assumed that the halo rotates differentially, $V(r) = constant$. The angular momentum distribution of the halo may then be determined, once a density profile is adopted. Hernquist (1990) provides an analytic model for an $r^{1/4}$ sphere which requires specification of a characteristic scale-length, a, such that $a = r_{eff}(1 + \sqrt{2})^{-1}$, and

$$\rho(r) = \frac{M}{2\pi} \frac{a}{r} \frac{1}{(r+a)^3}.$$

Adopting a half-light/mass radius for the stellar halo of 2.7 kpc (de Vaucouleurs & Pence 1978) gave the distribution of specific angular momentum, $M(h)$, shown in Figure 1, with

$$M(h) = \frac{h^2/(rV)^2}{(h/(rV) + a/r)^2}.$$

If there were no rearrangement of angular momentum, one would expect any gas which cooled from stellar halo formation to have a similar angular momentum distribution.

New rotational velocity data for the bulge have been derived recently by Ibata & Gilmore (1993). Their data, based on radial velocities for ~ 1000 stars, are consistent with differential rotation, with an amplitude rising to $V = 100$ km/s at a galactocentric distance of 3 kpc. Again using Hernquist's analytic density profile, with radial half-light radius $r_{eff} = 300$ pc, yields the angular momentum distribution for the bulge as shown in Figure 1. The angular momentum distributions of the central bulge and of the stellar halo are clearly very similar, showing that it is indeed possible for the present central bulge to be formed from the same gas as formed the oldest stellar populations in the Galaxy without requiring significant angular momentum loss.

The angular momentum distribution of the thick disk may be derived as above, once a density profile and rotation curve have been adopted. However, the global properties of the thick disk are yet poorly defined. The approximation of an infinitely thin disk (Freeman 1975) in differential rotation with constant rotation velocity, V, gives

$$M(h) = \frac{\pi \Sigma_o}{\alpha^2} \gamma(2, h\alpha/V),$$

with γ being an incomplete gamma function, and Σ_o is the central mass surface density. Adopting an exponential scale-length equal to that of the thin disk, $\alpha^{-1} = 4.5$ kpc (van der Kruit 1990), and a constant rotational velocity of amplitude $V = 180$ km/s (Ratnatunga & Freeman 1989), gives the angular momentum distribution plotted in Figure 1, where we have taken the edge of the disk at 4.5 scalelengths. That is, available chemical abundance models (Wyse & Gilmore) and dynamical constraints are both consistent

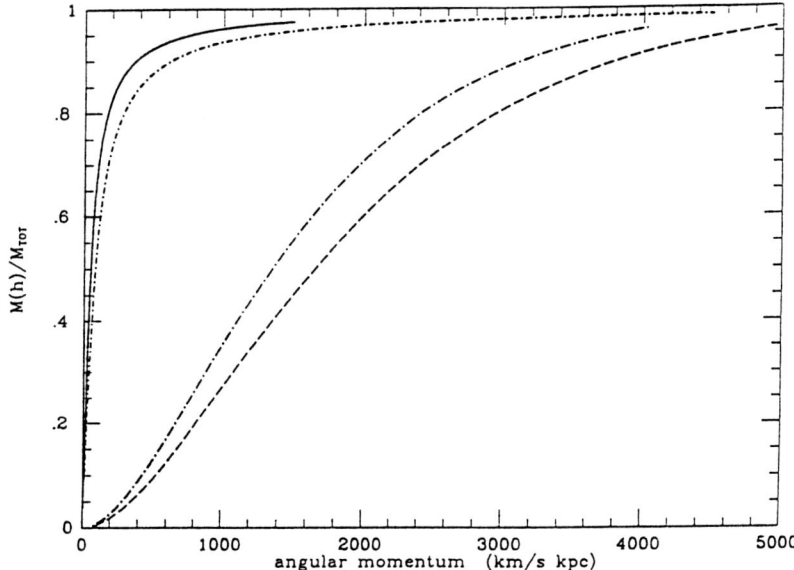

FIGURE 1. Normalized angular momentum distributions for the bulge (solid curve), the $r^{1/4} - law$ stellar halo (short dash-dot curve), the thick disk (long dash-dot curve), and the thin disk (long dash curve) from Wyse & Gilmore 1992.

with the Galactic bulge being part of that fragment of the early Universe about which the remainder of the present-day Milky Way Galaxy accreted.

The remaining important information required to test this is determination of the stellar velocity dispersion tensor in the bulge. The specific angular momentum distribution allows the bulge and the halo to be related, presumably with the bulge being the highly dissipated core of the halo. For this to be true one would expect a velocity ellipsoid in the bulge which is no more anisotropic than is that of the field subdwarf population. Determination of the velocity ellipsoid, and its variation with position in the bulge, is feasible with current astrometric capabilities. The result is awaited eagerly.

3. Kinematic evidence for mergers

The galaxy formation concepts outlined above suggest that a considerable amount of structure in the phase space distribution function for those stars (and DM particles) which inhabit the outer reaches of the galaxy is to be expected. This structure will be the remnant dispersion orbits occupied by the debris of former galactic satellites and near-neighbours which have now lost their former isolated identity. The existence of this structure provides a challenge in two ways: to devise dynamical analysis methods and/or sample selection methods which will still provide a 'fair sample' of the outer galaxy for dynamical studies; and to identify the fractional amount of phase space substructure, if any, and so test the (CDM) galaxy merger models.

The modelling limitations are minimized by ensuring that any kinematic sample which is to be analysed contains of order one star from any clump in phase space. That is, sparse sampling surveys over large areas of the sky are preferable to detailed pencil-beam

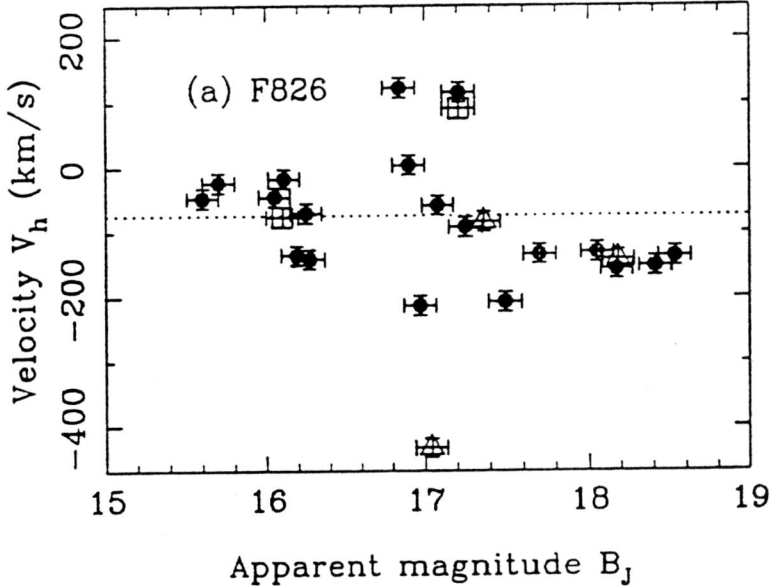

FIGURE 2. The distribution with apparent magnitude of the radial velocities of BHB stars in one field studied by Arnold & Gilmore (1992). The very small dispersion of the velocities at apparent magnitude 18.5 is suggestive of considerable cold structure in phase space, of the type expected if many stars occupy dispersion orbits.

surveys. Since extant data tend to have few stars down any one line of sight, it seems that current analyses which are designed to determine gross structural properties of the galaxy – such as the extent of the dark halo – benefit from patchy data. This is perhaps the only situation in astronomy where poor weather during observing allocations is beneficial, in that it prevents the accumulation of too much data on a single dispersion orbit

One would however like to prove that such orbits exist, and do contain much of the stellar mass of the outer halo. Since stars on a single orbit will be distributed over much of the sky, as seen by us in spatial projection, after only a few orbits, it is only very recent mergers which could be detected by spatial clumping, or in a spatial correlation function. Such remnants are likely to be rare, and may be restricted only to the Magellanic Stream. The most direct evidence is to determine the phase space distribution function, and to search for structure in that. Considerable efforts are being expended to do this for a variety of outer halo tracers. The most useful tracer objects studied to date are the metal poor globular clusters and distant horizontal branch HB stars.

A very considerable effort has been expended to derive proper motions for galactic globular clusters (and the Magellanic Clouds), and to complement the resulting space motions with age data. The present situation remains confused irreducibly by small number statistics: there simply are not very many clusters in the Galaxy. Nonetheless, recent efforts do suggest that perhaps 20 of the metal poor globular clusters have horizontal branch morphology such that they appear younger than the bulk of the population (Zinn 1993). A simple explanation of this result is that a fraction of order one-quarter of the very outer halo was accreted a few dynamical times after the bulk of the Galaxy became an identifiable gravitationally bound unit.

Astrometric surveys of halo field stars are underway in several groups, and are achieving levels of precision such that statistical analyses for significant substructure could be undertaken. Perhaps the most valuable survey in this regard in the near term is the Lick proper motion survey (Klemola, this meeting). Analysis of this substantial data set will be of considerable interest as a test of popular galaxy evolution models.

Until such astrometric data are available and have been analysed then one must rely on radial velocity data. Distant horizontal branch stars have been studied by several groups, with the current status of such work being well reviewed by Kinman (1993). A specific example of a survey of this type is that discussed by Arnold & Gilmore (1992; and 1993 in preparation).

An interesting feature of the Arnold & Gilmore results is the direct apparent evidence in their data for structure in phase space. They have studied two lines of sight. In one a group of BHB stars at galactocentric distance 30kpc is seen, with all four stars in their sample near this distance having indistinguishable velocities. This is discussed in Arnold & Gilmore (1992) and is shown in Figure 2.

In the second field studied by Arnold & Gilmore all eight stars in their sample with galactocentric distances between 15 kpc and 25 kpc have galactic orbits with the same sign of angular momentum, and again the local dispersion in the velocity distribution for those few stars is small. Each of these two results independently is of only marginal statistical significance. The interesting feature however is that the significance of these results has increased as the data set has increased, and that similar, again marginal when considered in isolation, results are seen in several other surveys. The existence of such phase space structure in stars of the thick disk and halo, in addition to the younger stellar populations, has been persuasively argued by Eggen for many years (Eggen 1987). Such moving groups are of course a specific high contrast example of the structure being considered here.

Analysis of new large samples of faint-star proper motion data promises to provide direct tests of currently popular CDM models of galaxy formation and evolution.

REFERENCES

Arnold, R. & Gilmore, G. 1992, MNRAS 257, 225

Dubinski, J. 1993, AJ (in press)

Eggen, O.J. 1987, *The Galaxy*, eds G. Gilmore & R. Carswell, Reidel, 211

Fich. M. & Tremaine, S. 1991, ARA&A 29, 409

Freeman, K. 1975, *Galaxies and the Universe* v 9????, 409

Frenk, C. & White, S. 1980, MNRAS 193, 295

Gilmore, G. & Wyse, R.F.G. 1993, *First Light in the Universe*, eds B. Rocca-Volmerange et al., Editions Frontieres in press

Gilmore, G., Wyse, R.F.G., & Kuijken, K. 1989, ARA&A 27, 555

Hernquist, L. 1990, AJ 356, 359

Ibata, R., & Gilmore, G. 1994, MNRAS (submitted)

Kinman, T. 1993, *Variable Stars and Galaxies*, ed B. Warner, ASP, 19

Kuijken, K. & Tremaine, S. 1993, ApJ (in press)

Ostriker, J. 1993, ARA&A 31, 689

Ratnatunga, K. & Freeman, K. 1989, ApJ 339, 126

Ryan, S. & Norris, J 1991, AJ 101, 1865

Silk, J., Wyse, R.F.G. 1993, *Physics Reports* 231, 293

Tremaine, S. 1993, *Back to the Galaxy* ed V. Trimble, (in press)

van der Kruit, P. 1990, *The Milky Way as a Galaxy*, (Gilmore, King & van der Kruit) (USB 1990)

Wyse, R.F.G., & Gilmore, G. 1992, AJ 104, 144

Wyse, R.F.G. & Gilmore, G. 1993, *Galaxy evolution: The Milky Way Perspective* ed S. Majewski, ASP, 209

Zinn, R. 1993, *The Globular Cluster-Galaxy connection*, eds G. Smith and J. Brodie, ASP, 38.

Stellar kinematics at the SGP

By JÜRGEN STOCK[1], RUI J. AGOSTINHO[2], JAMES A. ROSE[3]
AND ARTHUR R. UPGREN[4]

[1] Centro de Investigaciones de Astronomía, Mérida, Venezuela
[2] Departamento de Física da FCL-Univ. de Lisboa, Portugal
[3] University of North Carolina at Chapel Hill, USA
[4] Van Vleck Observatory, Middleton Ct 06457-6068, USA

A program of photographic objective prism spectroscopy is being carried out with the °10 prism on the CTIO Curtis Schmidt. From the digitized spectra, quantitative information is obtained for a complete and unbiased sample of stars in the magnitude range B < 13. In particular, information on T_{eff}, log g, [Fe/H], radial velocity, and proper motion is obtained for each star. Here we primarily discuss the radial velocity methods, which are based on the technique of opposed dispersion plate pairs developed by Stock (1992). Specifically, we find from a sample of 937 stars near the SGP that not only is the *rms* measuring error well-determined ($\sigma \sim 13$ km/s), but the error distribution function is as well. Consequently, we are able to derive accurate velocity dispersions for various populations. Our determination of the velocity dispersion of G dwarfs is in close agreement with that of Gilmore & Wyse (1987). Furthermore, we determine the velocity dispersion for late-F dwarfs with B=11 mean magnitude to be 22 km/s, and also find the velocity dispersion of 21 G giants to be 35 km/s.

1. Introduction

An understanding of the formation and evolution of our Galaxy requires a comprehensive inventory of the kinematics, chemical composition, and evolutionary states of its constituent stellar populations. To provide the necessary empirical data, a large, unbiased sample of stars must be examined. Objective prism spectroscopy offers the most efficient way to obtain such material when quantitative information is extracted from all usable spectra on the plate. The method of deducing physical stellar parameters, T_{eff}, log g, and [Fe/H] from a system of spectral indicators (Rose 1984) derived from digitized objective prism plates, has been developed by Rose (1991). Radial velocities and positions are obtained from opposed dispersion plates (the same plate material) following the methods developed by Stock (1992), Stock et al. (1984), and Stock & Osborn (1980). Both methods have been found to be suitable for this survey.

Using the Astrographic Catalog (AC) as a first epoch, proper motions for all stars in common with our survey are also determined. Although this is a long-term program covering a large range in Galactic coordinates, for the present time we are concentrating on the populations away from the Galactic plane. To reach a significant distance from the plane, we are working down to a limiting B magnitude of 13. For two fields near the SGP the data are already complete and will be discussed in this paper. For several additional fields, the positions and radial velocities, but not yet the spectral indices, are also available, and are presented here.

2. Observational material

All photographic plates are obtained with the Curtis-Schmidt telescope at CTIO† equipped with a 4°+6° prism combination which yields a dispersion of 86.3 Å/mm at

† Cerro Tololo Inter-American Observatory operated by AURA

Hδ. The use of an interference filter centered on $\lambda\lambda 4080$ Å with a band pass of 135 Å permits exposure times of 2 hr with a relatively low sky background. The IIa-O plates are hypersensitized in forming gas at 65° for 1 hr. The spectra are widened to $0.15-0.2$ mm, by moving the telescope in R.A. every 20 min. In most cases each field is covered by six plates, three with the dispersion in one direction and the other three with the dispersion in the opposite direction. For convenience we refer to the two plate sets as the "right" (R) and the "left" (L) set. For two fields near the SGP ($b = -86°$ and $b = -85°$ for Fields 1 and 2) the plates were scanned with the ZEISS PSK-2 Stereo Comparator at CIDA‡. The most recent plate material has been scanned with the Yale PDS microdensitometer. The development of computer programs for the determination of positions and radial velocitites from the same PDS-digitized data is now leading to a significant improvement in the final errors.

3. The radial velocities

3.1. Method of determination

Normally the prism is orientated such that the dispersion extends in the north-south direction. Hence the radial velocity will alter the apparent declination of the location of a spectrum, displacing it either north or south, depending on the orientation of the prism and, naturally, the sign of the velocity itself. The radial velocity then will show up as the difference between the declinations determined from the two plates. However, one has first to remove the field distortion introduced by the prism with a least-squares fit to the (x, y) data positions.

In our particular case, first the declinations from three plates with common dispersion direction are averaged (dec_L and dec_R), leading thus to the possibility of determining the error in the position $\Delta dec_{L,R}$ and hence in the radial velocity (Δv_r). The radial velocity is obtained from

$$v_r = 2 v_f (dec_L - dec_R), \qquad (3.1)$$

where the velocity factor v_f (=3.155 km/s/μ at Hδ, Agostinho 1992) is calculated from the dispersion curve produced by the prism (Stock & Osborn 1980). We should make it clear that the velocities determined in this manner are relative to the average velocity of all objects in the field. They should distribute like a gaussian around the average value, except for high velocity stars, as in Figure 3.

3.2. The error distribution.

The errors of the radial velocity are typically in the range $6-20$ km/s, but the tail of this distribution can have values up to ≈ 60 km/s. Since the goal is to derive velocity dispersions even for populations with $\sigma_z \approx 17$ km/s, we must have a correct description of the error distribution in order to subtract them quadratically from the calculated velocity dispersion.

The radial velocity error is obtained from Eq. 3.1 and also from considering that each measurement of the declination on a plate is a random sample (assumed to follow a normal probability distribution function, hereafter *pdf*) of its true value. Agostinho (1992) has shown that,

$$\Delta v_r = \frac{v_f \, \sigma_{dec}}{\sqrt{n(n-1)}} \sqrt{\chi^2_{2n-2}} \, ; \qquad (3.2)$$

‡ Centro de Investigaciones de Astronomia, Mérida, Venezuela

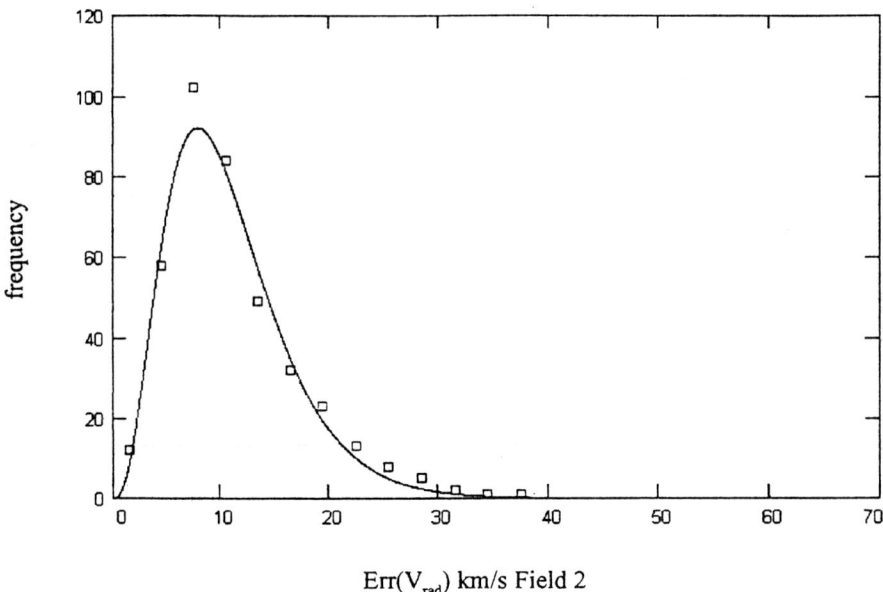

FIGURE 1. Error distribution for 385 stars in Field 2.

i.e. the radial velocity errors follow a $\sqrt{\chi^2}$ *pdf*, for a given n (number of plate pairs the star was measured on).

Most but *not all* stars have measurable spectra on all plates. Thus, the global error distribution is best described *in a simple form* by a gamma function $G(\mu, \eta)$†, since this is the most general case for this kind of *pdf* (Agostinho 1992),

$$G(\mu, \eta, t) = \frac{1}{\eta^\mu \, \Gamma(\mu)} \, t^{\mu-1} \, e^{-\frac{t}{\eta}}. \qquad (3.3)$$

Figure 1 shows the actual fit to the error distribution in Field 2. We obtained $\mu = 3.74 \pm 0.08$, $\eta = 2.91 \pm 0.07$ and an amplitude $A = 941 \pm 9$. The average value is $\langle \Delta v_r \rangle = 10.9$ km/s and the standard deviation $\sigma(\Delta v_r) = 5.62$ km/s. For Field 1 we obtained $\mu = 4.83 \pm 0.08$, $\eta = 3.20 \pm 0.06$ and an amplitude $A = 1347 \pm 10$. Its average value is $\langle \Delta v_r \rangle = 15.5$ km/s and the standard deviation $\sigma(\Delta v_r) = 7.04$ km/s. Stars with errors larger than 55 km/s were rejected since fewer than one were expected at these values. These fits are good but there is, however, a small deviation between the histogram peak and the fitted curve. This is related to the fact that the distribution is not *exactly* a gamma *pdf*.

We can conclude that $G(\mu, \eta)$ (Eq. 3.3) *gives a very good description of the error distribution*.

3.3. *External comparisons*

Radial velocities from slit spectra were obtained with the 1.5m telescope at CTIO for 25 randomly selected stars in each field. To assess the accuracy of the slit velocities, 21 radial velocity standards were also observed. A comparison between the objective prism radial velocities and those obtained from slit spectra, making proper allowance for the

† Note that $\sqrt{\chi^2_\nu(t)} = G(\frac{\alpha}{2}, 4, t)$ with $\alpha = \frac{\nu}{2} + 1$ and in our case $\alpha = n$

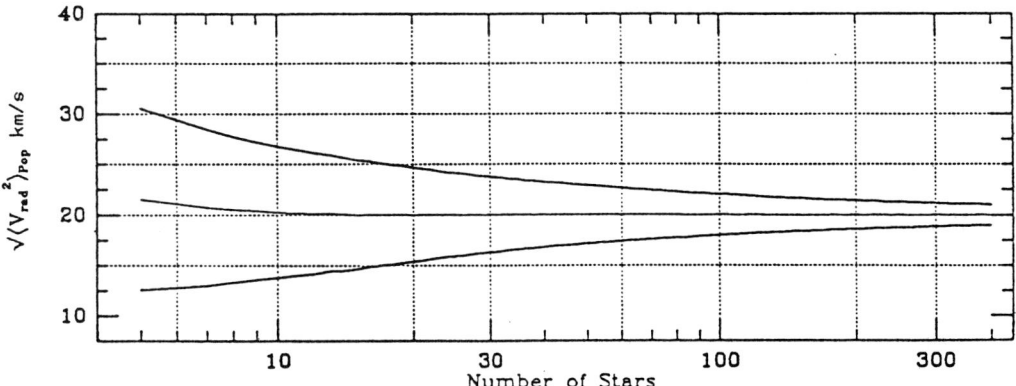

FIGURE 2. The error on the measured σ_z for a parent population with $\sigma_z = 20$ km/s and the error distribution of Field 2.

errors of the latter, yields an external error in the prism velocities of 12.0 km/s for Field 2. This agrees well with the value of 12.7 km/s arrived at from the discussion of the error distribution and the use of Eq. 4.5. For Field 1 we determine an external error of 21.8 km/s for the prism velocities, close to the 19.2 km/s obtained from the error distribution. We should point out that the match between the derived and the predicted values is also affected by the error in Eq. 4.5 (Figure 2).

4. The velocity dispersion σ_z

Consider that we are measuring velocities (v_p) drawn from an isothermal population; i.e. characterized by a gaussian distribution $\sim N(0, \sigma_z)$. Consider also that the error introduced by the measuring process is gaussian $\epsilon \sim N(0, \Delta v_r)$ — we have no direct access to it. Thus, the measured velocity is $v_r = v_p + \epsilon$ which implies that $v_r \sim N(0, \sigma_v)$ has variance,

$$\sigma_v^2 = \sigma_z^2 + \langle (\Delta v_r)^2 \rangle, \quad (4.4)$$

where $\Delta v_r \sim G(\mu, \eta)$, and we have also taken the average value of its square since it is a random variable (not a constant).

Knowing that $\langle X^2 \rangle = VAR(X) + \langle X \rangle^2$ we obtain for Eq. 4.4,

$$\sigma_z = \sqrt{\sigma_v^2 - \langle \Delta v_r \rangle^2 - \sigma^2(\Delta v_r)}, \quad (4.5)$$

which we use for small samples. The last two terms in Eq. 4.5 are the global errors, histogram average $\langle \Delta v_r \rangle$, and standard deviation $\sigma(\Delta v_r)$. The dispersion of the measured velocities is simply $\sigma_v^2 = \langle v_r^2 \rangle$.

For large samples one can fit a gaussian $N(0, \sigma_v)$ to the velocity distribution and a gamma pdf to the error distribution. Using the properties of a random variable $X \sim G(\mu, \eta)$ we can replace Eq. 4.5 by

$$\sigma_z = \sqrt{\sigma_v^2 - \mu(\mu + 1)\eta^2}. \quad (4.6)$$

For both fields at the SGP we should expect to obtain a similar σ_z since they are two uniform samples. For Fields 1 and 2 we have $\sigma_v = 31.47$ km/s, $\sigma_v = 29.70$ km/s and Eq. 4.6 yields $\sigma_z = 26.81 \pm 0.45$ km/s and $\sigma_z = 27.05 \pm 0.40$ km/s respectively, which shows the consistency of the method.

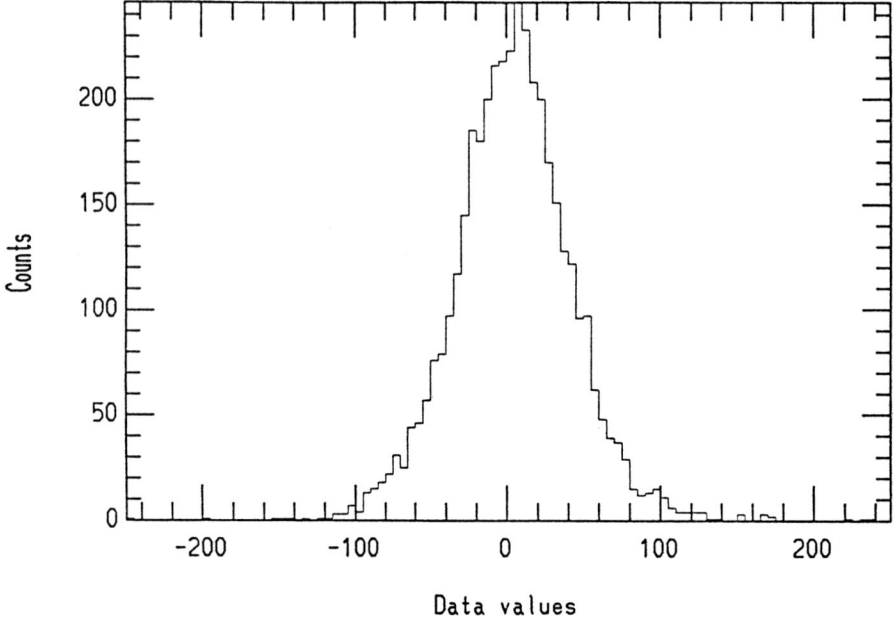

FIGURE 3. Radial velocity distribution for high and intermediate Galactic latitude.

To estimate the error in Eq. 4.5 we ran Monte Carlo simulations for several *parent* populations with $\sigma_z = 20-40$ km/s, combined with gamma *pdfs* consistent with the error distributions. In Figure 2 we have the result of the simulation for an initial population with $\sigma_z = 20$ km/s and an error distribution similar to that of Field 2. The x-axis is the number of stars, N, in a sample, and for a given N we performed 10 000 simulations, from which we computed the average value (thin line) of Eq. 4.5 and its 1-σ error (thick lines).

5. Results

5.1. *High velocity stars*

Figure 3 shows the distribution of v_r for fields at high and intermediate Galactic latitude. Similar work was carried out by J. Stock (1985) at low Galactic latitudes and the corresponding histogram is shown in Figure 4. Comparing both figures one finds a notorious absence of objects with velocities above 200 km/s at intermediate and high latitudes.

5.2. *Proper motions*

Positions with epochs around 1987 or 1988 are available for all fields from our plate material. S. Röser and U. Bastian of the *Astronomisches Rechen-Institut* at Heidelberg kindly made available the positions from the AC which has practically the same limiting magnitude as our plates. With these positions, proper motions with a time base of about 75 years were calculated. About 20% of our stars are in common with the PPM-Catalogue by Röser & Bastian (1991), Bastian et al. (1991). A comparison of the proper motions in R.A. for one field is shown in Figure 5.

We wish to draw attention to the appearance of a southward moving component of all

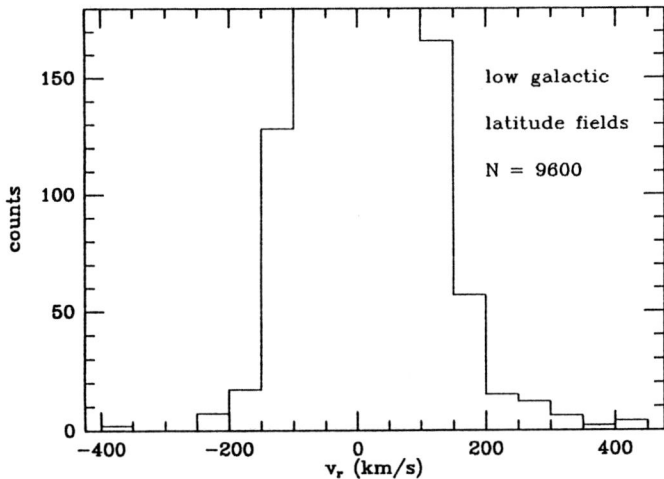

FIGURE 4. Radial velocity distribution for low Galactic latitude.

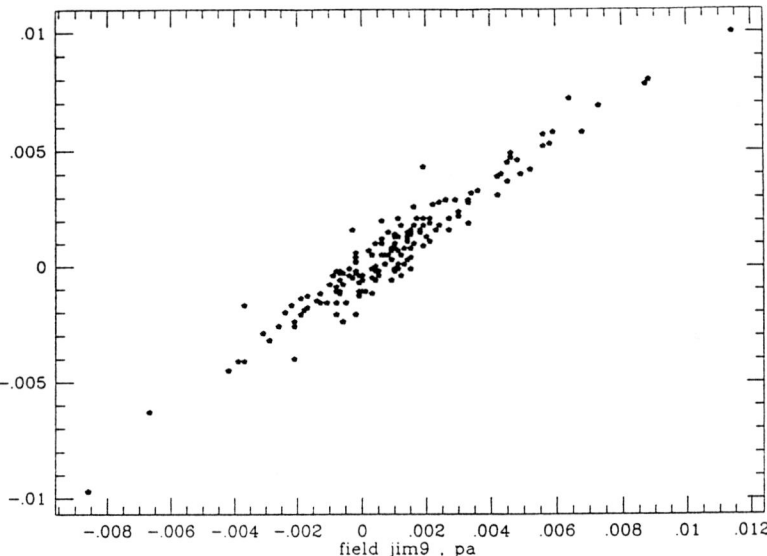

FIGURE 5. Comparison of our R.A. proper motion (x-axis) with the PPM-catalogue values.

stars which is apparent in all diagrams of the five fields under investigation. This is what one expects from the effect of Galactic rotation on the observed proper motions of halo stars.

5.3. Velocity dispersions

Combining the spectral indices with the 937 radial velocities obtained for the two fields at the SGP, we derived the following results. For the late-F stars characterized by $H\delta/FeI = 0.65 \approx F7$ and a scale height $h > 200$ pc, which Rose & Agostinho (1991) identified as the thick disk turnoff, we obtained $\sigma_z = 21.8 \pm 3.7$ km/s in the magnitude bin B=11 (50 stars), and $\sigma_z = 25.0 \pm 2.5$ km/s in the B=12 bin (109 stars). Assuming an absolute blue

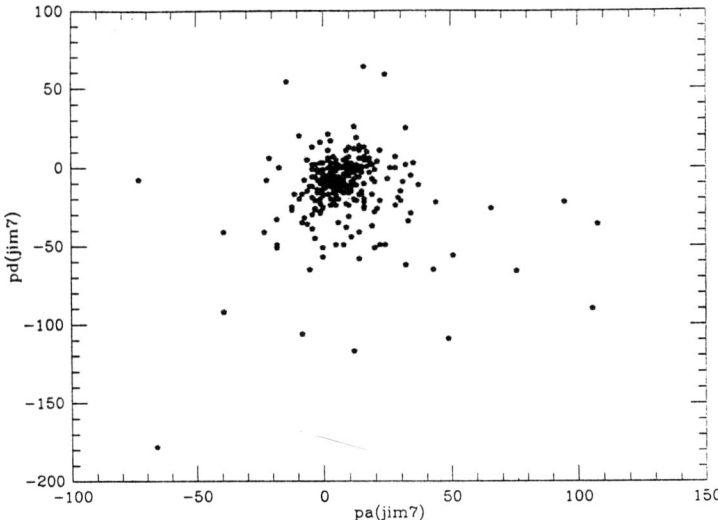

FIGURE 6. Proper motion diagram for data in one of our fields.

magnitude $M_B \approx 4.4$ one derives average distances of z=200 and 330 pc, respectively. These intermediate but rising values of σ_z between the standard 17 km/s for the thin disk (Sandage & Fouts 1987) and the ≈ 40 km/s of the thick disk, is due to a mixture of a cold (young thin) disk and an intermediate population, as our most recent data show.

The 112 G dwarfs in our sample have $\sigma_{zz} = 26.5 \pm 2.6$ km/s for B=12. With $M_B \approx 7.0$ and $M_B \approx 5.0$ for G5 and G0 stars, respectively, the distance above the Galactic plane is in the range $z = 100 - 250$ pc. Gilmore & Wyse (1987) obtained $\sigma_{zz} = 27.5 \pm 2.5$ km/s for G dwarfs at $z \approx 200$ pc, which is an almost identical result.

In our sample there are 21 G giants characterized by a $\langle[Fe/H]\rangle \approx -0.7$. Assuming an absolute magnitude $M_B \approx 1.8$ one derives a distance z=1.1 kpc above the Galactic plane. We computed a velocity dispersion $\sigma_z = 35.3 \pm 6.3$ km/s, which is comparable to the 37 km/s assigned by Carney et al. (1989) to the thick disk. These results show the *excellent* agreement between σ_z obtained from objective prism data and slit velocities.

5.4. *PDS reductions*

With the help of a strictly linear coordinate transformation spectral scans of plates of identical prism orientation can be co-added. Figure 7 shows the scan of a single spectrum. The result of adding to it the other two spectra with the dispersion in the same sense as the first, and co-adding the spectra of the same object on the plates with opposite dispersion (after cross-correlation), is given in Figure 8, which is composed of six individual scans. This process can be applied to any object, even if it's individual spectra are too weak to show any recognizable feature. In both the directions of the dispersion and perpendicular to it, the precision of the final coordinates is found to be considerably better than the measurements obtained with a Zeiss PSK-2 stereo-comparator used so far. It is hoped that the final positions (hence radial velocities as well) and proper motions based on the PDS scans will show a comparable improvement.

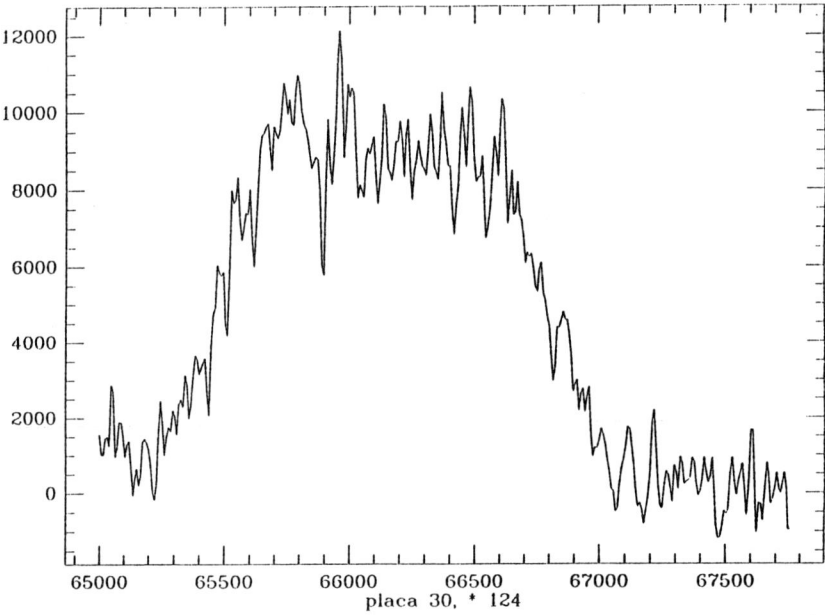

FIGURE 7. Spectrum of star #124 in one of our fields.

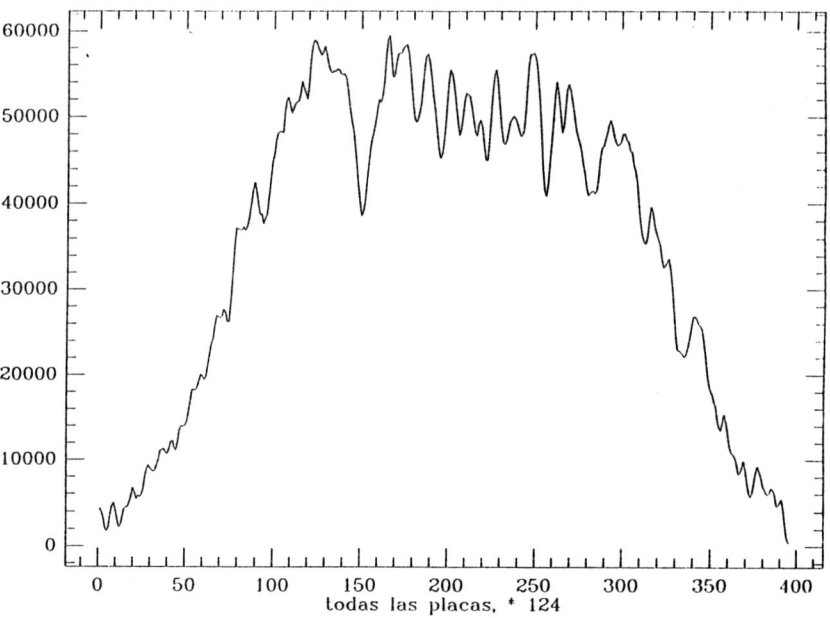

FIGURE 8. Co-added spectrum of star #124 composed of six individual scans.

Acknowledgements

This work was supported in part by grant BIC 3/87 from Junta Nacional de Investigacão Científica e Tecnológica (JNICT) from Portugal, and the NSF grant AST-8919455 to the University of North Carolina.

REFERENCES

Agostinho, R. J. 1992 *The Use of Objective Prism Radial Velocities for Kinematic Studies of Galactic Populations* Ph. D. Thesis, Univ. of North Carolina at Chapel Hill, USA

Bastian, U., Röser, S., Nesterov, V. V., Polozhentsev, D. D., Potter, Kh. I., Wielen, R., Yagudin, L. I. & Yatskiv, Ya. S. 1991 AAS, 87, 159

Carney, B. W., Latham, D. W., & Laird, J. B. 1989 AJ, 97,423

Gilmore, G., & Wyse, R. 1987, *The Galaxy*, Eds. G. Gilmore and B. Carswell, Reidel, 247

Green, J. R. & Margerison, D. 1978 *Statistical Treatment of Experimental Data* ed. Elsevier Scientific Pub. Comp.

Rose, J. A. 1984 AJ, 89, 1238

Rose, J. A. 1985 AJ, 90, 787

Rose, J. A. 1991 AJ, 101, 937

Rose, J. A. & Agostinho, R. J. 1991 AJ, 101, 950

Röser, S. & Bastian, U., 1991 *The PPM Catalogue* Astronomisches Rechen-Institut, Heidelberg

Sandage, A. & Fouts, G. 1987 AJ, 93, 592

Stock, J. 1992 Rev. Mexicana A&A, 24, 45

Stock, J., MacConnell, D. J., Osborn, W. & Alvarez, H. 1984 AJ, 89, 1897

Stock, J. & Osborn, W. 1980 AJ, 85, 1366.

Galactic kinematics from the Cambridge APM Proper Motion Project

By DAFYDD WYN EVANS

Royal Greenwich Observatory, Madingley Road, Cambridge CB3 0EZ, UK

Values for the Solar motion and differential Galactic motions are presented using data obtained from Schmidt plates scanned by the APM at Cambridge. Absolute proper motion measurements of the red stars yield values of $A = 19 \pm 6$ and $B = -13 \pm 6$ km/s/kpc for the Oort constants. A rotational shear of -21 ± 7 km/s/kpc in the z direction for the disc stars was also observed. This work is part of the APM Proper Motion Project (Evans 1988) which covers a total sky area of almost 100 square degrees and has astrometry and photometry for $\sim$170 000 stars.

1. Introduction

For this project four widely separated fields in areas of low galactic obscuration were chosen. Plates from the Palomar Observatory Sky Survey (POSS) were used for the first epoch positions and UK Schmidt telescope plates for second epoch. This provided a baseline of $\sim$ 25 years.

Colour and magnitude information was measured from the red and blue POSS images. The photometric accuracy of the final data was 0.09 magnitude for both O and E and 0.12 for colour, (O−E). The error in the colour zero points was estimated to be 0.05 magnitude.

An investigation of the high proper motion stars in this project can be found in Evans (1992).

2. Systematic Astrometric Errors and their Removal

These errors broadly fall into two groups (a) astrometric errors that are solely dependent on the position of the image on the plate and (b) those dependent on the physical properties of the images in question.

The first group includes those errors due to plate deformations and various systematic x-y table errors which are introduced during scanning. The biggest error in this category is due to the effects of differential refraction by the atmosphere. Most of those errors can be removed by a simple 'measurement and subtraction' technique, the basic assumption of which is that the proper motion of a star is independent of its position on a plate.

Among the second group of errors are the shifts due to the variation of atmospheric refractivity with image colour and the magnitude dependent error due to an incorrect alignment of the APM laser. Generally these errors are smaller than the first sort but since they can mimic shifts which are astronomical in origin, great care must be taken when trying to account for them. In some cases these are the limiting factor in the accuracy of the project.

One of these errors is caused by the vertical misalignment of the laser beam, which is in error by typically 1°. To correct this error each plate is scanned twice, the second time rotated by 180°. When the positions are averaged this error is removed.

A similar magnitude effect was measured (after 0°/180° averaging) between plate pair combinations which had little epoch difference. The most probable explanation is that the emulsion has sheared, either during processing or storage. For an emulsion 25μm

	σ/μm
APM (0° vs. 180° scan)	0.7
UKST (2 plates)	1.2
POSS (2 plates)	2.2

TABLE 1. This table shows the contributions to the overall proper motion error from the various sources. Note (i) that these errors are $\sqrt{2}$ larger that the positional errors and (ii) the APM contribution has already been subtracted from the plate values.

	Error/mas y^{-1}
(a) Refraction estimate error	0.11
(b) Misclassification error	0.09
(c) Intrinsic error	0.14
(d) Emulsion shear	0.51

TABLE 2. This table gives the various contributions to the overall proper motion zero point error in milliarcseconds per year.

thick a shear of around 2° could account for this error. No correction is possible for this error since the subtraction of magnitude dependent terms will remove astronomical information from the data cf. Chiu (1980).

The absolute value of the average proper motion of all stars on the plate is then determined by using the positions of all the galaxies as an absolute reference frame. Note that (i) no magnitude equation is used and (ii) the **positions** are not corrected and will still be in error by typically $\sim 1''$.

3. Overall accuracy

The overall accuracy of these measurements can be made by measuring the width of the residuals between scans of different plates with little epoch difference. The error estimate is calculated from the contributions due to the APM, UKST plate and POSS plate (given in Table 1). This is equal to 1.9 μm. For a 25 year baseline this is equivalent to **5mas/y**.

The various contributions to the proper motion zero point errors are given in Table 2 These errors are caused by (a) errors in estimating the average galaxy colour causing an error in the theoretical estimate of the refraction correction (b) inaccuracies caused by the misclassification of stars and galaxies (c) intrinsic random errors in using galaxies to set the zero point (d) the emulsion shear affecting the average stars which are $\sim 2^m$ brighter than the faintest ones. The overall effect is a zero point error of **0.55mas/y**.

To confirm the error estimates and test the zero point accuracy, the proper motions of known QSOs were measured in the 8 and 12 Hour Fields. For the 8 Hour Field the mean proper motions (expected to be zero) of these QSOs were found to be $\langle\mu_\alpha\rangle = 0.6 \pm 0.7$ and $\langle\mu_\delta\rangle = 0.0 \pm 0.7$ mas/y with dispersions of $\sigma_\alpha = 5.0 \pm 0.6$ and $\sigma_\delta = 4.8 \pm 0.5$ mas/y. The 12 Hour Field QSO results were $\langle\mu_\alpha\rangle = 1.6 \pm 1.3$, $\langle\mu_\delta\rangle = -1.5 \pm 1.3$, $\sigma_\alpha = 3.8 \pm 1.0$ and $\sigma_\delta = 3.8 \pm 1.0$ mas/y. These results are consistent with a zero average proper motion and proper motion errors as quoted above.

U	7.3	±	1.5	km/s
V	13.9	±	2.3	km/s
W	8.8	±	2.2	km/s
l	62°	±	6°	
b	29°	±	7°	
A	19	±	6	km/s/kpc
B	−13	±	5	km/s/kpc
M	−21	±	7	km/s/kpc

TABLE 3. This table shows the results of the kinematical analysis. (U, V, W) is the Solar velocity, in Galactic coordinates, with respect to the stars considered, (l, b) is the corresponding position of the Solar apex and A, B and M are the differential Galactic rotation parameters.

4. Solar Motion and Differential Galactic Motion

The two main systematic motions that can be detected in proper motion data are due to Solar motion and differential Galactic motions. The first causes the measured proper motion to increase linearly with parallax, π, while the effect of the second is a constant zero point shift in proper motion.

The data in each field is binned by parallax and is restricted to the range $0''.001 < \pi < 0''.005$ and stars with (O−E)> 2.0 (redder than about K5). In this colour range the contamination by giants would be negligible and the parallax range would restrict the stars to mainly those of the thin disc. The data was also restricted to a central 2° radius area so that systematic errors from photometric field effects would be kept to a minimum.

The parallax of each star was calculated using the (O−E) photometry, a colour-colour transformation given in Evans (1989) and a colour-magnitude relation from Chiu (1980). Biases in these parallaxes were allowed for using a Lutz-Kelker type correction. The data for the four fields are given in Figure 1. Note that μ_l is considered in terms of an angular measure on a great circle i.e. $\mu_l \not\equiv \Delta l/t$. This is so that the units of μ_l and μ_b are consistent with each other. The error bars in these diagrams do not show the systematic errors and show only the random measuring errors. Note that the systematic errors are about 0.5 mas/y for each coordinate in each field.

The gradients of these plots are used to calculate the Solar motion and the intercepts for the differential Galactic motion terms. Only the Oort constants, A and B, and a perpendicular velocity shear in the direction of Galactic rotation, M (Murray 1986), were included in these differential terms. The differential expansion terms, C and D, were not included since there were insufficient proper motion offsets to provide a reliable result. They are also expected to be small (Balona & Feast 1974).

The approximate formulae for the differential terms are

$$\mu_l \approx (A \cos 2l + B) \cos b + M \sin |b| \cos l \qquad (4.1)$$

$$\mu_b \approx -\frac{1}{2} A \sin 2l \sin 2b - M \sin |b| \sin b \sin l, \qquad (4.2)$$

where l and b are the Galactic coordinates of the sample. A, B and M in these formulae are in units of $''/y$.

For the analysis the full, unapproximated, formulae (Evans 1988) are used since the approximation used in Equations 4.1 and 4.2 breaks down at distances of ~ 1 kpc for some of the fields. The full results are given in Table 3. The errors quoted include the

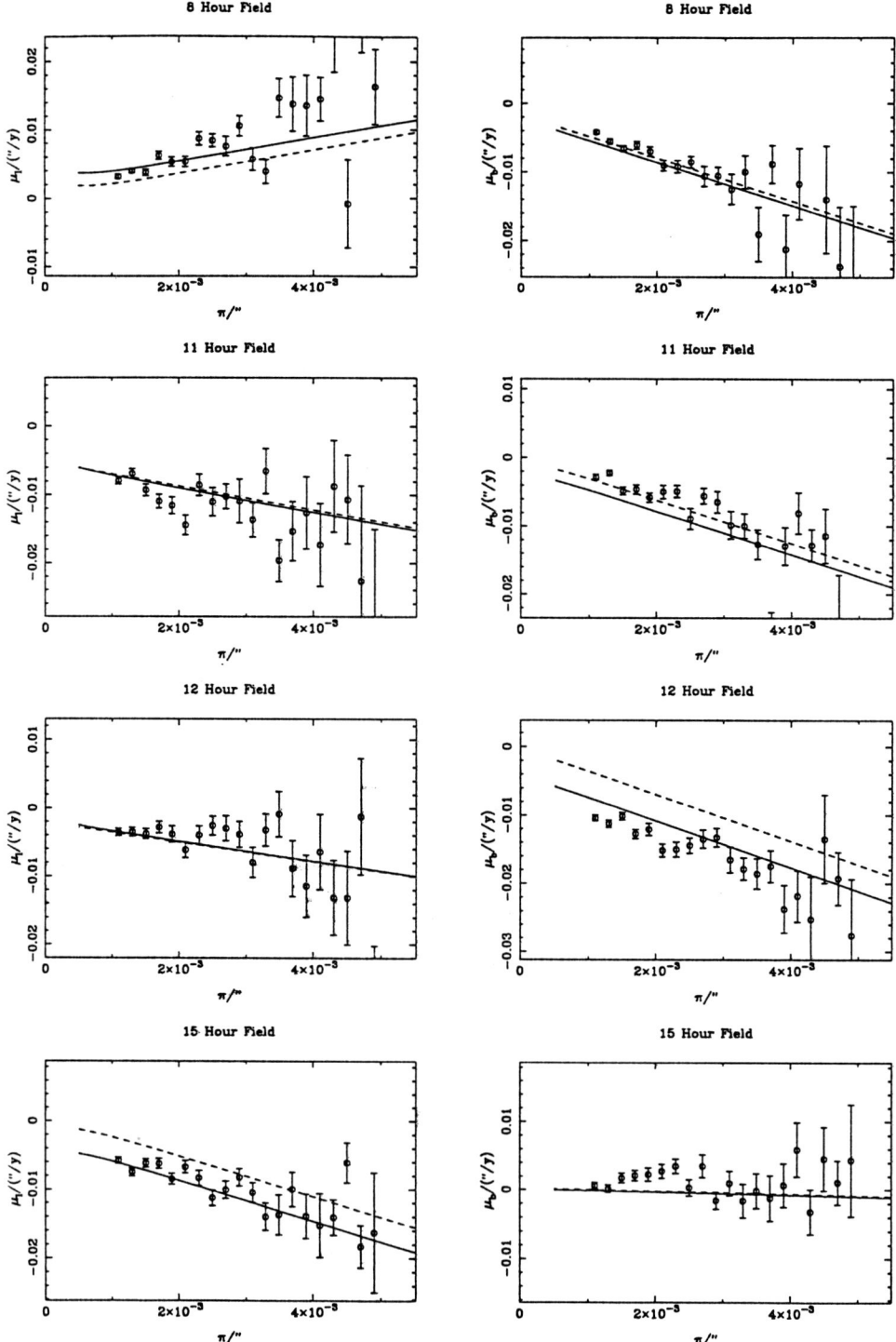

FIGURE 1. These diagrams show the proper motion versus parallax plots for the four fields of the project. The stars considered in this analysis had (O−E)> 2.0, $0''.001 < \pi < 0''.005$ and were in a central 2° radius area of the field. The solid line is the Lindblad-Oort model plus the velocity shear in the z direction and the dashed line is the same model without this shear. The error bars just represent the random measuring errors.

effects of systematic errors and have been calculated using a Monte Carlo method. The value of the Solar velocity derived for this sample of stars is in reasonable agreement with that of the Basic Solar motion, (9,11,6) km/s. The values derived for the Oort constants A and B are consistent with most recent determinations, but due to the large errors it is not possible with this data to discriminate between other peoples' results. The value for the velocity shear in the z direction of $M = -21 \pm 7$ km/s/kpc is in moderate agreement with that found originally by Murray (1986) in the Southern Galactic hemisphere of $M = -36 \pm 5$ km/s/kpc. The validity of this additional term to the Lindblad-Oort model can be seen in the 12 Hour Field μ_b versus π plot in Figure 1.

Using measurements of the velocity dispersions as a function of height, it was found that this shear is accompanied by an increase in the velocity dispersion similar to that seen with the Strömberg drift.

REFERENCES

Balona, L.A. & Feast, M.W., 1974, MNRAS, 167, 621
Chiu, L.-T.G., 1980, ApJS, 44, 31
Evans, D.W., 1988, PhD thesis, Cambridge University
Evans, D.W., 1989, A&AS, 78, 249
Evans, D.W., 1992, MNRAS, 255, 521
Murray, C.A., 1986, MNRAS, 223, 649.

Structure and kinematics of the Galaxy with Schmidt plates

By N. V. KHARCHENKO[1] E. SCHILBACH[2] AND R.-D. SCHOLZ[2]

[1]Main Astronomical Observatory, Ukrainian Academy of Sciences, Goloseevo, Kiev 252127, Ukraine

[2]WIP-Astronomie, Universität Potsdam, Sternwarte Babelsberg, An der Sternwarte 16, 1590 Potsdam, Germany

Absolute proper motions and stellar B, V magnitudes are determined in two sky regions near the North Galactic pole (NGP) by means of Tautenburg Schmidt plates. The limiting magnitude is $B = 20$, and the overall sky area investigated is 16 square degrees. Kinematical characteristics in the galactocentric and rotational directions, eccentricities of galactic orbits, parameters of spatial distribution and their changes with galactic Z distance are obtained. Four subsystems distinguished in the NGP direction have spatial boundaries at $Z = 0.3, 1.0$ and 4.5 kpc and age boundaries of 0.1, 0.9 and 0.95 of the Galaxy age. For these subsystems the semi-thicknesses are 0.15, 0.3, 0.65 and 1.5-2.5 kpc, respectively. In these subsystems there are four different relations between the velocity in the galactic rotation direction with respect to the Local Standard of Rest and the velocity dispersion: $\sigma = 2v, 16.0 - 1.3v, 53.2 - 0.4v, 150$. These are the consequence of different dynamical conditions in the Galaxy during the formation of these subsystems.

1. Introduction

The importance of studying fields close to the galactic meridian and poles has been long recognized. Due to the axial symmetry of the galactic structure, these fields are well suited for the study of gradients of galactic parameters, and then for dynamics and evolution of the Galaxy. Investigations in the frame of the programme of the meridional section (programme MEGA) have been carried out (Einasto et al. 1985) for the last ten years. The programme includes 64 selected sky areas for the determination of astrometric and physical observed stellar characteristics; 47 fields are the Kiev part (Kharchenko 1983) and 17 fields are treated in Potsdam (Schilbach 1987) by means of plates obtained with the largest Schmidt telescope in Tautenburg (D = 2 m, F = 4 m). Kharchenko (1993) has already obtained kinematical and spatial characteristics down to 2 kpc on the basis of the 47 Kiev fields. Here we present the first stellar statistical results down to 10 kpc and more obtained on the basis of Tautenburg Schmidt plates.

2. Observations

We have selected two fields near the North Galactic Pole; the 14th field of the MEGA programme (F14) and the field with the globular cluster M3. Table 1 gives the data for these fields.

In these fields we have determined the absolute proper motions of about 10 000 stars with respect to 3000 galaxies and their stellar magnitudes in the B, V Johnson system (the area of the cluster M3 itself was excluded). The rms errors of these data are ±3 mas/yr and ±0.1 mag., respectively. The magnitude equation of the proper motions was studied by means of galaxies (F14) and of cluster members (M3). There is no magnitude equation in F14, but in the M3 field we found a large error in the y proper motions. After correcting for this error, the mean proper motions $(\overline{\mu}_x, \overline{\mu}_y)$ and the orientation

	F14	M3
$\alpha_{1950.0}$	$12^h 48.8^m$	$13^h 40.3^m$
$\delta_{1950.0}$	$+30° 28'$	$+28° 41'$
(l,b)	$(80°,+88°)$	$(42°,+79°)$
Area	8.9 sq. degrees	7.5 sq. degrees
B_{lim}	19.0	20.5
No. of plate pairs	2	5
Measuring machine	MAMA, Paris	APM, Cambridge

TABLE 1. Field characteristics.

$\overline{B}$	field	$\overline{\mu}_x$	$\overline{\mu}_y$	σ_{μ_x}	σ_{μ_y}	ϕ
13.5	F14	-6.0 ± 1.4	-6.2 ± 1.2	18.6	16.0	144°
	M3	-8.0 ± 1.5	-5.3 ± 1.5	16.8	11.0	178°
14.5	F14	-8.0 ± 1.0	-8.7 ± 0.9	16.3	14.4	151°
	M3	-4.9 ± 0.9	-6.7 ± 0.9	13.8	11.3	162°
15.5	F14	-7.2 ± 0.7	-5.6 ± 0.6	14.4	12.7	161°
	M3	-5.7 ± 0.7	-6.8 ± 0.7	12.8	10.4	165°
16.5	F14	-7.4 ± 0.4	-6.3 ± 0.4	12.4	11.2	168°
	M3	-4.9 ± 0.5	-6.2 ± 0.5	10.7	9.1	167°
17.5	F14	-6.3 ± 0.3	-5.9 ± 0.3	11.2	9.9	169°
	M3	-5.2 ± 0.3	-6.3 ± 0.4	9.1	7.9	171°

TABLE 2. Proper motion characteristics in two fields (in mas/yr).

V	$B-V < 0.2$	$0.2 < B-V < 1$	$B-V > 1$
< 14	16 (2.8%)	433 (83%)	80 (14.1%)
$14 - 15$	17 (3.3%)	455 (88%)	45 (8.7%)
$15 - 16$	17 (4.9%)	756 (87%)	96 (11.0%)
$16 - 17$	49 (3.4%)	1179 (82%)	203 (14.2%)
$17 - 18$	92 (4.6%)	1477 (73%)	430 (21.5%)

TABLE 3. Colour-magnitude distribution.

angle (ϕ) of the dispersion ellipse in two fields are identical in accordance with the t-test at the level of 0.95 (see Table 2. The reason for the different values of proper motion dispersions ($\sigma_{\mu_x}, \sigma_{\mu_y}$) is the different number of plate pairs in two fields. This fact was taken into account by giving different weights to the proper motions when stellar kinematical characteristics were calculated.

3. Kinematical and spatial stellar characteristics

The distances were determined by estimating absolute stellar magnitudes which were obtained on the basis of a joint analysis of the colour–magnitude diagram (see Table 3), proper motions and reduced proper motions, the colour–absolute magnitude diagrams of Population I (Straizis 1977) and of the globular cluster M92 (Sandage 1970).

The value of interstellar light absorption near our fields was found to be 0.03–0.06 mag. from the results of spectral classification (Andruk et al. 1992). For computing the

Z [kpc]	β [kpc]	a/b	T/T_G	σ [km/s]
	0.15	0.1		$2v$
0.3			0.10	
	0.30	0.25		$16.0 - 1.3v$
1.0			0.90	
	0.65	0.4		$53.2 - 0.4v$
4.5			0.95	
	1.5-2.5	0.5		150

TABLE 4. Subsystem characteristics in the NGP direction.

velocity components and spatial distribution, the influence of proper motion rms errors, solar velocity with respect to the LSR, as well as limited distances of stars with different absolute magnitudes, were taken into account.

From the analysis of the kinematical and spatial distributions of stars in our fields we have concluded that four subsystems are observed in the NGP direction. The spatial boundaries (Z) of these subsystems are given in Table 4.

The relative age boundaries (T/T_G) have been derived from the age calibration of the relationship between the rotation velocity around the galactic Z-axis (v) and velocity dispersion (Einasto 1973). This relationship is given in the last column. The different coefficients in the relationship for each subsystem are the consequence of different dynamical conditions in the Galaxy during their formation. The density ellipsoid axis ratios (a/b) have been calculated on the basis of the quantitative connection between (a/b) and velocity dispersion (σ) for long-period variables stars (Kharchenko 1993). From the value of the relationship between the semi-thickness (β) and the axis ratio of the density ellipsoid, we have concluded that the bulk of the stellar population of the Galaxy is concentrated within Z distances of 5 kpc and galactic radius of 20 kpc.

REFERENCES

Andruk, V.N., Bartashiute, S.A., Kharchenko, N.V., Malyuto, V.D., & Shvelidze, T.D., 1992, in *The feedback of chemical evolution on the stellar content of galaxies*, ed. D. Alloin & G. Stasinska, pp. 189–192, Observatoire de Paris

Einasto, J., 1973, *Astron. circular*, No. 790, 3

Einasto, J., Malyuto, V.D. & Kharchenko, N.V., 1985, *Astron. circular*, No. 1394, 1

Kharchenko, N.V., 1983, *Astrometrija i astrofizika*, **49**, 61

Kharchenko, N.V., 1993, in *Developments in astrometry and their impact on astrophysics and geodynamics*, ed. I.I. Mueller & B. Kolaczek, 231, Kluwer Academic Publishers, Dordrecht

Kharchenko, N.V., 1993, *Kinematics and physics of celestial bodies*, in press

Sandage, A. 1970, Main-sequence photometry, colour-magnitude diagrams, and ages for the globular clusters M3, M13, M15, and M92, ApJ, **162**, 841

Schilbach, E., 1987, *Tartu Astrofuusika observ. Teated*, No. 84, 8

Straizis, V., 1977, *Multicolour photometry of stars*, 312 pp Mokslas, Vilnius.

Evolution of old populations of our Galaxy

By P. E. NISSEN

Institute of Physics and Astronomy, University of Aarhus, DK-8000 Aarhus C, Denmark

New accurate data on the chemical composition and ages of F and G main sequence stars in the solar neighborhood based on high resolution spectroscopy and Strömgren photometry is discussed and compared with kinematical data. The results point to the existence of three old, kinematically discrete populations:

(i) The old disk stars with ages between 3 and approximately 10 Gyr, [Fe/H] between -0.6 and $+0.3$, and α-element/iron ratios close to the solar ratio.

(ii) The thick disk stars with ages >10 Gyr, [Fe/H] between -1.2 and -0.4, and non-solar α-element/iron ratios that depend on the galactocentric distance of the birthplace of the star.

(iii) The halo stars with ages >10 Gyr, [Fe/H] less than about -1.0, and a non-solar α-element/iron ratio that seems to be independent of [Fe/H].

The three populations overlap in [Fe/H] and age. For each population the velocity dispersions and $<V>$ are nearly independent of age and [Fe/H]. The large majority of the thick disk stars in the solar neighborhood appear to have been formed in the inner parts of the Galaxy.

Some implications of these results for evolutionary scenarios of the Galaxy are discussed.

---

1. Introduction

F and G main sequence stars in the solar neighborhood ($d < 200$ pc) are bright enough that the detailed chemical composition of their atmospheres can be determined from high resolution spectroscopy. Hence, the abundances of elements representing various sites and time scales of nucleosynthesis can be studied, e.g. (i) the α-elements (O, Mg, Si, Ca and Ti) produced by massive stars exploding as SNeII on a short time scale, (ii) the iron-peak elements produced by SNeII *and* less massive stars exploding as SNeI on a time scale of typically 1 Gyr, and (iii) the s-process elements synthesized in AGB stars on a time scale of 3-5 Gyr.

The F and G stars have deep outer convection zones. This probably prohibits significant changes of the abundances of the heavier elements in their atmospheres as a function of time. The observed composition of a star therefore represents the composition of the matter in the Galaxy at the time and at the place of the birth of the star. The very good agreement between solar photospheric and meteoritic abundances and the constancy of abundances of various elements as a function of T_{eff} in star clusters like Hyades and Coma support this assumption. Furthermore, the F and G stars have ages ranging over the whole lifetime of the Galaxy. Thus, by studying the relations between chemical composition, ages and kinematics for F and G stars in the solar neighborhood we may learn a lot about the chemical and dynamical evolution of the Galaxy. In the following some new studies in this field are briefly reviewed and discussed.

2. The Edvardsson et al. sample of disk stars

A survey of the chemical composition, ages and kinematics of 189 F and G stars has recently been completed by Edvardsson et al. (1993a). Details about this work may be found in their rather extensive paper. The stars were selected from the Olsen (1988) catalogue of $uvby$-β photometry, which contains nearly all F and early G-type stars brighter than $V = 8.3$. Using the β and c_1 indices, stars with $5600 < T_{\text{eff}} < 7000$ K and somewhat evolved from the ZAMS were first selected and then divided into nine [Fe/H] intervals ranging from -1.0 to $+0.3$ by the aid of the m_1 index. High resolution, high S/N spectroscopy were obtained for the ~ 20 brightest stars in each metallicity range. Hence, the stars are evenly distributed in [Fe/H] but the sample is without any kinematical bias.

¿From a model atmosphere analysis of the observed spectra the abundances of 13 different elements were determined. As discussed by Edvardsson et al. the error of [Fe/H] is 0.05 dex, whereas the error of $[\alpha/\text{Fe}] \equiv \frac{1}{4}([\text{Mg/Fe}]+[\text{Si/Fe}]+[\text{Ca/Fe}]+[\text{Ti/Fe}])$ is 0.03 dex only. The error of T_{eff} as determined from $b-y$ is less than 100 K.

The age of a star is determined from comparing its position in the $\log T_{\text{eff}}$-δM_V diagram with isochrones computed by VandenBerg (1985). Here δM_V denotes $M_V(ZAMS) - M_V(star)$ as determined from the Balmer discontinuity index, c_1. The error of the *differential* ages of the stars is estimated to be about 25%, corresponding to an error of the logarithmic age, $\sigma(\log Age) = 0.10$. The absolute values of the ages are, however, more uncertain.

The kinematical parameters are computed from accurate radial velocities (± 1 km s^{-1}) and proper motions based on second epoch observations with the Carlsberg Automatic Meridian Circle for the large majority of the stars. The distances of the stars follow from absolute magnitudes as determined with the $\delta c_1 - \beta$ method of Crawford (1975). The accuracy of the distance determinations is estimated to be about 15%. Hence, the largest contribution to the error of the space velocity ($\sigma < 5$ km s^{-1}) comes from the error of the distance determination.

Binaries for which the light from the secondary component affects the photometric or spectroscopic data by a significant amount were excluded. Each star was checked for interstellar reddening by comparing the β and $b-y$ indices of the star, but in all cases the amount of reddening was negligible.

As the distances are critical for the accuracy of the space velocities we have compared the photometric distances with distances from the new General Catalogue of Trigonometric Parallaxes by van Altena et al. (1993). Fig. 1 shows the comparison for stars having $\sigma(\pi)/\pi < 0.20$. Taking into account the errors the agreement is quite satisfactory although a few stars deviate by more than 3σ.

3. The age-abundance-kinematics relations in the disk

The radial velocity, distance and proper motion data were used to calculate velocity components U, V, and W relative to the sun. Furthermore, galactic orbits were computed as described in Edvardsson et al. (1993a). The mean value of the apogalactic and the perigalactic distance of a star is called R_m. As discussed by Grenon (1987) R_m only changes by about ± 0.5 kpc as a result of 'orbital diffusion' (Wielen 1977). Hence, it seems reasonable to adopt R_m as a best estimate of the distance from the galactic center at which the star was originally formed in a nearly circular orbit.

The Edvardsson et al. paper contains a detailed discussion of the relations between various abundance ratios. Here we limit ourselves to a few of the more important results. In Fig. 2 $[\alpha/\text{Fe}]$ is plotted versus [Fe/H] with different symbols for stars formed in (i) the inner regions of the Galaxy, $R_m \leq 7$ kpc, (ii) the 'solar cylinder', $7 < R_m < 9$ kpc, and (iii) the outer regions, $R_m \geq 9$ kpc. For [Fe/H] > -0.4 $[\alpha/\text{Fe}]$ is close to zero, and the scatter in $[\alpha/\text{Fe}]$ is 0.04 dex only. For [Fe/H] < -0.4 both $[\alpha/\text{Fe}]$ and the scatter increase. It is, however, interesting that the degree of 'overabundance' of α-elements with respect to iron is correlated with R_m. At a given value of [Fe/H] stars formed in the inner regions of the Galaxy tend to have higher $[\alpha/\text{Fe}]$ values than stars formed in

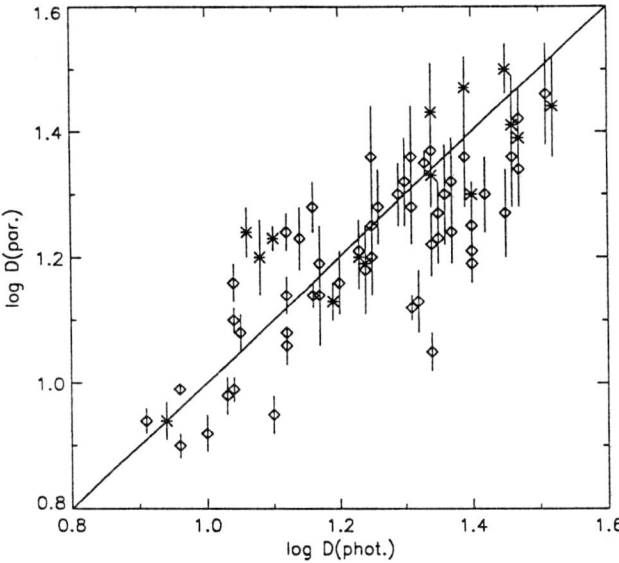

FIGURE 1. Comparison of logarithmic distances (in pc) determined from Strömgren photometry and from parallaxes in the new catalogue of van Altena et al. (1993). Error bars indicate ± the standard error. ◇ refers to stars with $-0.4 < [Fe/H] < +0.3$ and ∗ to stars with $-1.0 < [Fe/H] < -0.4$.

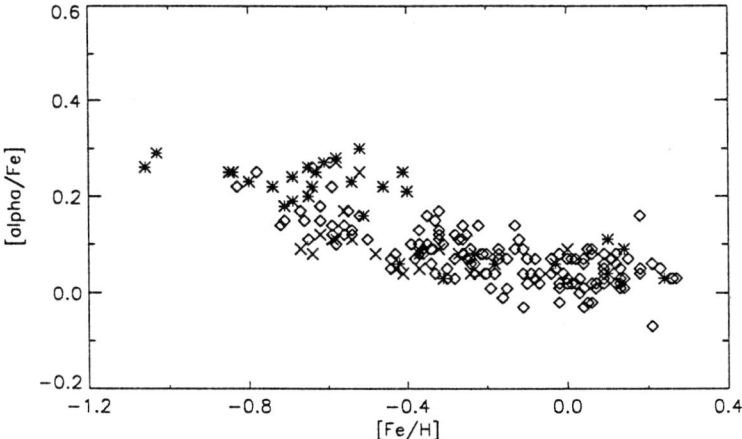

FIGURE 2. $[\alpha/Fe]$ versus $[Fe/H]$. The symbols refer to the following mean galactocentric distances of the stars: ∗, $R_m \leq 7$ kpc; ◇, $7 < R_m < 9$ kpc; ×, $R_m \geq 9$ kpc.

the outer regions. Assuming that the transition from a high value of $[\alpha/Fe]$ to a solar value is due to the appearance of supernovae of type I we conclude that the the chemical evolution has proceeded faster in the inner regions of the galactic disk. This is consistent with models of disk formation by Burkert et al. (1992) implying that the disk formed from inside out.

Another interesting result from the Edvardsson et al. survey is the large scatter in

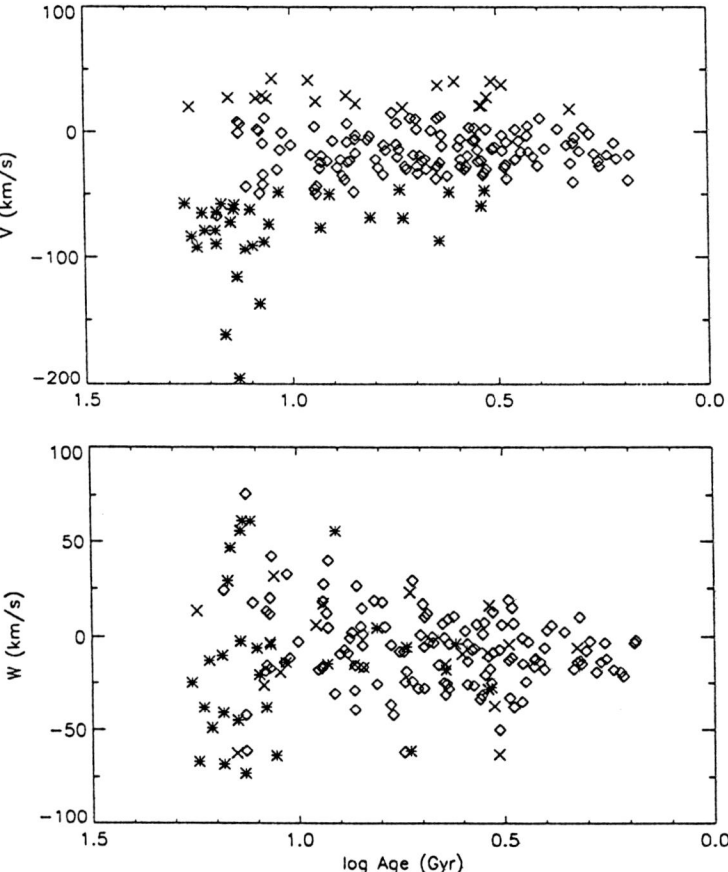

FIGURE 3. The velocity components V, and W versus the logarithm of the stellar age. The symbols refer to the following mean galactocentric distances of the stars: $*, R_m \leq 7$ kpc; $\diamond, 7 < R_m < 9$ kpc; $\times, R_m \geq 9$ kpc.

the age-metallicity relation as shown in their figure 14. At a given age and R_m the rms scatter in [Fe/H] is about 0.20 dex, *i.e.* four times larger than the estimated error of [Fe/H]. The corresponding very small scatter in [α/Fe] makes it difficult to explain the large scatter in the age-metallicity relation by local bursts of star formation, because such bursts would produce an even larger scatter in [α/Fe] (Gilmore & Wyse 1991). As discussed by Edvardsson et al. (1993a) there is no obvious explanation of the large scatter in the age-metallicity relation as due to observational errors or 'orbital diffusion'.

In Fig. 3 the V and W velocity components of the stars in the Edvardsson et al. sample are plotted versus the logarithm of the age with different symbols according to R_m. As expected from simple dynamical considerations there is a tight relation between R_m and V. In particular, we note that the majority of stars older than 10 Gyr appear to have been formed in the inner regions of the Galaxy.

The distribution of W velocities shows an interesting structure. For ages between 3 and 10 Gyr, *i.e.* $0.5 < \log Age < 1.0$ the velocity dispersion in W is approximately constant (± 21 km s^{-1}), whereas it is nearly a factor of two higher (± 38 km s^{-1}) for the group of stars with $Age > 10$ Gyr. Freeman (1991) has analyzed the same data by plotting

$\Sigma |W|$ versus the age rank of the star. He found evidence for a division of the stars in three kinematically discrete populations: (i) the thin disk with $\sigma(W) \simeq 10$ km s^{-1} and $Age < 3$ Gyr, (ii) the old disk with $\sigma(W) \simeq 20$ km s^{-1} and $3 < Age < 12$ Gyr, and (iii) the thick disk with $\sigma(W) \simeq 40$ km s^{-1} and $Age > 12$ Gyr. Edvardsson et al. (1993b), on the other hand, show that if the analysis is limited to stars actually formed in the 'solar cylinder' ($7 < R_m < 9$ kpc), then the data is consistent with a heating mechanism, which increases the velocity dispersions as $(Age)^{0.4}$. We conclude that the thick disk stars now observed in the solar neighborhood have primarily been formed in the inner regions of the Galaxy with a few coming from the outer parts. Surprisingly few, if any, of the thick disk stars have been formed in the 'solar cylinder'.

4. The halo stars

Several extensive studies of the kinematics and metallicities of high proper motion stars have been completed recently. Carney et al. (1990a, 1990b) used high resolution, low S/N spectra and (mainly) UBV photometry to study metallicities and kinematical parameters for a sample of 740 stars. Nissen & Schuster (1991) and Schuster et al. (1993) used $uvby$-β photometry to make a similar study of 1214 stars. In both works the distribution of stars in the V-[Fe/H] diagram indicates the existence of three discrete populations: the old disk, the thick disk and the halo. It is unlikely that this structure of the V-[Fe/H] diagram is due to the kinematical bias of the samples selected.

The most important result from these works is that there is no evidence of a radial or vertical chemical gradient in the halo. Furthermore, there is no significant change in $<V>$ as a function of [Fe/H] for the halo stars.

Schuster & Nissen (1989) also determined ages of the turnoff stars from their position in the c_1-$(b-y)$ diagram (the HR-diagram of the Strömgren system). A mean age of about 16 Gyr was found with indication of a cosmic scatter (± 2.5 Gyr) and a weak trend with [Fe/H].

The detailed chemical composition of halo stars has been the subject of several studies, most recently by Nissen et al. (1993). A significant overabundance of α-elements is found ($[\alpha$/Fe$] = 0.4$) with a very small scatter, ± 0.06 dex. Considering that the stars analyzed are likely to have been formed in widely different parts of the Galaxy, the small dispersion in $[\alpha$/Fe$]$ suggests that the IMF was similar in different regions of the halo and that the mixing of nucleosynthesis products was very efficient, like that in the disk.

5. Discussion and future work

It seems fairly well established that the old stars in the solar neighborhood consist of at least three discrete populations: the old disk, the thick disk and the halo stars, with entirely different explanations of their kinematical properties. The velocity dispersion of the old disk is probably due to a 'heating' mechanism that saturates in about 3 Gyr. The thick disk stars have either been formed with their high velocity dispersion, mostly inside the solar orbit, or have been accelerated by a more violent event, e.g. a merger of a satellite galaxy. The halo stars have a negligible net angular momentum and insignificant gradients in their kinematical parameters as a function of metallicity. All of this supports the scenario by Searle & Zinn (1978), according to which the outer galactic halo was formed independent of the disk by accretion of smaller satellite galaxies, rather than the dissipational collapse model of Eggen, Sandage & Lynden-Bell (1962).

Much work is, however, needed before a clear picture of the formation and evolution of the Galaxy can be reached. Ages, metallicities and kinematics are needed with the same high accuracy as obtained by Edvardsson et al. for thousands of stars. Work in this direction is in progress based on the big catalogues of Strömgren photometry of F and G stars. In particular it will be interesting to look for structure in the age-metallicity-kinematics space that relates to merger events and maybe explains the puzzling large spread in the age-metallicity relation. Accurate dating of the thick disk stars relative to the halo stars is another fundamental problem.

For the halo stars it would be particular important to extend the studies of the detailed chemical composition and ages to a sample of say 200 stars, including a fair sample of stars with metallicities in the range $-4.5 < [Fe/H] < -2.5$. Such work is in progress with the very metal poor stars being selected from the Beers et al. (1992) survey. In particular, it would be interesting to see if $[\alpha/Fe]$ is strictly constant or slightly dependent on $[Fe/H]$, and if an age-$[Fe/H]$ relation is present among the halo stars. A search for clustering in the abundance-kinematics plane as a relic of satellite accretion should also be carried out.

Acknowledgements

The Danish Natural Science Research Council is thanked for support.

REFERENCES

van Altena W.F., Lee J.T., Hoffleit E.D. 1993, General Catalogue of Trigonometric Parallaxes, Yale Univ. Obs., New Haven.
Beers T.C., Preston G.W., Shectman S.A. 1992, AJ 103, 1987.
Burkert A, Truran J.W., Hensler G. 1992, ApJ 391, 651.
Carney B.W., Aguilar L., Latham D.W., Laird J.B. 1990a, AJ 99, 201.
Carney B.W., Latham D.W., Laird J.B. 1990b, AJ 99, 572.
Crawford D.L. 1975, AJ 80, 955.
Edvardsson B., Andersen J., Gustafsson B., Lambert D.L., Nissen P.E., Tomkin J. 1993a, A&A, in press.
Edvardsson B., Gustafsson B., Nissen P.E., Andersen J., Lambert D.L., Tomkin J. 1993b, in preparation.
Eggen O.J., Lynden-Bell D., Sandage A.R. 1962, ApJ 136, 748.
Freeman K.C. 1991, in 'Dynamics of Disc Galaxies', ed. B. Sundelius, Gothenburg University, 15.
Gilmore G., Wyse R.F.G. 1991, ApJ 367, L55.
Grenon M. 1987, JA&A 8, 123.
Nissen P.E., Schuster W.J. 1991, A&A 251, 457.
Nissen P.E., Gustafsson B., Edvardsson B., Gilmore G. 1993, A&A, submitted.
Olsen, E.H. 1988, A&A 189, 173.
Schuster W.J., Nissen P.E. 1989, A&A 222, 69.
Schuster W.J., Parrao L., Contreras Martinez M.E. 1993, A&AS 97, 951.
Searle L., Zinn R. 1978, ApJ 225, 357.
VandenBerg D.A. 1985, ApJS 58, 711.
Wielen R. 1977, A&A 60, 263.

Galactic structure with the GSC plate archive − Colors and proper motions in the main meridional section of the Galaxy: first results.

By M. G. LATTANZI[1] †, A. SPAGNA[2], B. M. LASKER[1], G. MASSONE[2], B. McLEAN[1], B. BUCCIARELLI[1] ‡ AND M. POSTMAN[1]

[1] Space Telescope Science Institute, 3700 San Martin Drive, Baltimore, MD21218, U.S.A.

[2] Torino Astronomical Observatory, Strada Osservatorio 20, I-10025 Pino Torinese, Italy

A study of absolute proper motions, colors, and magnitudes in 12 fields along the main meridional section of the Galaxy ($l_{II} \approx 180°$) has been initiated. This paper presents results obtained for field N081, a region of high star density in the galactic plane (with $l_{II} \approx 140°$).

1. Introduction

Large field Schmidt plate collections may be processed with modern scanning machines and analysis software to produce catalogs of colors and proper motions that extend to faint magnitudes. In pursuit of the opportunities that such cataloging offers for galactic structure and stellar statistics, we have initiated a pilot project along the main meridional section of the Galaxy ($l_{II} \sim 180° \pm 40°$). The data are from 12 fields of the *Guide Star Catalog* Schmidt plate archive (Lasker et al. 1990). These plates support a magnitude-limited survey to V=17.5 everywhere and probably fainter (V=19.5) at Southern declinations. The photometric goal is an average precision in V of $0\overset{m}{.}11$ and $0\overset{m}{.}15$ in at least one color. The astrometric goal is a proper motion precision limited only by the plate material.

Magnitude-limited surveys, being free of biases introduced by kinematic or spectroscopic selection, are especially suited for detailed analysis of the kinematic properties of the Galactic disk elements and the spheroid (Wyse & Gilmore 1986). In the context of Galactic models, it is possible to derive, eg., a new estimate of the asymmetric drift of the POP II stellar component (rotation and z-differential rotation of the halo), while data from fields close to the galactic plane may be used to provide an independent measurement of the K_z component of the local galactic potential.

A primary tool for stellar statistics is the Reduced Proper Motion diagram (RPM, cf, eg, Luyten 1922, Jones 1972), wherein the quantity $H = m + 5\log\mu + 5 = M + 5\log T + A_m - 3.38$ (m is apparent magnitude, μ total observed proper motion, M absolute magnitude, T tangential velocity in km/sec, and A_m extinction) is plotted versus color. For fields with little extinction, H is distance independent.

This tool has been shown to be prolific for the discovery of POP I white dwarfs (WD; Evans 1988), and many are expected from our survey. However, a more exciting catch may be POP II WD. Data for the (only!) five presently known (Liebert, Dahn & Monet 1988) show that they belong to a distinct region of the RPM. Their sample also shows that POP II WD will be at the faint end (V=18–20) of our survey. The existence of statistically large sample of POP II WD is of considerable relevance to the history of

† Affiliated with the Astrophysics Division, SSD, ESA; on leave from Torino Observatory
‡ On leave from Torino Observatory

the halo; cf, Flemming, Liebert & Green (1986) for an analogous WD study of the disk population.

In this first publication from the pilot program, we report on a region of 2° square, centered at $\alpha = 2^h11^m.4$ and $\delta = 59°11'$ (1950; l, b =133°, −1.7°), and located about 2.5° South-West of the center of Quick-V plate N081 (V_{pg} = IIa-D + W12; epoch=1983; Lasker et al. 1990). The center lies on the open cluster STOCK 2 (St2), which we use to control our calibrations with an independent catalog to V=15.5; this contains proper motions good to 0.5 mas/yr and photometry to $0^m.1$ (Lattanzi et al. 1993).

For the derivation of survey proper motions and colors, plate N081 was matched with the two deep POSS-I exposures (Minkowski & Abell 1963), XO081 (B_{pg} = unfiltered 103a-O, epoch=1952) and XE081 (R_{pg} =103aE + red plexiglass; epoch=1952).

Given the relatively high star density and given that none of the algorithms utilized for the derivation of both colors and proper motions have been specialized for crowded regions, we expect the results to provide a worst–case example of what can (and cannot) be accomplished with the survey material and algorithms presently at our disposal.

2. Photometric reductions

As the focus here is validation of the astrometric data and procedures, the photometric effort was accordingly limited, with primary attention to the blue band (O) and the yellow (V), and with differences between the photographic and the sequence passbands being treated with some approximations. The available photometric sequences lie about 2° North-East of the St2 field, and they have B–V colors to V$\simeq 15^m$ from the *Guide Star Photometric Catalog* (GSPC; Lasker et al. 1988) and V–R colors to V$\simeq 20^m$ from the GSPC extension (see, eg., Postman et al. 1992 and Ferrari et al. 1993).

The BVR colors are transformed to the natural bandpass of the plates before the photometric calibration. The photographic photometry is based on instrumental magnitudes linearized by the Bunclark (1982) point-spread-function (PSF) method, with the zero points set from the photometric sequences. This method, of course, requires that the PSF be identical in the plate areas of the sequence and of the program stars. Deviations from this condition will contribute to photometric error.

The internal errors, as estimated by Bunclark's code, are $0^m.15$ for N081 (V) and $0^m.19$ for XO081 (O). External validation of our photometry in this field is presently limited to the differences *Schmidt plate magnitudes − St2 magnitudes*, together with the first and second moments of their distribution. This subtraction *does not* take into account the contribution of the color terms in transforming reduced plate magnitudes to B and V magnitudes (Russell et al. 1990; Humphreys et al. 1991), and corresponding offsets are to be expected.

The V-plate differences give $< V_{pg} - V > \simeq -0^m.09$ (derived from stars fainter than 12^m) with an *rms* about the mean of $\simeq 0^m.24$. The results for the V plate are consistent with Bunclark's errors, combined with those of the GSPC, and the color term expected for this bandpass (Lasker et al. 1990). Similarly, O-plate gives an offset of $\simeq 0^m.36$ (derived from stars fainter than 14^m) and a rms close to $0^m.23$. The intrepretation of this error analysis is similar to the V, except that the contributions from the color equation are much larger. Humphreys et al. (1991) provide the relation O–B = f(B–V), from which we estimate, with the typical B–V color for our field set to 1^m (see Figure 1), an offset of about $0^m.3$. The offsets above were applied to the tranformed plate data; clearly future work will make explicit use of the color equations.

The R plate has yet to provide reliable photometry; nevertheless, it was retained in

Plate	Epoch	Objects	Selection
N081	1983.85	23 617	$V < 18.0, C = 0, e < 0.25$
XO081	1952.71	43 382	$O < 19.5$
XE081	1952.71	81 623	$E < 17.0$

TABLE 1. Number of objects available on the plates within the 2° × 2° St2 field.

this program, as astrometry of objects therefrom correctly classified as stars contributes usefully to the proper motions.

3. Astrometric reductions

The major steps in our derivation of proper motions are as follows:

• *Object selection.* Selection of objects found in plate processing is based on classification and shape parameters therefrom. Only stars and galaxies are retained, and magnitude limits are set.

• *Plate matching.* The first task is *plate orientation*. Stars from existing catalogs define the transformations (one for each axis) to convert from secondary plates to the reference plate. Next is the *global matching* task. After selection of a search radius, the objects on the master list are matched against all objects from the other plates. Single (successful) and multiple (failed) matches are generated. Last is the *selection of Reference Objects*. A subset of successful matches is flagged as possible reference objects (RO).

• *Accurate coordinate transformations.* The first step is *overlapped subplates tessellation*. The plate area is divided into smaller overlapping sections (subplates). Second is *plate coordinate transformation*. For each subplate, an accurate coordinate transformation (usually a cubic) is fit to the local RO. Third comes *collocation* improvement, to detect and remove systematic effects left in the RO residuals (Bucciarelli, Lattanzi & Taff 1993). At this stage we prepare for a possible *iteration*. Large RO residuals are flagged and eliminated. The coordinate transformation task is then repeated until convergence is obtained. Usually, this happens in two iterations.

• *Proper motion reduction.* Once all positions are in a consistent frame, proper motions μ_x, μ_y and their errors are computed via linear least squares.

Plate N081 is the reference. It contains 23 617 objects brighter than 18, classified as stellar (C=0), and with image $e < 0.25$ (see Table 1). Stars with higher e are rejected, as they appear systematically brighter than the comparison photometry.

The alignment solution, good to $\sigma_{x,y} \simeq 0.62$ pixels (1 pixel$\simeq 1\farcs 7$) for the XO081-to-N081 (*blue*) transformation and to $\sigma_{x,y} \simeq 1.4$ pixels for the XE081-to-N081 (*red*) transformation, uses 45 PPM stars. The search radius was set to $r \simeq 14\farcs 4$, i.e., four times the mean error of the red transformation; this is close to the mean separation among objects ($d \sim 25''$) on the E plate. As reference stars (no galaxies are available in this field) for the accurate transformations, we selected ~ 650 stars in V=[16,18].

Before the matching, pairs closer than r on the master plate (N081) were discarded to minimize ambiguous matches. The number of objects common to all three plate is 10 269 (52% of the sample originally available on the V plate). Plots of star counts and color distributions show that no significant bias has been introduced in the final sample.

4. Proper motion accuracy

Proper motion errors principally depend on two factors: (a) plate measurements and (b) coordinate transformations, which remove the geometrical and optical distortions. We can evaluate the expected proper motion error from the fit-error of the transformations to the RO (assumed to be at rest), i.e. $\sigma_{BV} \simeq 0.14$ pixels (each coordinate) for the blue transformation, and $\sigma_{RV} \simeq 0.18$ pixels for the red. Taking the mean proper motion $<\mu> = (\mu_{BV} + \mu_{RV})/2$ as best estimate, the application of the error propagation formula (the correlation between μ_{BV} and μ_{RV} is $\simeq 0.5$ because of the common V measurement) yields $\sigma_{<\mu>} \simeq 7.6$ milli-arcseconds (mas) per year.

As internal check, we also consider proper motions obtained with the plate pairs XO081–N081 and XE081–N081: $\Delta\mu_x = \mu_{RV} - \mu_{BV} = (x_R - x_B)/\Delta t$, and similarly in y. Because the XO081 and XE081 plates have the same epoch, the $\Delta\mu$ are simply proportional to the position errors associated with both centroider and transformation. For our 10 269 objects, $\Delta\mu_x = 0.6 \pm 10.0$ mas/yr and $\Delta\mu_y = -1.1 \pm 10.3$ mas/yr (1σ). As expected, this is slightly larger than the individual σ_μ's because the transformation (and its error) is now applied on both XO081 and XE081 objects. However, these estimates confirm that the errors are dominated by centroiding rather than the transformation.

External validation was done against our independent catalog in St2. Its proper motions are on the FK4 system as defined by the AGK3U catalog (Bucciarelli et al. 1992). After converting the survey μ_x, μ_y to $\mu_\alpha \cos\delta$, μ_δ, we compared the proper motions of 838 stars in common with the comparison catalog. Then, $\Delta\mu_\alpha \cos\delta(Schmidt - Catalog) = 3.72 \pm 7.30$ mas/yr and $\Delta\mu_\delta = 5.46 \pm 7.29$ mas/yr (1σ error); these dispersions are consistent with the expected value. The positive biases in $\Delta\mu_\alpha \cos\delta$ and $\Delta\mu_\delta$ represent the effect of a residual offset between the proper motion system defined by our 'distant' stars and the inertial frame (offsets of the same order of magnitude are obtained even after transformation of the AGK3U proper motions to the FK5 system). Also, the values of the biases are consistent with the 'distant' stars being between 1 and 2 kpc away, as explained below.

5. Results and discussion

Present knowledge of the structure and kinematics of the Galaxy is largely based on fields with $|b| \geq 30°$, and investigations of low-latitude fields such as the present may be expected to contribute to more general understanding. These fields, however, require considerable attention to effects of reddening and extinction. For the St2 field, Krzeminski & Serkowski (1967) give a mean interstellar extinction rate of 4.0 mag/kpc and an average reddening at the cluster ($\simeq 300$ pc away) of $E_{B-V} \simeq 0.35$.

Figure 1 shows the color–magnitude diagram (CMD) for all reliable stellar images ($e < 0.25$, §3). The circles for V<14.5 are from the catalog of Lattanzi et al. (1993). As noted earlier, the present photometry is not reliable for objects brighter than V=12. However, the systematic error decreases for $V > 12$ and the two sequences practically overlap at V$\approx$ 13, well above the limit of reliable photometry in the comparison catalog (V$\simeq$ 14.5). The expected location of POP I WD derived from Allen (1973) is drawn on the bottom left of Figure 1. The unreddened line is corrected for interstellar absorption at the assumed distance of 200 pc ($A_V \simeq 0.8$ and $E_{B-V} \simeq 0.2$). The dashed line on the red side of Figure 1 marks the survey limit, as imposed by the magnitude limits in Table 1.

The RPM associated with the CMD of Figure 2 (but excluding the bright stars, i.e. V<12 and B<14) is shown in Figure 2. The loci for POP I WD and for red giants

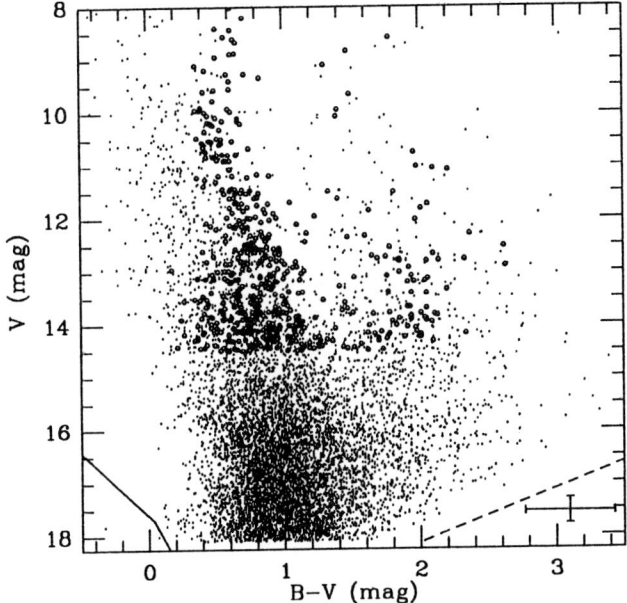

FIGURE 1. Color-magnitude diagram of all single-matched stars from our survey (*dots*). For comparison, data from the reference catalog of Lattanzi et al. (1993) are also shown (*circles*). Typical error bars for faint stars are drawn near the lower right corner.

with $(B-V)_0 = [1,2]$, drawn as solid lines in Figure 2, are based on (1) the WD color and magnitude data in Allen (1973), (2) an adopted absolute magnitude of $<M_V> = 0$ and a typical distance of 1 kpc (i.e., $A_V = 4$) for giants with intrinsic color of $\geq 1^m$, and (3) a tangenital velocity appropriate for the young disk population (eg., Jones 1972), T=30 km/sec. The natural dispersion and errors are such that 90% of the population lies within 1-2 units of H about the given loci.

WD, which are close and therefore little reddened (to the limit of our survey), and intrinsically very red giants, which have even more extreme colors if reddened, are easily identifed on the observed RPM, even in the presence of strong interstellar absorption. Hence the fair agreement Figure 2 shows between predictions and the appearance of the RPM is particularly interesting as a validation. The lack of points close to the WD track is not surprising as the expected number is between 1 and 4 WD per square degree. (We count only one candidate WD at B-V$\simeq$0.55 and H$\simeq$16 in our high proper motion sample, $|\mu| > 0\rlap{.}''04$ /yr.)

While visually inspecting the initial set of 18 candidate WD, we noted three interesting objects. An unsuccessful match is likely to be a large amplitude variable. We also noted a faint common proper motion pair (B-V$\simeq$1.4, H$\simeq$16, and $|\mu| \simeq 0\rlap{.}''12$ /yr), which could belong to the old disk population of subdwarfs. The pair is not in Luyten (1981) nor in the WDS catalog (Worley & Douglas 1984).

We are certainly encouraged by these results, and do expect further improvements after enhancing the crowded-field capabilities of the software. We are also processing plates in the region of SA57, which is expected to provide best-case results.

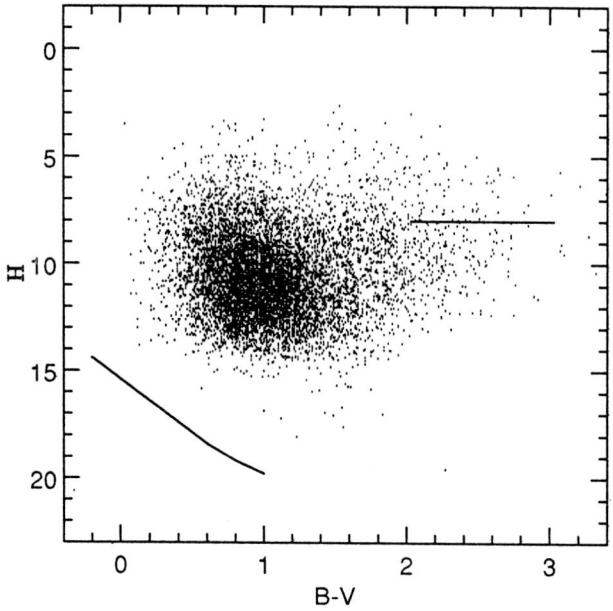

FIGURE 2. Reduced proper motion diagram. Only single–matched stars fainter than V=12 are drawn.

Acknowledgements

We wish to thank L. Siciliano for providing us with the deep photometric sequences and A. Ferrari for encouraging this collaboration between Torino and STScI. AS acknowledges the assistance of M. Sarasso and R. Morbidelli with the DIRA2 database.

Space Telescope Sc. Institute is operated by AURA, Inc. for NASA under contract No.NAS5–26555. The Italian work in this project is supported by the Italian Space Agency (ASI) under contract ASI 1992 RS 78.

REFERENCES

Allen, C.W. (1973), *Astrophysical Quantities*, The Athlone Press, Univ. of London, third edition.
Bucciarelli, B. et al. (1992), *Astron. J.* **103**, 1689.
Bucciarelli, B., Lattanzi, M.G. & Taff, L.G. (1993), *Astrophys. J. Suppl.* **84**, 91–99.
Bunclark, P.S. (1983), *Occasional Reports of the Royal Observatory, Edinburgh*, No.10, p.149 (R.S. Stobie & B. McInnes, eds.).
Evans, D.W. (1988), Cambridge University D. Phil. thesis.
Ferrari, A., et al. (1993), *The Messenger* **72**.
Flemming, T.A., Liebert, J. & Green, R.F. (1986), *Astrophys. J.* **308**, 176.
Humphreys, R.M., et al. (1991), *Astron. J.* **102**, 395.
Jones, E.M. (1972) *Astrophys. J.* **173**, 671–676.
Krzeminski, W. & Serkowski, K. (1967), *Astrophys. J.* **147**, 988.
Lasker, B.M. et al. (1990), *Astron. J.* **99**, 2019–2058.
Lattanzi, M.G., Massone, G. & Munari, U. (1993), in preparation.
Liebert, J., Dahn, C.C. & Monet, D.G. (1988), *Astrophys. J.* **882**, 891.
Luyten, W.J. (1922), *Lick Obs. Bull.* No. 336.

Luyten, W.J. (1981) *The NLTT Catalogue*, Univ. of Minnesota, Minneapolis.

Minkowski, R.L. & Abell, G.O. (1963), *Stars and Stellar Systems*, K.Aa. Strand, Ed. (Univ. of Chicago, Chicago), Vol. 3, p. 481.

Postman, M., et al. (1992), *Digitised Optical Sky Surveys*, MacGillivray & Thomson (eds.), 61–63.

Russell, J.L. et al. (1990), *Astron. J.* **99**, 2059–2081.

Wyse, R.F.G. & Gilmore, G. (1986), *Astron. J.* **91**, 855.

Worley, C.E. & Douglas, D. (1984), *Washington Double Star Catalog 1984.0*, USNO, Washington, D.C.

Galactic internal motions derived from proper motion surveys

By MASANORI MIYAMOTO

National Astronomical Observatory, Mitaka, Tokyo 181, Japan

The overall pattern analysis of the proper motions given by the modern astrometric catalogue ACRS is reviewed. We have determined the luni-solar precessional correction and the equinoctial motion correction with reference to K–M giants, which are considered to be well-relaxed in the Galaxy. Then, based on the FK5 proper motion system thus improved, we have detected a delicate systematic motion of young stars, which is inherent in the Galactic warp. The prospect for Galactic kinematics is also described, taking into consideration the huge number of modern proper motions now available and in the quite near future.

1. Prologue

The astrometric catalogues SAOC and AGK3, which have been used in the astronomical community for years, have been recently replaced with the modernized ones, ACRS (Corbin & Urban, 1991) and PPM (Röser & Bastian, 1989; Bastian & Röser, 1992). The astrometric data supplied by these catalogues are of great value not only in the astrometry, but also for Galactic kinematics: These catalogues contain a huge number of stars with a large variety of the spectral type and luminosity class, and moreover a large subset of the proper motions in these catalogues is more reliably determined than those in previous catalogues, with the benefit of observations at more than two different epochs and the up-to-date reference frame, FK5 (Fricke et al., 1988). These proper motions combined with modern astrophysical data provide the best available material for global Galactic kinematics.

The encouraging support from the modern proper motion surveys will be further strengthened, because the proper motions of a huge number of stars is being determined by the northern and southern astrographs (Klemola et al., 1987, van Altena et al., 1986), HIPPARCOS, and the photoelectric meridian circles (Carlsberg Automatic Meridian Circle – La Palma, Bordeaux Meridian Circle, USNO Meridian Circles, and Tokyo Meridian Circle) will, additionally, be available in the quite near future. Thus, we are now entering the second era of "Stellar Astrometry" after half a century. Our tractable Galactic volume with transverse velocity data of stars will extend to more than 5 kpc from the sun, even at a modest estimate. This situation is a marked improvement on the past.

I hope that the present review of the overall pattern of analysis of the proper motions, which has been carried out recently together with Sôma and Yoshizawa on the basis of the ACRS, gives an encouraging example of the application of proper motion analysis in Galactic kinematics.

2. Background of the present analysis

The equinox and the celestial pole as the origin of the celestial coordinate system move continually on the celestial sphere due to the luni-solar precession, planetary precession, and observational errors of the equinox. Therefore, the precessions and the fictitious (erroneous) equinox motion play a primary role in the reference frame. Unless the motion of the origin of the reference frame is known precisely, the proper motions given in the astrometric catalogues will not provide the precise projection of the stellar spatial motions on the celestial sphere.

In a series of papers Fricke (1967a,b, 1977a,b) determined the luni-solar precessional correction to Newcomb's value and the fictitious motion of the equinox, which formed the basis of the "IAU System of Astronomical Constants (1976)" and the fundamental reference frame, FK5 System (Fricke et al., 1988). However, the geodetic VLBI (McCarthy & Luzum, 1991) and LLR (Williams et al., 1991) observations have been suggesting an additional correction of $\Delta p \approx -0\rlap{.}''3/\text{cy}$ to the luni-solar precessional constant of the IAU System. Thus, the IAU value of the luni-solar precessional constant $p = 5038\rlap{.}''7784/\text{Julian century (J2000.0)}$ is still problematical beyond the decimal point.

If the precessional correction and possibly also an equinox correction to the FK5 system are really necessary, then, the proper motions of the modern astrometric catalogues given in the FK5 system should apparently demonstrate two systematic flows of stars around the ecliptic and celestial poles in addition to the familiar flow along the Milky Way (Galactic rotation), corresponding to the erroneous determination of the precession constant and the spurious equinox motion. Anyway, the above tiny uncertainties in the rotation of the FK5 system cannot be disregarded in the context of Galactic kinematics, because even the most basic quantity, the Galactic rotation $B - A$, amounts to only $\sim 0\rlap{.}''5/\text{cy}$, which corresponds to the proper motion of the Galactic center seen from the sun.

Thus, in the light of modern astrometric data, the rotations still left in the FK5 system should be re-examined together with the Galactic rotation and the solar motion (Miyamoto & Sôma, 1993, hereafter referred to as Paper I). Then, the modern proper motion data in the FK5 system thus improved can be used for measuring Galactic internal motions, and especially for revealing a delicate systematic motion of young stars inherent in the Galactic warp (Miyamoto et al., 1993, hereafter referred to as Paper II).

3. Analysis models of the systematic stellar velocity field and the rotations of the reference frame

In performing the overall pattern analysis of the proper motions μ_α and μ_δ given in the ACRS Part 1, we adopt the three-dimensional Ogorodnikov-Milne model (Ogorodnikov, 1932; Milne, 1935) for the systematic stellar velocity field at the solar neighbourhood, and permit two systematic rotations Δp and $(\Delta e + \Delta \lambda)$ of the FK5 system around the ecliptic and celestial poles, respectively. The first rotation, Δp, denotes the luni-solar precession correction and the second, $(\Delta e + \Delta \lambda)$, the equinox motion correction Δe and the planetary precession correction $\Delta \lambda$.

Then, we have the following equations of condition for least-squares

$$\begin{pmatrix} \mu_\alpha \cos \delta \\ \mu_\delta \end{pmatrix} = M\,X, \tag{3.1}$$

with the unknown vector

$$X^T = (S_1\ S_2\ S_3\ \omega_1\ \omega_2\ \omega_3\ D^+_{12}\ D^+_{13}\ D^+_{23}), \tag{3.2}$$

and the 2×9 known matrix M of the trigonometric functions of star positions, where (S_1, S_2, S_3) are the solar motion components in Galactic rectangular coordinates, and $(\omega_1, \omega_2, \omega_3)$ the components of the systematic rotations of stars in equatorial rectangular coordinates given by

$$
\begin{aligned}
\omega_1 &= D_{32}^- \cos\alpha_{\rm GC} \cos\delta_{\rm GC} + D_{13}^- \cos\alpha_{\rm GR} \cos\delta_{\rm GR} \\
&\quad + D_{21}^- \cos\alpha_{\rm GP} \cos\delta_{\rm GP} + \Delta p \cos\alpha_{\rm EP} \cos\delta_{\rm EP} \\
\omega_2 &= D_{32}^- \sin\alpha_{\rm GC} \cos\delta_{\rm GC} + D_{13}^- \sin\alpha_{\rm GR} \cos\delta_{\rm GR} \\
&\quad + D_{21}^- \sin\alpha_{\rm GP} \cos\delta_{\rm GP} + \Delta p \sin\alpha_{\rm EP} \cos\delta_{\rm EP} \\
\omega_3 &= D_{32}^- \sin\delta_{\rm GC} + D_{13}^- \sin\delta_{\rm GR} \\
&\quad + D_{21}^- \sin\delta_{\rm GP} + \Delta p \sin\delta_{\rm EP} - (\Delta e + \Delta\lambda)
\end{aligned} \quad (3.3)
$$

with the directions $(\alpha_{\rm GC}, \delta_{\rm GC})$, $(\alpha_{\rm GR}, \delta_{\rm GR})$, $(\alpha_{\rm GP}, \delta_{\rm GP})$, and $(\alpha_{\rm EP}, \delta_{\rm EP})$ of the Galactic rectangular coordinate axes and the ecliptic pole, respectively.

The rate-of-strain components $2D_{ij}^+$ and the vorticity components $2D_{ij}^-$ at the sun which are to be determined are given in galactocentric cylindrical coordinates by

$$
\left.\begin{aligned}
D_{12}^+ &= \frac{1}{2}\left(\frac{\partial V_\theta}{\partial R} - \frac{V_\theta}{R} + \frac{1}{R}\frac{\partial V_R}{\partial \theta}\right) \\
D_{21}^- &= \frac{1}{2}\left(\frac{\partial V_\theta}{\partial R} + \frac{V_\theta}{R} - \frac{1}{R}\frac{\partial V_R}{\partial \theta}\right)
\end{aligned}\right\}, \quad (3.4)
$$

$$
\left.\begin{aligned}
D_{13}^+ &= -\frac{1}{2}\left(\frac{\partial V_R}{\partial z} + \frac{\partial V_z}{\partial R}\right) \\
D_{13}^- &= -\frac{1}{2}\left(\frac{\partial V_R}{\partial z} - \frac{\partial V_z}{\partial R}\right)
\end{aligned}\right\}, \quad (3.5)
$$

and
$$
\left.\begin{aligned}
D_{32}^+ &= -\frac{1}{2}\left(\frac{1}{R}\frac{\partial V_z}{\partial \theta} + \frac{\partial V_\theta}{\partial z}\right) \\
D_{32}^- &= -\frac{1}{2}\left(\frac{1}{R}\frac{\partial V_z}{\partial \theta} - \frac{\partial V_\theta}{\partial z}\right)
\end{aligned}\right\}. \quad (3.6)
$$

It should be noted that the elements D_{12}^+ and D_{21}^- are identical to the familiar Oort constants A and B, respectively, under the proviso that the stellar velocity field is axisymmetric, or $V_R = 0$. Another point should be noted. Since equations (3.3) give only three conditions for the five unknowns D_{32}^-, D_{13}^-, D_{21}^-, Δp, and $(\Delta e + \Delta\lambda)$ of the systematic rotations of stars and the reference frame, we can determine at most only three unknowns. Therefore, in the classical proper motion analysis (Fricke, 1977a; Schwan, 1988), special attention was paid to determining only three unknowns Δp, $(\Delta e + \Delta\lambda)$, and the Oort constant $B = D_{21}^-$, putting implicitly the other unknowns D_{32}^- and D_{13}^- equal to zero (this case is called the Oort-Lindblad model).

4. Precession and equinox corrections to the FK5 system

On the basis of the Oort-Lindblad model, Fricke assumed implicitly that the vorticity vector of a group of his selected stars (512 FK4 and FK4 Sup stars in the heliocentric distance interval 0.1 – 1.3 kpc) had only a component (the Oort constant B) perpendicular to the Galactic plane. However, if the assumption of complete perpendicularity of the vorticity vector does not hold for the group of stars, the precession correction and the equinox motion thus determined might be affected by the other vorticity components lying in the Galactic plane.

In fact, Fricke's material is an inhomogeneous mixture of stars, biased to young ages; so that these stars are not guaranteed to be well-relaxed, and as a result do not always follow only the familiar Galactic rotation. The nearer stars in his material are disturbed by localized velocity fields (star streams, Gould-belt, etc.), and the motion of the more distant young stars are, on the other hand, liable to be disturbed by the Galactic warp (see Paper II).

Now, the main constitution of the Galaxy can be considered to be in a steady-state, with some geometrical symmetries, so long as drastic dynamical behaviour of the Galaxy is not introduced. The stellar system in such a state exhibits only the familiar plane-parallel Galactic rotation described by the Oort constants A and B, and as a consequence, the vorticity vector of the system is perpendicular to the Galactic plane. Thus, it is natural to define the luni-solar precession motion and the equinox motion with reference to a representative group of stars in a steady-state, since the overall pattern analysis of the proper motions cannot determine both the three components of the vorticity vector of the stellar motion and the rotations left in the reference frame, as mentioned above.

The old and well-relaxed population of stars such as K–M giants are supposed to have already reached a steady-state. Thus, we have chosen from ACRS Part 1 a group of about 30 000 K–M giants with heliocentric distance greater than 500 pc, avoiding the Gould-belt region.

In Paper I, taking first the luni-solar precession correction $\Delta p = -0\rlap{.}''3/\mathrm{cy}$ indicated by the VLBI and LLR observations as an initial guide, we have shown in the three-dimensional Ogorodnikov-Milne model that the generally accepted idea that the K–M giants are a steady-state constituent of the Galaxy is compatible with the luni-solar precession correction proposed by the VLBI and LLR observations. Next, applying the two-dimensional Oort-Lindblad model to these stars, we have determined the corrections Δp and $(\Delta e + \Delta\lambda)$ to the FK5 system, together with the Oort constants A and B and the solar motion. The results are given in the second column of Table 1, where the first line gives the solution and the second line the standard error. The precession correction thus obtained confirms the one proposed by the VLBI and LLR observations.

Some remarks need to be made. The present result gives the Galactic rotational velocity, $V_\theta = -177$ km/s, which is considerably smaller than the recommended IAU (1985) value (-220 ± 20 km/s). The present sample may be a mixture of K–M giants with a wide range of metallicity and contain a large amount of metal-deficient K–M giants which are members of the slowly rotating thick-disk and halo of the Galaxy. But, the slow Galactic rotation of the present sample does not in itself distort the present determination of Δp and $(\Delta e + \Delta\lambda)$, since it is the perpendicularity of the vorticity vector to the Galactic plane which is of prime importance. In the next section, we shall show that young stars chosen from the ACRS Part 1 give a result for the Galactic rotational velocity which is almost identical to that of the IAU recommendation.

5. Kinematics of the Galactic Warp

It has been known since the early H I surveys of our Galaxy that the H I gas layer in the outer part of the Galaxy is distorted systematically above the Galactic plane on the northern hemisphere and below it on the southern hemisphere. The distortion of the H I layer (Galactic Warp) starts near the solar neighbourhood, and the nodal line of the warp happens to be quite close to the Galactic center–sun–anticenter line (see Fig. 2 of Henderson et al., 1982). More recently, it has become clear that such non-planar distortions in the outer H I gas layers of spiral galaxies are more the rule than the exception. However, almost no information has not yet been obtained on the kinematics of the warp. In fact, in order to delineate the geometric form of the warp, the radial velocity data of the H I gas have usually been used for locating the gas with pure circular rotation around the galactic center, not only in our Galaxy, but also in nearly face-on galaxies (Rogstad et al., 1974 and 1976; Bosma, 1981; Bottema et al., 1987).

Unknowns	K–M Giants	Young Stars
Δp	$-0\overset{''}{.}267$ / cy $\pm 0\overset{''}{.}028$	$-0\overset{''}{.}267$ / cy (fixed)
$\Delta e + \Delta \lambda$	$-0\overset{''}{.}116$ / cy $\pm 0\overset{''}{.}026$	$-0\overset{''}{.}116$ / cy (fixed)
S_1	$+13.6$ km/s ± 0.3	$+ 8.7$ km/s ± 0.8
S_2	$+23.3$ km/s ± 0.3	$+15.9$ km/s ± 0.8
S_3	$+11.9$ km/s ± 0.3	$+ 9.1$ km/s ± 0.7
S_{total}	29.5 km/s	20.3 km/s
D_{12}^{+} (A)	$+0\overset{''}{.}263$ / cy $\pm 0\overset{''}{.}012$	$+0\overset{''}{.}285$ / cy $\pm 0\overset{''}{.}019$
D_{21}^{-} (B)	$-0\overset{''}{.}176$ / cy $\pm 0\overset{''}{.}010$	$-0\overset{''}{.}260$ / cy $\pm 0\overset{''}{.}015$
D_{13}^{+}	$0''$ / cy	$-0\overset{''}{.}059$ / cy $\pm 0\overset{''}{.}011$
D_{13}^{-}	$0''$ / cy	$+0\overset{''}{.}059$ / cy $\pm 0\overset{''}{.}011$
D_{23}^{+}	$0''$ / cy	$+0\overset{''}{.}039$ / cy $\pm 0\overset{''}{.}010$
D_{32}^{-}	$0''$ / cy	$+0\overset{''}{.}039$ / cy $\pm 0\overset{''}{.}010$
V_θ	-177.1 km/s ± 6.2	-219.9 km/s ± 9.8

TABLE 1. Results for the two-dimensional Oort-Lindblad model

Since young optical objects such as O–B stars, supergiants, bright giants, etc., have a very close relation to the H I gas, these objects can be considered as an optical counterpart of the H I gas, and it is expected that these young stars share the same kinematic behaviour as the H I gas. Thus, in order to detect the warping motion inherent in the Galactic warp, we have chosen about 3000 O–B5 stars, supergiants, and bright giants in the domain $0.5\,\text{kpc} \leq r \leq 3.0\,\text{kpc}$ and $|z| \leq 0.25\,\text{kpc}$ (avoiding the Gould-belt region) from the ACRS Part 1.

Now, we have already determined the values of the corrections Δp and $(\Delta e + \Delta \lambda)$ to the FK5 system. Fixing these corrections in equations (3.3), we can determine the nine kinematical parameters S_1, S_2, S_3 and $D_{ij}^{\pm}$ for these young stars, in principle, on the basis of the Ogorodnikov-Milne model (see Paper II). However, we should pay attention to the fact that the derivatives of the systematic velocity with respect to z are hardly detectable, since the layer of the young stars is very thin ($h_z < 100\,\mathrm{pc}$) and the tilt of the layer is very small. Therefore, in the present model of the warping motion we assume that the derivatives with respect to z are zero. This means that

$$D_{13}^+ = -D_{13}^- \quad \text{and} \quad D_{23}^+ = +D_{32}^-. \tag{5.7}$$

Under these conditions, we have found the kinematic parameters. The results are given in the third column of Table 1. The standard errors of $D_{13}^+ = -D_{13}^-$ and $D_{23}^+ = +D_{32}^-$ are even smaller than those of the Oort constants D_{12}^+ and D_{21}^-, so that these parameters are determined fairly well. The results given in Table 1 provide new kinematic information on the young stars. We have

$$2D_{13}^- = -2D_{13}^+ = \frac{\partial V_z}{\partial R} = 5.6 \pm 1.0 \text{ km/s/kpc}, \tag{5.8}$$

and

$$2D_{32}^- = 2D_{23}^+ = -\frac{1}{R}\frac{\partial V_z}{\partial \theta} = 3.7 \pm 0.9 \text{ km/s/kpc}, \tag{5.9}$$

in addition to the solar motion 20.3 ± 1.3 km/s and the classical Galactic rotation -219.9 ± 9.8 km/s at $R = 8.5$ kpc. These results imply that the thin sheet of young stars is performing a rotation around the axis pointing to the Galactic center in a positive sense with the angular velocity 4 km/s/kpc and simultaneously another rotation around the axis pointing to $l = 90°$ and $b = 0°$ with an angular velocity of 6 km/s/kpc. In order to simplify the apparent view of the warping motion, we introduce the concept of a kinematic warp, defining the inclination ϵ of the kinematic warp with respect to the Galactic plane by

$$\epsilon = \tan^{-1}|D_{13}^-/D_{21}^-|, \tag{5.10}$$

where $\epsilon = 12°\!.7$. Then, as is shown in Figure 1, the young stars as the optical counterpart of the H I warp are streaming around the Galactic center in the plane of the kinematic warp with the velocity -225 km/s, and simultaneously the plane is rotating around the Galactic center–sun–anticenter line in the sense of increasing the tilt of the plane with an angular velocity of 4 km/s/kpc. But this tilting may simply be an oscillation in a secular sense.

6. Epilogue

In the present work, the modern proper motion data are fitted to a simple model of the Galactic internal motions by using the least-squares method. Considering that our tractable Galactic volume with the spatial velocity of stars will expand to more than 5 kpc from the sun in the near future, it is very interesting to apply the maximum likelihood method to fitting a more sophisticated model of the internal motions, which includes the velocity ellipsoid, the shear with respect to the Galactic height of the Galactic rotation, etc.

We have not yet performed a similar overall pattern analysis of the proper motions taken from PPM. Prior to that analysis, we should know the systematic difference between ACRS and PPM proper motion systems, although these catalogues have been compiled on the same FK5 system. The systematic difference should be represented objectively by the method developed by Brosche (1966). However, a gross systematic difference can be found by an easy way (as for the cross-identification between ACRS and PPM, see Sôma and Miyamoto, 1993). Forming the differences of the proper motions $\Delta\mu_\alpha$ and $\Delta\mu_\delta$ between ACRS and PPM, and then applying the least-squares method to

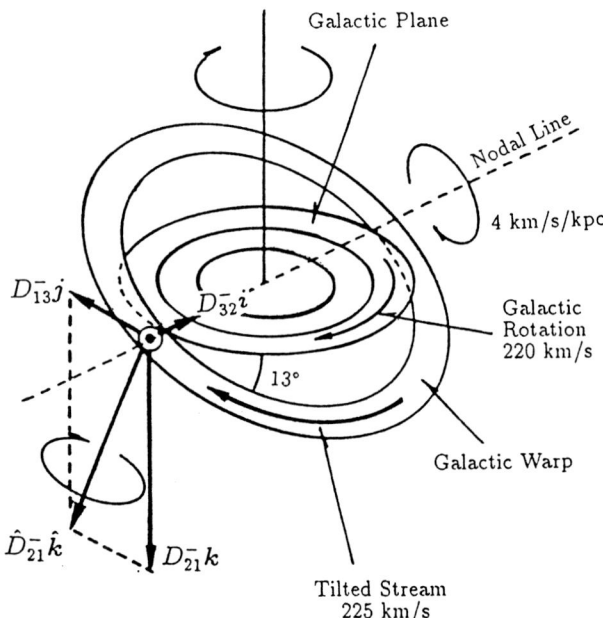

FIGURE 1. A schematic view of the Galactic warping motion

the equations of condition (Sôma, 1993)

$$\left.\begin{aligned}\Delta\mu_\alpha \cos\delta &= \omega_1 \cos\alpha \sin\delta + \omega_2 \sin\alpha \sin\delta - \omega_3 \cos\delta \\ \Delta\mu_\delta &= -\omega_1 \sin\alpha + \omega_2 \cos\alpha\end{aligned}\right\}, \quad (6.11)$$

we can express the systematic differences as the rotations ω_1, ω_2, and ω_3 around the three axes of the equatorial rectangular coordinates. Thus, we found that the total difference between ACRS- and PPM-proper motion systems amounted to 0″.1/cy, which seems to be rather large from the point of view of determining accurately Galactic internal motions. The cause of this difference should be carefully investigated.

REFERENCES

Bastian, U. & Röser, S. 1992 PPM South Star Catalogue, Astron. Rechen-Institut Heidelberg.
Bosma, A. 1981 ApJ, **86**, 1825.
Bottema, R., Shostak, G. S. & Van der Kruit, P. C. 1987 Nature, **328**, 401.
Brosche, P. 1966 Veröff. Astron. Rechen-Institut Heidelberg, No. 17.
Corbin, T. E. & Urban, S. E. 1991 Astrographic Catalogue Reference Stars (ACRS), U. S. Naval Observatory.
Fricke, W. 1967a AJ, **72**, 642.
Fricke, W. 1967b AJ, **72**, 1368.
Fricke, W. 1977a Veröff. Astron. Rechen-Institut Heidelberg, No. 28.
Fricke, W. 1977b A&A, **54**, 363.
Fricke, W., Schwan, H. & Lederle, T. 1988 Veröff. Astron. Rechen-Institut Heidelberg, No. 32.
Henderson, A. P., Jackson, P. D. & Kerr, F. J. 1982 ApJ, **263**, 116.

Klemola, A. R., Jones, B. F. & Hanson, R. B. 1987 AJ, **94**, 501.

McCarthy, D. D. & Luzum, B. J. 1991 AJ, **102**, 1889.

Milne, E. A. 1935 MNRAS, **95**, 560.

Miyamoto, M. & Sôma, M. 1993 AJ, **105**, 691 (Paper I).

Miyamoto, M., Sôma, M. & Yoshizawa, M. 1993 AJ, **105**, 2138 (Paper II).

Ogorodnikov, K. F. 1932 Z. Astrophys., **4**, 190.

Rogstad, D, H., Lockhart, I. A. & Wright, M. C. H. 1974 ApJ, **193**, 309.

Rogstad, D. H., Wright, M. C. H. & Lockhart, I. A. 1976 ApJ, **204**, 703.

Röser, S. & Bastian, U. 1989 Catalogue of Positions and Proper Motions (PPM) North, Astron. Rechen-Institut Heidelberg.

Schwan, H. 1988 A&A, **198**, 116.

Sôma, M. 1993 private communication.

Sôma, M. & Miyamoto, M. 1993 Publ. Nat. Astron. Obs. Japan (in press).

van Altena, W. F., Girard, T., López, C. E., Klemola, A. R., Jones, B. F. & Hanson, R. B. 1986 Highlights Astron. **6**, 89.

Williams, J. G., Newhall, X. X. & Dickey, J. O. 1991 A&A, **241**, L9.

Kinematics of the stellar populations from a proper motion survey

By CAROLINE SOUBIRAN

Observatoire de Paris, 6a Avenue de l'Observatoire, 75014 Paris, France

This paper presents a kinematical study in a 7 square degree field at high galactic latitude, including 2500 stars. The data are absolute proper motions, and B and V magnitudes measured from Schmidt plates. The distributions of U and V components of the galactic space motions with respect to the Sun have been analysed in several bins of distance up to 2.5 kpc above the plane. Using a stochastic ML algorithm, the old disk, thick disk and halo populations have been deconvolved, and the following parameters of their velocity ellipsoid have been deduced : mean velocity ($\bar{U}, \bar{V}$), dispersions (σ_U, σ_V), rotational velocity V_{rot}, vertex deviation. No vertical gradient is observed in the kinematics of the thin disk.

1. Introduction

The present project is conducted with the MAMA machine and Schmidt plates, and also with spectroscopic observations. The goal is to describe the stellar populations in a few kpc around the Sun and to study their properties in terms of spatial distribution, kinematics and metallicity. There are two main difficulties in this kind of study. The first one concerns the separation in a clean way of the different populations which overlap considerably in their distributions. The second one is the paucity of thick disk and halo stars in the solar neighbourhood. This latter problem can be resolved in selecting these stars on a chemical or kinematical criterion, or in observing at large distances. The second way of proceeding has been chosen for this project. Schmidt plates are used because of their wide field and faint limiting magnitude, producing large samples of stars reaching a few kpc, with the possibility of multicolour photometry and absolute proper motion for each object detected in the field. This point is important because it means that no selection is made on the size of the proper motion. The second part of the project is just starting : it is the measurement of metallicities and radial velocities for a part of this proper motion survey.

2. The sample

The methods concerning the astrometric reduction, zero point of proper motions, distance estimation, deconvolution of velocity distributions have already been tested on a small field near the NGP (Soubiran, 1992, Soubiran, 1993).

In order to minimize the random errors on proper motions a set of eleven plates has been used for the astrometric reduction and five plates for the photometric reduction, which has been performed with an accuracy of about 0.08 mag. The plates are glass copies of the first POSS and original plates from Tautenburg and OCA. They cover a period of 40 years. For the preliminary study only a central part of 7 square degrees has been digitized. Nearly 5000 objects have been measured down to V=18.5. In the following the sample has been used only down to V=17.5 which is the limit of good proper motions and the limit where the separation between stars and galaxies is complete. The proper motions of all these objects has been computed in a differential way by comparing

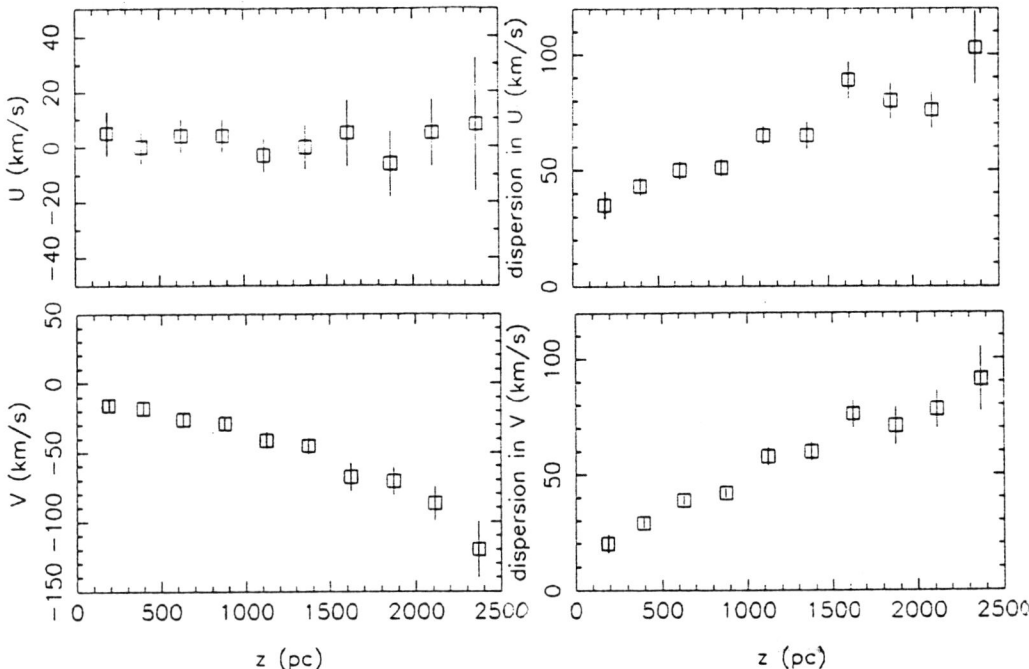

FIGURE 1. Kinematics of the total sample (2370 stars) with standard errors versus z distance above the plane

the (x,y) positions at the different epochs, with respect to the background stars. The resulting accuracy in proper motions is better than 2 mas/yr.

The zero point of proper motions is an important step. It is defined by the mean proper motion of galaxies with respect to the background stars. Extragalactic sources have to be numerous enough to define a good transformation between relative and absolute proper motions, but not all the galaxies can be used for that goal because of large random errors on their position. Here 136 galaxies and quasars have been selected to define the zero point with a low dispersion of 0.3 mas/yr.

Photometric distances have been assigned to stars with $0.3 \leq B-V \leq 1.3$ using a colour-magnitude relation of main sequence stars. From proper motions and distances, (U, V) velocities with respect to the Sun have been computed without knowing the radial velocities as the studied field is near the NGP. This sample includes 2370 stars.

3. Deconvolution of velocity distributions

The kinematics of the sample versus z distance above the plane is presented on Figure 1. The U velocity is clustered around zero at all distances while the asymetric drift has a gradient of about −40 km/s/kpc. An increase of the dispersions with z is also observed, which is an expected feature and corresponds to the rising proportion of old stars with hotter kinematics as one looks further above the plane. The sample is a mixture of different populations with different kinematics and as their relative proportion varies with z, it produces a gradient in the mean kinematics of the sample.

Figure 2 shows what one expects to find for the relative proportion of the three populations at a given distance z. The thin and thick disk should be in equal proportion at

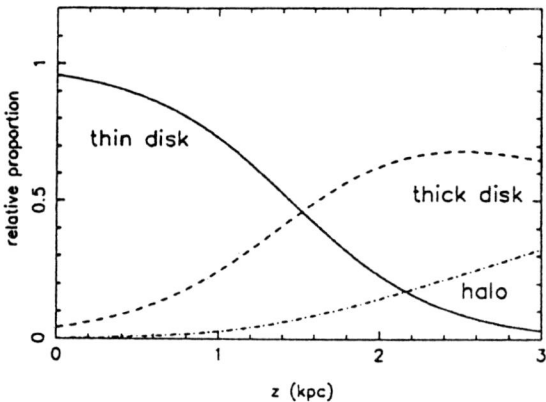

FIGURE 2. Expected relative proportion of each population adopting an exponential density for the thin and thick disks with scale heights of 350 pc and 1.3 kpc respectively, a radial density for the halo, and local densities of 4% and 0.2% for the thick disk and halo (Gilmore et al., 1989, Mihalas & Binney, 1981)

about 1.5 kpc. This proves that the sample has to be divided in several bins of distance to analyse the contribution of each population. The velocity distributions of the thin and thick disks and of the halo can be deconvolved in the different bins of distance to deduce the mean velocity and dispersions, with a possible vertical gradient, for each population. The SEM algorithm has been used for that goal. It is a noninformative method which solves iteratively the ML equations, with a stochastic step, in the case of a multivariate mixture of gaussian distributions (Celeux & Diebolt, 1986). The separation of different components is shown in Figure 3, in three bins of distance, for the V velocity. As the sample has been divided in six bins of distance, it has been possible to see for each population the evolution of its estimated parameters with the distance z above the plane.

3.1. *The thin disk*

A population with a low dispersion, corresponding to the thin disk, has been identified in five bins of distance up to 1.5 kpc. The estimated parameters are presented in Figure 4. The most obvious feature is that there is no vertical gradient in these parameters. The kinematics of the thin disk appears to be the same at 0.5 kpc and at 1.5 kpc. In particular, there is no increase of the asymetric drift. In Figure 4 the values which are expected in the immediate solar neighbourhood (Delhaye, 1965 and Wielen, 1974) are indicated. There is no major difference, except on the mean U velocity which is expected to be around 10 km/s. The weighted averages of the estimated parameters are : $\bar{U} = 3 \pm 4$ km/s, $\bar{V} = -17 \pm 3$ km/s, $\sigma_U = 41 \pm 4$ km/s, $\sigma_V = 23 \pm 3$ km/s.

3.2. *The thick disk*

This population has also been recognized in five bins of distance with no clear vertical gradient. The estimated parameters have been averaged : $\bar{U} = 0 \pm 15$ km/s, $V_{rot} = 179 \pm 16$ km/s, $\sigma_U = 56 \pm 11$ km/s, $\sigma_V = 43 \pm 6$ km/s. A comparison has been made with a similar proper motion study at the NGP (Spaenhauer, 1989) where the parameters of the thick disk have been computed from an unique gaussian fit at a mean distance of 2 kpc : $\bar{U} = 24 \pm 13$ km/s, $V_{rot} = 140 \pm 12$ km/s, $\sigma_U = 89$ km/s, $\sigma_V = 80$ km/s. As no deconvolution of the stellar populations has been performed on these data, the results probably suffer from halo contamination and that may be why the circular velocity is

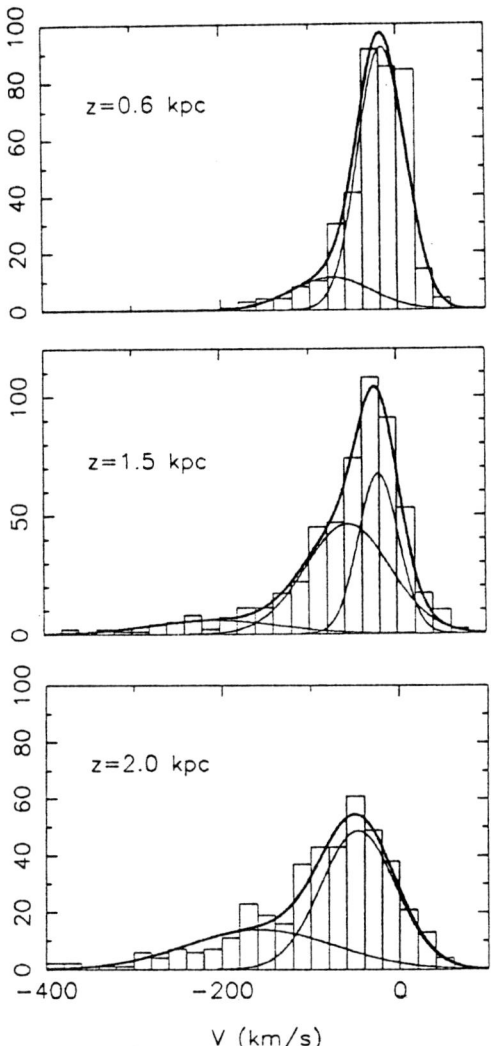

FIGURE 3. SEM deconvolution of gaussian components overlaid on the V histogram for three bins of distance

found to be so low and with high dispersions. On the contrary, Yoss et al. (1991) found for the thick disk a higher circular velocity $V_{rot} = 207$ km/s with small dispersions : $\sigma_U = 31$ km/s, $\sigma_V = 38$ km/s. In this latter study of close stars, the separation between the 2 disks has been performed by a cut in metallicity and the results may suffer from thin disk contamination. For most of the other studies (Wyse & Gilmore, 1986, Sandage & Fouts, 1987, Carney et al., 1989, Ratnatunga & Freeman, 1989, Morrison et al., 1990, Majewski, 1992 (at a mean distance of 2 kpc), Schuster et al., 1993) there is a good agreement for the circular velocity with values ranging between 160 km/s and 190 km/s.

3.3. The halo

This population is not well represented in the sample because of the limiting distance of 2.5 kpc, but the solutions of SEM have been very steady: $\bar{U} = -4 \pm 8$ km/s, $V_{rot} = 58 \pm 12$ km/s, $\sigma_U = 137 \pm 13$ km/s, $\sigma_V = 79 \pm 6$ km/s.

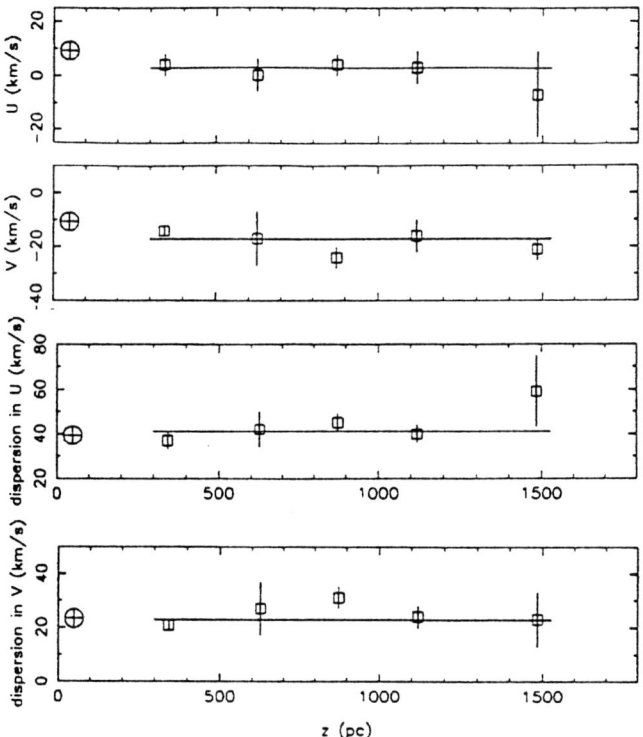

FIGURE 4. Kinematics of the thin disk versus z distance. The horizontal lines indicate the weighted average for each parameter. The crossed circles indicate the values which are usually adopted for the immediate solar neighbourhood (Delhaye, 1965, Wielen, 1974).

The rotational velocity which has been found is at the upper limit of other determinations computed from kinematically or spectroscopically selected samples of halo stars (Norris et al., 1985, Carney & Latham, 1986, Norris, 1986, Morrison et al., 1990, Ryan & Norris, 1991). This result together with the low dispersion in V may indicate that some disk stars have not been well separated from the halo.

4. Vertex deviation

An analysis of the bivariate distribution of (U,V) allows study of the vertex deviation. It is interesting to know what the vertex deviation becomes for old stars. Figure 5 has been taken from Kuijken & Tremaine (1991). There are not many determinations of the vertex deviation for kinematically hot stars. This study has allowed the addition of two points on this diagram, and indicates a tendancy toward positive vertex deviation for old stars.

5. Future work

For the present sample, spectra have already been obtained for more than 100 stars and we expect to obtain 500 spectra by next year. These spectra are measured at Observatoire de Haute Provence for the brightest part (V $\leq$ 14), and with the Multi-

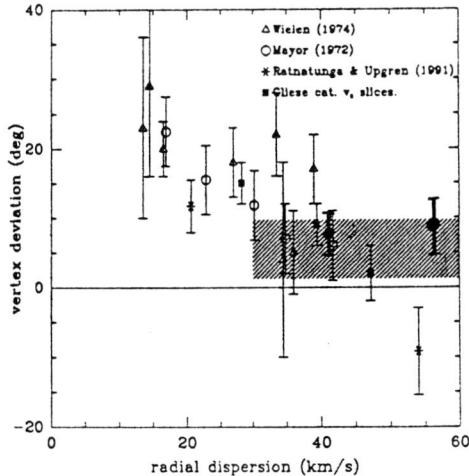

FIGURE 5. Diagram of the vertex deviation versus U dispersion taken from Kuijken & Tremaine (1991). The 2 full circles are the results of the present study for the thin disk and thick disk.

Object Spectrograph at CFHT (V ≤ 17). These observations are made without any preselection on a kinematical criterion. The radial velocities will be computed with an accuracy of 8 to 15 km/s. By comparison with synthetic spectra (Cayrel et al., 1991), metallicities will be deduced as well as effective temperatures and gravities to give better distance estimates. We will then have a significant sample with no bias of selection to allow an SEM deconvolution of the stellar populations in the 4-dimensional space of $(U, V, W, [Fe/H])$.

REFERENCES

Carney B.W., Latham D.W., 1986, AJ 92, 60.
Carney B.W., Latham D.W., Laird J.B., 1989, AJ 97, 423.
Cayrel R., Perrin M.N., Barbuy B., Buser R., 1991, A&A 247, 108.
Celeux G., Diebolt J.,1986, Rev. Statistique Appliquée, 34, 35.
Delhaye J., 1965, in Galactic structure, eds A. Blaauw & M. Schmidt, Chicago.
Gilmore G., King I., van der Kruit P., 1989, The Milky Way as a Galaxy, eds R. Buser & I.King, Geneva Observatory.
Kuijken K., Tremaine S., 1991, in Dynamics of disk galaxies, Varberg, Sweden.
Majewski S.R., 1992, ApJS 78, 87.
Mihalas D., Binney J., 1981, Galactic Astronomy, eds W.H. Freeman and Co, San Fransisco.
Morrison H.L., Flynn C., Freeman K.C., 1990, AJ 100, 1191.
Norris J., 1986, ApJS 61, 667.
Norris J., Bessell M.S., Pickles A.J., 1985, ApJS 58, 463.
Ratnatunga K.U., Freeman K.C., 1989, ApJ 339, 126.
Ryan S.G., Norris J.E., 1991, AJ 101, 1835.
Sandage A., Fouts G., 1987, AJ 92, 74.
Schuster W.J., Parrao L., Contreras Martínez M.E., A&AS 97, 95.
Soubiran C., 1992, A&A 259, 394.

Soubiran C., 1993, A&A (in press).

Spaenhauer A., 1989, in The gavitational force perpendicular to the galactic plane, eds A.G.D. Philip & P.K. Lu, L.Davis Press, New York, p 45.

Wielen R., 1974, Highlights in Astronomy (3), ed. Contopoulos G., Reidel, Dordrecht, p. 395.

Wyse R.F.G., Gilmore G., 1986, AJ 91, 855.

Yoss K.M., Bell D.J., Detweiler H.L., 1991, AJ 102, 975.

Dynamical analysis of local kinematical data

By J. J. BINNEY

Department of Theoretical Physics, University of Oxford, Keble Road, Oxford, OX1 3NP, UK

By Occam's razor, we should first compare data with steady-state, axisymmetric models. The strong Jeans' theorem provides the right framework for this exercise. Two techniques now exist for the construction of models with three-integral distribution functions $f(\mathbf{J})$. In principle each stellar type has its own DF $f(\mathbf{J})$, while the potential $\Phi(\mathbf{x})$ is common. The DFs and Φ can be suitably parameterized and the values of these parameters chosen by optimizing the fit between observations and the model densities in the observational domain.

Only limited information can be gleaned if one is confined to stars in the solar neighbourhood; such data should be combined with data from pencil-beam surveys that penetrate to distances $\sim R_0$.

1. Introduction

HIPPARCOS will soon deliver good trigonometric parallaxes for tens of thousands of stars uniformly distributed around the Sun. Many of these stars will also have reliable proper motions from a combination of ground-based and HIPPARCOS data. With modern spectrographs it will not take ground-based astronomers long to obtain good radial velocities for those of these stars which do not already have them. Hence, within a few years we shall have space velocities for a large and homogeneous sample of stars near the Sun.

At distances greater than that ($d \approx 100 \,\mathrm{pc}$) to which HIPPARCOS can get good parallaxes, we already have an impressive and rapidly growing body of reliable proper motions. When these data are combined with photometric parallaxes, one obtains good estimates of five of the six phase-space coordinates for stars up to several kiloparsecs away. It should be possible to determine the sixth coordinate for quite large samples of stars.

These observational developments need now to be matched by a renaissance in dynamical studies of the Milky Way. The challenge is to integrate all available kinematic data into a global dynamical model of the Milky Way. If successful, this enterprise will reveal the structure of the Galaxy's potential, and thus of the matter, much of it probably non-baryonic, which generates it. It will also provide important clues to the origin and evolution of the Galaxy.

In practice it will be important to model stellar data in parallel with kinematic data on the distribution of gas within the Galaxy. In the interest of brevity I here concentrate on stellar data. Kuijken & Tremaine (1993) have recently discussed in some detail the interpretation of the kinematics of gas in the Milky Way.

2. Steady-state, axisymmetric models

Although the Milky Way is probably neither axisymmetric nor in a steady state, we should start from axisymmetric, steady-state models. There are several reasons for this. First, the older, higher-velocity stars (classical Population II stars) are the stars with the most to teach us about dark matter in the Galaxy, since they are the stars which explore the potential away from the Plane. Since their motions are insensitive to many non-axisymmetric components of the Galactic potential, the latter may be assumed axisymmetric. Since these stars are old, they are likely to be dynamically well mixed, and can thus be represented by a steady-state model. Second, a natural way to look for evidence of non-axisymmetry is to look for discrepancies between the predictions of axisymmetric models and observation. Such discrepancies should be most prominent for younger, lower-velocity stars.

2.1. *The potential*

The first step in the construction of a galaxy model is the choice of potential. The Milky Way's circular-speed curve $v_c(R)$ is thought to be either flat or slowly rising in the vicinity of the Sun. Moreover, at small radii the $2\,\mu$m brightness profile (Kent, Dame & Fazio 1991) and models of the gaseous disk (Binney et al. 1991) suggest that $v_c(R) \propto R^{0.1}$ in the range $2\,\text{pc} \lesssim R \lesssim 2\,\text{kpc}$. We have little information on the potential's vertical structure beyond what can be deduced from the Oort-limit problem studied by Bahcall (1984) and by Kuijken & Gilmore (1989). Hence at this stage an adequate representation of $\Phi(R, z)$ is as a sum of disk and bulge–halo components, where the latter is generated the two-power mass distribution

$$\rho_{\text{bh}}(R, z) = \rho_b \frac{a_b^\beta}{a^\gamma (a + a_b)^{(\beta - \gamma)}} \quad \text{where} \quad a^2 \equiv R^2 + z^2/q^2. \tag{2.1}$$

Here $\rho_b, a_b, \beta, \gamma$ and q are constants; $q < 1$ is the flattening of the mass distribution, ρ_b gives its normalization and β and γ determine is radial variation at $R \gg a_b$ and $R \ll a_b$, respectively. The potential of an elliptical mass distribution such as (2.1) can be readily obtained (e.g. §2.3 of Binney & Tremaine 1987). Dehnen (1993) has given DFs that generate (2.1) in the case $q = 1$, $\beta = 4$.

The disk's potential is conveniently recovered by one of the techniques described by Cuddeford (1993).

2.2. *The integrals*

By the strong Jeans theorem, the steady-state assumption enables us to take the distribution function (DF) of stars of type α to be a function $f^{(\alpha)}(\mathbf{I})$ of the potential's isolating integrals I_1, I_2, I_3. The first integral is conveniently taken to be either energy or radial action J_r, while I_2 is naturally taken to be the azimuthal component of angular momentum L_z, which is identical with the azimuthal action J_a. The third integral is well known to be problematical. Some orbits may not even respect an I_3. When I_3 exists, it may usually be identified with the latitudinal action J_l. In the case of a star that does not stray far from the plane, v_z is approximately independent of R and we have

$$J_l \simeq \frac{1}{2\pi} \oint v_z \, dz. \tag{2.2}$$

In the epicycle approximation, $J_l = E_z/\nu$ is just the ratio of the vertical energy to the vertical frequency ν. In the case of a high-velocity star, J_l is approximately equal to $L - L_z$, the difference between the star's approximately constant total angular momentum L and L_z.

Analytic expressions for $J_r(\mathbf{x}, \mathbf{v})$ and $J_l(\mathbf{x}, \mathbf{v})$ are unavailable for any suitable potential. However, in the case of a Stäckel potential, it is possible to obtain $J_r(\mathbf{x}, \mathbf{v})$ and $J_l(\mathbf{x}, \mathbf{v})$ from a single numerical quadrature each (de Zeeuw 1985). Gerhard (see Gerhard & Saha 1991) has shown how first-order perturbation theory can be used to obtain approximate expressions for $J_r(\mathbf{x}, \mathbf{v})$ and $J_l(\mathbf{x}, \mathbf{v})$ for potentials that may be regarded as perturbed spherical potentials. It may be possible to adapt this technique to the case of orbits in a nearly planar potential, such as that of a thin disk. An alternative, non-perturbative approach, employs numerically fitted canonical transformations to map tori of constant

J in a toy potential, such as the spherical isochrone, into approximate invariant tori of any given galactic potential (McGill & Binney 1990, Binney & Kumar 1993, Kaasalainen & Binney 1994).

2.3. The distribution functions

A great deal of work has been done over the last decade on the problem of choosing a distribution function which is compatible with a given body of observational data. Most of this work has concentrated on systems with either spherical potentials, or perfectly planar disks, since in these cases all the necessary integrals are known analytically. Recent advances include the development by Hunter & Qian (1993) of a new procedure for deriving DFs of the form $f(E, L_z)$ from a given density distribution $\rho(R, z)$, and the derivation by Collett & Evans (1993) of exact DFs for planar exponential disks. Cuddeford (1991), Gerhard (1991), Dejonghe & Merritt (1992), Louis (1993), Dehnen (1993) and others have further expanded the range of DFs available for spherical galaxies.

While all these studies contribute to our general understanding of the relation between a DF and the galaxy it generates, they are none of them directly relevant to work on the Milky Way; they all ignore the third integral, which strongly affects solar neighbourhood kinematics.

Bishop (1987) and de Zeeuw & Hunter (1990) have presented self-consistent systems with oblate Stäckel potentials in which the third integral is dominant. In these models only shell orbits are populated, that is, all stars have $J_r = 0$. Most of these models are unstable (Merritt & Stiavelli 1990).

Recently Dehnen & Gerhard (1993) have explored a much wider class of self-consistent models with three-integral DFs. Their models have flattened isochrone potentials, which are very similar to oblate Stäckel potentials, but their third integrals are obtained from perturbation theory rather than separation of the Hamilton-Jacobi equation, and therefore it should be possible to generalize them to potentials with realistically flat circular-speed curves. They show that widely differing velocity structures are compatible with given $\rho(R, z)$ and have discussed how long-slit observations may be used to choose which velocity structure best fits an external galaxy.

Considerably insight into the workings of three-integral galaxies can be obtained by studying non-spherical systems in spherical or Stäckel potentials. In this approach one either makes no effort to ensure self-consistency (Binney & Petrou 1985), or one seeks approximate self-consistency (May & Binney 1986, Binney & Petit 1989). Since real galaxies contain many stellar populations, each with a different DF, and probably contain a great deal of non-baryonic matter as well, there really is little point in making a single population self-consistent. On the other hand, it is very desirable to work with a realistic potential. For different reasons, neither spherical nor Stäckel potentials can be deemed realistic.

The integral space associated with orbits in a given potential, 'action space', enables one to think about DFs in a way that is independent of specific choices of $\Phi(R, z)$. In particular, it enables one to take advantage of our extensive knowledge of DFs for spherical systems when building an axisymmetric model. Specifically, one starts by recovering a DF of the form $f(E, L)$ for the spherical anologue of the problem in hand, and then expresses it in the form $f(J_r, |J_a| + J_l)$ by evaluating $J_r(E, L)$ by simple quadratures and using the relation $L = |J_a| + J_l$. In a flattened version of the original spherical potential, $f_0(\mathbf{J}) \equiv f(J_r, |J_a| + J_l)$ will generate a slightly flattened system similar to the original model. A more flattened model of the same mass will be obtained with $f_\alpha(\mathbf{J}) \equiv f(J_r, \alpha|J_a| + J_l/\alpha)$, where $\alpha < 1$.

Perfectly flat disks with DFs of the form $f(E, L_z)$ can be fattened into disks of finite thickness by a different technique: one first expresses f as $f(J_r, J_a)$ and then works with the DF

$$f_z \equiv \frac{\exp\left(-J_l/\beta(J_a)\right)}{\beta(J_a)} f(J_r, J_a). \tag{2.3}$$

Here $\beta(J_a)$ determines the vertical dispersion ($\simeq (\beta/\nu)^{1/2}$) as a function of radius. So long as the disk is reasonably thin, its surface density will be independent of β.

3. What can modelling teach us?

Since f is a function of three variables, it should be possible to obtain a perfect fit to the local velocity distribution $n(\mathbf{v})$ of a stellar population no matter what potential Φ one employs. In fact, it is not hard to see that, given $\Phi(R, z)$, $f(\mathbf{I})$ is uniquely determined by $n(\mathbf{v})$ in that portion of integral space in which stars visit the Sun. Obviously, $n(\mathbf{v})$ cannot constrain $f(\mathbf{I})$ in the rest of action space. May & Binney (1986) estimated that about two thirds of the spheroid's mass lies in the portion of integral space that is not determined by $n(\mathbf{v})$.

Hence it is essential to complement solar-neighbourhood data with 'in situ' observations, that is studies of the stellar density and radial velocities of stars in cones that reach deep into the Galaxy from the Sun. Along some lines of sight, especially those nearly perpendicular to the Plane, one will be seeing stars on orbits that pass through the solar neighbourhood. Consequently, the statistics of the stars will have already been determined, and agreement between theory and observation will constitute a test of the potential.

Along other lines of sight one will be predominantly studying stars that never visit the Sun. The same orbits will, however, intersect each line of sight at more than one distance, and intersect several different lines of sight. The densities of stars of a given population will fit together in a coherent way only if the potential is right.

These considerations suggest the following procedure:

1. Choose a potential and calculate for it a complete set of isolating integrals $\mathbf{I}(\mathbf{x}, \mathbf{v})$. This can be done either by perturbation theory or by the torus-fitting technique.

2. Choose a distribution function $f^{(\alpha)}(\mathbf{I})$ for each stellar population. The general form of $f^{(\alpha)}$ would be motivated by toy models such as those of Dehnen & Gerhard (1993), but its detailed structure would depend on a number of undetermined parameters. Each $f^{(\alpha)}$ should be normalized such that $\int d^3\mathbf{x}\, d^3\mathbf{v}\, f^{(\alpha)} = 1$.

3. For each star i with a space velocity, evaluate the value $\mathbf{I}_i$ taken by $\mathbf{I}(\mathbf{x}, \mathbf{v})$ at the star's position in phase space, and thus evaluate the probability density

$$P_{\alpha i} \equiv \psi^{(\alpha)}(d_i) f^{(\alpha)}(\mathbf{I}_i), \tag{3.4}$$

where $\psi^{(\alpha)}(d)$ is the distance-dependent selection function for population α. Uncertainties in the phase-space coordinates $\mathbf{w} \equiv (\mathbf{x}, \mathbf{v})$ may be allowed for by defining $P_{\alpha i}$ to be

$$P_{\alpha i} \equiv \int d^6\mathbf{w}\, P(\mathbf{w}|\mathbf{x}_i, \mathbf{v}_i) \psi^{(\alpha)}(d(\mathbf{w})) f^{(\alpha)}(\mathbf{I}(\mathbf{w})), \tag{3.5}$$

where $P(\mathbf{w}|\mathbf{x}_i, \mathbf{v}_i)$ is the probability that $\mathbf{w}$ gives the true coordinates given that $\mathbf{x}_i$, $\mathbf{v}_i$ are observed.

If only the star's transverse velocity is known, we define

$$P_{\alpha i} \equiv \int dv_{\text{los}}\, \psi^{(\alpha)}(d_i) f^{(\alpha)}(\mathbf{I}_i), \tag{3.6}$$

where the integral is over the all radial velocities. If only the star's radial velocity is known, we define $P_{\alpha i}$ to be

$$P_{\alpha i} \equiv \int \mathrm{d}^2 \mathbf{v}_i \, \psi^{(\alpha)}(d_i) f^{(\alpha)}(\mathbf{I}_i), \tag{3.7}$$

where the integral is over the plane in velocity space perpendicular to the line of sight. Clearly, when account is taken of observational errors, the definitions 3.6 and 3.7 should be replaced by definitions of the type 3.5.

4. Calculate the likelihood

$$\mathcal{L} \equiv \sum_{\substack{\text{stellar} \\ \text{populations } \alpha}} \sum_{\text{stars } i} \log(P_{\alpha i}). \tag{3.8}$$

5. Maximize $\mathcal{L}$ by adjusting the parameters in the DFs.

6. Repeat steps 1–5 for different potentials, finally choosing the potential with the largest peak likelihood.

4. Non-axisymmetries

The Milky Way's potential is surely not perfectly axisymmetric. Three potential sources of non-axisymmetry are: (i) spiral structure; (ii) a bar in the visible component; (iii) triaxiality of the dark halo. The essential differences between these possibilities are that (i) has radius-dependent phase, while for (ii) and (iii) the potential's phase is radius-independent. Also in cases (i) and (ii) the potential will have non-negligible pattern speed, while in case (iii) the pattern speed will be very small.

Spiral structure affects only gas and stars on nearly circular orbits, while a bar will significantly perturb the orbits of all stars that are confined to its corotation radius. Consequently, solar-neighbourhood observations in principle suffice to distinguish a barred from a spiral potential perturbation.

It has long been suspected that a bar lurks at the centre of the Milky Way (de Vaucouleurs 1964, Listz & Burton 1980). Evidence for a central bar has recently been accumulating (Binney et al. 1991, Blitz & Spergel 1991b, Weinberg 1992, Whitelock & Catchpole 1992), and Binney et al. have claimed that its pattern speed is higher than previously suggested; corotation is placed at $R = 2.4\,\mathrm{kpc}$ rather than near R_0 as has been traditional (e.g. Yuan 1984). If the bar's corotation really lies as far in as $2.4\,\mathrm{kpc}$, the bar is unlikely to perturb solar-neighbourhood stars detectably.

The existence of a bar at $R < 2.4\,\mathrm{kpc}$ does not exclude the existence of a larger bar with corotation beyond the Sun, such as that advocated by Blitz & Spergel (1991a) to explain the envelope of the observed HI (l,v) plot. However, the arguments presented by Blitz & Spergel for the this bar have been strongly contested by Kuijken & Tremaine (1993), who argue that both the kinematics of solar-neighbourhood stars and the neutral-hydrogen (l,v) plot are best explained by the Galaxy's dark halo being triaxial at about the level predicted by numerical simulations of halo formation (Binney 1978, Frenk et al. 1987).

4.1. The vertex deviation

Kuijken & Tremaine (1993) discuss in detail how non-axisymmetry of the potential will betray itself. An obvious diagnostic is the radial component $\bar{v}_R$ of the velocity of the local standard of rest with respect to the galactic centre. However, this is hard to determine accurately; Kuijken & Tremaine conclude $\bar{v}_R = -1 \pm 9\,\mathrm{km\,s^{-1}}$. A more useful diagnostic is the vertex deviation; the angle between the long axis of the local velocity ellipsoid and the direction to the galactic centre. An accurate calculation of this angle is difficult,

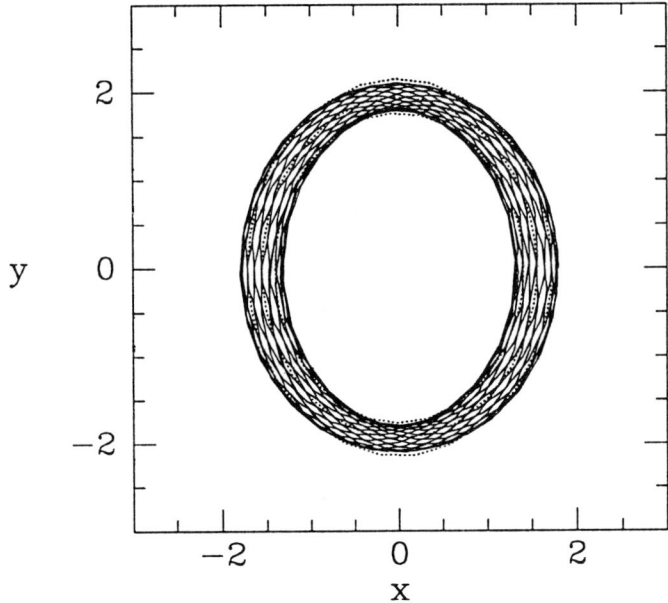

FIGURE 1. A non-closed orbit (full curve) and three closed orbits (dashed curves) in the potential $\Phi = v_0^2 \log[x^2 + y^2/(0.8)^2]$. At any given point in a Stäckel potential, the two velocity vectors with which a non-closed orbit cuts the local closed orbit always make equal angles with the closed orbit. This ensures that in a Stäckel potential one principal axis of the velocity ellipsoid is always parallel to the local closed orbit. One consequence of this equal-angle property, is that in a Stäckel potential non-closed orbits are bounded by closed orbits. Here the non-closed orbit is not bounded by closed orbits and the equal-angle property holds only for the central closed orbit. Consequently, the velocity ellipsoid is not necessarily aligned with the local closed orbit. In particular, if only the non-closed orbit shown were populated (because f were a δ-function), the vertex deviation would be smaller than if the equal-angle property held near the outside of the orbit, and larger than predicted by the equal-angle property near the inside of the orbit. These two effects will tend to cancel in a typical disk, since the DF typically dependends on J_a much more weakly than on J_r, and, at a given point, the numbers of stars near peri- and apo-centre are approximately equal.

but a simple argument conveys the essence of the matter. In a Stäckel potential the principal axes of velocity ellipsoids are always aligned with the separating coordinate system. Moreover, closed orbits coincide with coordinate curves. Consequently, in a cool disk immersed in a Stäckel potential, the long axis of the velocity ellipsoids are aligned with the elliptical closed orbits. Figure 1 shows that this state of affairs approximately holds in more general barred potentials.

4.2. The Oort equation

Standard texts (e.g. Binney & Tremaine 1987) derive Lindblad's equation

$$\frac{\sigma_\phi^2}{\sigma_R^2} = \frac{1}{2}\left(1 + \frac{\mathrm{d}\ln\overline{v}_\phi}{\mathrm{d}\ln R}\right) \qquad (4.9)$$

under the assumptions that the potential is axisymmetric and that the three third moments,

$$S_{RR\phi} \equiv \overline{v_R^2(v_\phi - \overline{v}_\phi)},$$

$$S_{\phi\phi\phi} \equiv \overline{(v_\phi - \overline{v}_\phi)^3}, \qquad (4.10)$$
$$S_{R\phi z} \equiv \overline{v_R(v_\phi - \overline{v}_\phi)v_z},$$

all vanish. Moreover, for a low-velocity population, the mean-streaming speed $\overline{v}_\phi$ will differ little from the circular speed v_c. Equating the two speeds, we obtain Oort's relation

$$\frac{\sigma_\phi^2}{\sigma_R^2} = \frac{-B}{A-B} \qquad (4.11)$$

between the dispersions and the Oort constants A and B. Determinations of A and B from either proper motions (Hanson 1987) or radial velocities (Rohlfs & Kreitschmann 1988, Fich Blitz & Stark 1989) lead to the conclusion that $|B| > A$ and thus that v_c is outwards-rising. On the other hand, the values of σ_ϕ and σ_R of old stellar populations (Jahreiss 1974) imply with (4.11) that $|B| < A$ and thus that v_c is outwards-declining.

A partial resolution of this puzzle is that (4.11) is invalid because the neglected third moments S_{ijk} and their derivatives are non-negligible (Cuddeford & Binney 1993). Since the S_{ijk} are exceedingly difficult to measure, Cuddeford & Binney have sought an equation analogous to (4.11), which would enable $-B/(A-B)$ to be reliably determined from second moments of local velocities. By studying this problem in the context of disks with simple DFs, they show that $-B/(A-B)$ *can* be successfully predicted if one takes *modified* moments, that is, the expectation of $v_i v_j w(\mathbf{v}; u, \sigma)$, where

$$w(\mathbf{v}; u, \sigma) \equiv e^{-((v_\phi - u)^2 + v_R^2 + v_z^2)/2\sigma^2}, \qquad (4.12)$$

and u and σ are suitably chosen constants roughly equal to the mean-streaming speed and the radial velocity dispersion of the population under study.

However, when one applies the new method to Jahreiss' (1974) data for the McCormick K & M dwarfs in the Gliese catalogue, one finds that $|B|$ is still predicted to be smaller that A; the best-fitting circular-speed curve is $v_c(R) \propto R^{-0.2}$. Evans & Collett (1993) and Kuijken & Tremaine (1993) have reached similar conclusions by working directly with models based on simple DFs.

4.3. A triaxial halo?

Kuijken & Tremaine (1993) argue that the likely resolution of the conflict between the shape of the local velocity ellipsoid and direct determinations of $v_c(R)$ is that the Milky Way's potential is significantly non-circular near the Sun because the halo is triaxial. They show that ellipticity in the potential of order $\epsilon \simeq 0.1$, with the Sun near the potential's minor axis (the major axis of the disk), suffices to bring all the data into agreement. An ellipticity of the order 0.1 is what one expects *a priori* from consideration of the violent relaxation of a halo of collisionless matter (Binney 1978, Frenk et al. 1987).

One advantage of using actions as integrals, is that a model characterized by a set of DFs $f^{(\alpha)}(\mathbf{J})$ can be readily generalized to the case of a slowly rotating, weakly barred potential, such as that of a triaxial halo. Such generalization is possible because (i) well inside the potential's inner Lindblad resonance, the azimuthal action J_a of short-axis tube orbits is a natural generalization of L_z, and becomes identical with L_z in the axisymmetric limit, and (ii) if the potential's core is not too small, it is possible to regard J_a as an extension of L_z even for the box and long-axis tube orbits (e.g. Binney & Spergel 1982). (The case of a coreless potential is more complex; Miralda & Schwarzschild 1989.)

The technology required to obtain expressions of the form $\mathbf{I}(\mathbf{x}, \mathbf{v})$ or $\mathbf{x}(\mathbf{J}, \boldsymbol{\theta})$, $\mathbf{v}(\mathbf{J}, \boldsymbol{\theta})$ for a triaxial potential is not yet in place. But from the work of Kaasalainen (1994) it appears that at least for short-axis tube orbits it should be straightforward to obtain $\mathbf{x}(\mathbf{J}, \boldsymbol{\theta})$, $\mathbf{v}(\mathbf{J}, \boldsymbol{\theta})$ by applying perturbation theory to an axisymmetric potential. Moreover,

5. Conclusions

I believe that we shall soon be in possession of sufficient astrometric data to determine the galactic potential between, say, $R_0/3$ and $5R_0$. Moreover, we should be able to determine the DFs of several stellar populations. Since dark matter is almost certainly a major contributor to the potential, a knowledge of the latter will immediately tell us the disposition of the former. The DFs of the Galaxy's populations probably contain significant clues to the formation of the Milky Way.

The available information promises to be highly redundant. So we shall not only determine Φ and the $f^{(\alpha)}$, but obtain either ample reassurance that our model is a good one, or clues as to why it is imperfect.

The redundancy will be greatest and the modelling process most straightforward if we assume that the Milky Way is axisymmetric and in a steady state. Then the $f^{(\alpha)}$ are functions of three integrals, and we know how to calculate the latter for any given potential. I have argued that the way forward is to parameterize Φ and the $f^{(\alpha)}$ and then to seek the values of the parameters which maximize the likelihood of any given body of observational data. Over the next three years a concerted attempt to carry through this strategy will be made in Oxford.

It is to be expected that significant discrepancies between theory and observation will remain at the end of this period. Two candidate explanations are (i) that the Galaxy is not axisymmetric, and (ii) that it is not in a steady state. Both non-axisymmetry and deviations from equilibrium are likely to be most pronounced for low-velocity disk populations. Indeed, recent work on the Oort ratio suggests that the effects of triaxiality in the dark halo are already apparent in the dynamics of disk stars. Fortunately, the easiest step to take beyond steady-state, axisymmetric models is that towards slowly-rotating triaxial models. So within the next decade we may be studying non-equilibrium effects by comparing our data with models of the Milky Way in which the halo is a triaxial object. Only then will galactic dynamics have fulfilled the promise of the classic papers published by Jeans & Eddington in 1915–16.

REFERENCES

Bahcall, J.N., 1984, ApJ, 276, 169

Binney, J.J., 1978, MNRAS, 183, 779

Binney, J.J., Gerhard, O.E., Stark, A.A., Bally,J & Uchida, K.I., 1991, MNRAS, 252, 210

Binney, J.J. & Kumar, S., 1993, MNRAS, 261, 584

Binney, J.J. & Petit, J-M., 1989. In Dense Stellar Systems, ed. D. Merritt (Cambridge University Press) p. 43

Binney, J.J. & Petrou, M., 1985, MNRAS, 214, 449

Binney, J.J. & Spergel, D.N., 1984, MNRAS, 206, 159

Binney, J.J. and Tremaine, S., 1987 Galactic Dynamics, Princeton University Press

Bishop, J.L., 1987, ApJ, 322, 618

Blitz, L. & Spergel, D. N., 1991a, ApJ, 370, 205

Blitz, L. & Spergel, D. N., 1991b, ApJ, 379, 631

Collett, J.L. & Evans, W., 1993, MNRAS, (in press)

Cuddeford, P., 1991, MNRAS, 253, 414

Cuddeford, P., 1993, MNRAS, 262, 1076
Cuddeford, P. & Binney, J.J., 1993, MNRAS, (in press)
Dehnen, W., 1993, MNRAS, (in press)
Dehnen, W. & Gerhard, O.E., 1993, MNRAS, 261, 311
Dejonghe, H. & Merritt, D., 1992, ApJ, 391, 531
de Vaucouleurs, G., 1964. In IAU Symposium 20: The Galaxy and the Magellanic Clouds, eds Kerr, F. J. & Rodgers, A. W. (Sydney: Australian Academy of Science), p. 195
de Zeeuw, P.T., 1985, MN, 216, 273
de Zeeuw, P.T. & Hunter, C. 1990, ApJ, 356, 365
Evans, N.W. & Collett, J.L., 1993, MNRAS, (in press)
Fich, M., Blitz, L., Stark, A.A., 1989, ApJ, 342, 272
Frenk, C., White, S.D.M, Davis, M. & Efstathiou, G., 1987, ApJ, 327, 507
Gerhard, O.E., 1991, MNRAS, 250, 812
Gerhard, O.E. & Saha, P., 1991, MNRAS, 251, 449
Hanson, R.B., 1987, 94, 409
Hunter, C. & Qian, E., 1993, MNRAS, 262, 401
Jahreiss, H., 1974. Ph.D. thesis, Heidelberg
Kaasalainen, M., 1994, MNRAS, (in press)
Kaasalainen, M. & Binney J.J., 1994, MNRAS, (in press)
Kent, S.M., Dame, T.M. & Fazio, G., 1991, ApJ, 378, 131
Kuijken, K. and Gilmore, G., 1989, MNRAS, 239, 571, 605, 651
Kuijken, K. and Tremaine, S., 1993, ApJ, (in press)
Liszt, H.S. & Burton, W.B., 1980, ApJ, 236, 779
Louis, P.D., 1993, MNRAS, 261, 283
May, A. & Binney, J.J., 1986, MNRAS, 221, 857
McGill, C., Binney, J., 1990, MNRAS, 244, 634
Merritt, D. & Stiavelli, M., 1990, ApJ, 358, 399
Miralda-Escudé, J. & Schwarzschild, M., 1989, ApJ, 339, 752
Rohlfs, K. & Kreitschmann, J., 1988, 20 A&A, 151
Weinberg, M.D., 1992, ApJ, 384, 81
Whitelock, P., & Catchpole, R.M., 1992. In The Center, bulge and disk of the Milky Way, ed. L. Blitz (Kluwer: Dordrecht)
Yuan, C., 1984, ApJ, 281, 600.

Absolute proper motions and tangential velocities to B=22.5 at the SGP

By XINJIAN GUO[1], TERRENCE M. GIRARD[1], WM. F. VAN ALTENA[1] AND CARLOS E. LÓPEZ[2]

[1]Yale University Observatory, P. O. Box 6666, New Haven, Connecticut 06511, U.S.A.

[2]Yale Southern Observatory and Felix Aguilar Observatory, University of San Juan, San Juan, Argentina

Absolute proper motions, referred to a few hundred galaxies, are derived for ∼5000 stars to B=22.5 in a field near the South Galactic Pole and the Galactic globular cluster NGC 288 based on a series of photographic plates exposed on the prime focus of the CTIO 4-meter telescope. The optical field angle distortion center and coefficients were accurately determined by comparing the 4-m plates with a flat field astrographic plate of the same field. The magnitude equation introduced by guiding error and non-cosmic color equation were differentially corrected for each plate relative to a standard plate. The absolute proper motions are therefore free of these systematic errors. The photographic photometry was carefully calibrated into the UBV system using our CCD photometry with the CTIO 0.9-meter telescope and other photometric data for NGC 288 from the literature. Photometric parallaxes for ∼3000 field stars and the corresponding tangential velocities or U, V components were determined with an adopted metallicity deficiency correction. Preliminary results for the stellar spatial distribution and kinematics are presented. The absolute proper motion and space motion of the Galactic globular cluster NGC 288 are also calculated using an adopted radial velocity and distance modulus.

1. Introduction

The spatial, kinematic, and chemical structure of the Galaxy are of great importance to our understanding of the formation and evolution of the Galaxy. Many observations have been or are being successfully carried out to determine these basic properties. However, due to the limitation of observing ability and relatively low speed and accuracy of the old types of measuring machines, it was very difficult to accurately determine the various properties of the Galaxy. Only in recent years, with the availability of high precision measuring machines and more reliable data reduction algorithms, does the tuning of a much finer picture of the Galaxy become possible. The existence of the Thick Disk proposed by Gilmore & Reid (1983) has been confirmed by more and more different types of observations. But the general characteristics of such a disk component, its relationship with the old Thin Disk and Halo (Spheroid component), as well as its formation and evolution history are not well known. The motion of the Halo is still controversial in the aspect of prograde motion (Norris 1986; Morrison et al. 1990; Ryan & Norris 1991) or retrograde motion (Reid 1990; Majewski 1992).

To answer these questions, the most valuable kinds of observations are the multi-color photometry, spectroscopy and absolute proper motion surveys for a complete sample of stars to a very faint magnitude limit in a selected area. This kind of *in situ* deep survey is only apparent magnitude limited, therefore it is free of metallicity and kinematic biases. If such a small area is chosen close to the two Galactic Poles, one can derive the *(U, V)* velocity components (toward the Galactic center and rotation, respectively) directly from the photometric parallaxes and absolute proper motions without necessarily knowing the radial velocities. So the difficulty of obtaining measurable spectra for a large number of very faint stars can be avoided. Good examples of this kind survey are those already carried out by Majewski (1992) and Soubiran (1993) for two selected areas close to the North Galactic Pole (NGP). We are conducting a similar survey toward the South Galactic Pole (SGP) using photographic plates exposed on the CTIO 4-m prime focus and CCD photometry with the CTIO 0.9-m telescope. Considerable attention has been paid to the study of the optical field angle distortion (OFAD), magnitude and color equation corrections. We found that these systematic errors were large that they would translate into significant systematic errors in the final stellar kinematics. A detailed discussion has been made by Guo et al. (1993). A summary of that study is presented here, which concentrates on the recent work on photometry and absolute proper motions as well as some preliminary results of stellar space distribution and kinematics.

2. OBSERVATIONS AND MEASUREMENTS

The selected area has an effective radius of $\sim 25'$ or an area of ~ 0.5 square degrees centered at $(\alpha, \delta) = (00^h 49^m, -27°00'; 1950.0)$. The Galactic globular cluster NGC 288 is in a quadrant of the field and can be used to provide photometric standards, reference stars to define the magnitude and color equations, and to determine its space motion. The plates were taken at the prime focus of the CTIO 4-m telescope with the UBK-7 coma corrector between 1981 and 1992. Eight of the 25 IIIa-F/GG495 plates taken between 1981 and 1986 are of better quality because of the newer emulsion and better seeing, while the 5 new plates (2 IIIa-F/GG495 and 3 IIIa-J/GG385) taken in 1992 are of lower quality mainly due to the bad seeing and possibly also due to the very old emulsion (8 years old). The magnitude limit of the new plates is about 1-1.5 lower than the old plates, which decreases our sample from ~ 7000 to ~ 5000 with reasonable measuring errors.

To determine the OFAD coefficients and center positions, two flat field plates (emulsion 103a-O) were taken in 1988 using the Yale Southern Observatory (YSO) astrograph in El Leoncito, Argentina. Although NGC 288 can provide some photometric standards for the calibration of the photographic photometry into the UBV system, we still made CCD BVR photometry with the CTIO 0.9-m to obtain more standards across the field to increase the calibration accuracy by incorporating coordinates into the transformation, since the OFAD affects the plate background and changes the photometry zero-point with distance from the center. However, on the new plates we found that a non-OFAD background variation caused by old emulsions, uneven baking or developing, or other unknown reasons dominated the background variation.

All the photographic plates were measured with the laser encoded Yale PDS microdensitometer. Measured coordinates, photometric indices and other image parameters such as radius, peak density, etc. were obtained with the 2–Dimensional Bivariant Gaussian Fitting routine (Lee & van Altena 1983). The initial PDS scan input catalogue contained 8377 entries (field and cluster stars, galaxies). But the final number of images with reasonable centroiding errors for each plate was between 4500 and 7000 depending on the quality and seeing of the plate. All CCD frames (object fields and standard star fields) were reduced using IRAF and DAOPHOT packages. BVR data for 746 field and NGC 288 stars were obtained with an internal error in the range $\pm 0.02 - 0.04$ mag. for $V \leq 20$.

3. Photographic photometry and photometric parallaxes

Besides the 746 stars from our CCD photometry, 527 stars in the same field, mostly belonging to NGC 288, were also selected from the literature to serve as photometry standards to calibrate the photographic photometry indices into the UBV system. The system transformation between UBV and UJF was adopted from Majewski (1992). Eleven IIIa-F/GG495/UBK-7 plates were used to derive f photometric indices, but only 3 IIIa-J/GG385/UBK-7 plates were used to derive the j photometric indices. The internal errors of the final V magnitudes and B–V colors for 4449 stars and galaxies could not be much better than ± 0.04 and ± 0.06 mag. due to the small number of available j plates and their lower quality as mentioned before.

To derive photometric parallaxes, we start by making the simplifying assumption that all the stars are single (not in binary or multiple systems), main-sequence stars and adopt a relation for M_V/B–V. Reduced proper motion techniques could be used to deconvolve the different stellar populations and luminosity classes (Chiu 1980; Soubiran 1993), but not very reliably. Other ways will be explored to consider the contamination due to other stars, including multiples.

We adopted the magnitude-color relation of the Hyades open cluster (Mermilliod 1992; Hanson 1980) as the basic relation to derive absolute magnitudes from the colors for the field dwarfs and subdwarfs. The dependence of the M_V correction on the ultraviolet excess $\delta_{0.6}$(U-B) was adopted from Hanson (1979). Since we do not yet have plates in the ultraviolet passband, ultraviolet excesses are not available for individual stars and only a statistical mean correction can be made at this time. We have assigned a metallicity value to stars at the same vertical distance by adopting the derived dependence of [Fe/H] on z (Gilmore & Reid, 1983). The dependence of the ultraviolet excess on [Fe/H] derived by Carney (1979) was adopted. Before deriving distances z from the distance modulii, we can reasonably assume the extinction and reddening of this near SGP field are both zero (Gilmore & Reid 1983).

4. Proper motions

The principle of dealing with the systematic effects such as the differential atmospheric refraction, OFAD, magnitude and color equations is to pre-correct for them separately and iteratively for the measuring coordinates before proceeding to derive relative proper motions. This is done because, first, they introduce systematic errors which can be removed beforehand; second, more terms in the final solutions with a limited number of reference stars give less accurate solutions; and, third, some effects often can not be represented by analytical functions, e.g. the magnitude equation.

Differential atmospheric refraction affects the plate scale as a function of position on the plate. It thus affects the relative proper motions derived from different plates. The correction can be made for each object relative to the plate center using the formulae given by Taff (1981).

OFAD also changes the plate scale and the higher order coefficients and is believed to be another large systematic error source in proper motion work using 4-m plates, if the OFAD coefficients and center are not very well known.

After having derived the OFAD coefficients, one has to define the OFAD center with considerable accuracy relative to which the corrections to measuring coordinates are to be made. The OFAD center position on a 4-m plate can be located with an accuracy better than ± 0.03 mm by minimizing the χ^2 of the plate transformation between the 4-m plate and an astrograph flat field plate (Guo et al. 1993). This error only leads to an error in position smaller than $\pm 0\overset{''}{.}04$.

The proper motion magnitude equation is complicated by the mixing of the cosmic one due to the solar motion and the non-cosmic one due mainly to the guiding error. In principle, the two kinds of magnitude equations can be separated if a group of stars of about the same distance and tangential velocity with a large magnitude range can be found. The members of a globular cluster or an open cluster are ideally suited for defining the non-cosmic magnitude equation. Like the correction for differential atmospheric refraction, it is also easier to correct the differential magnitude equation of one plate relative to a common plate.

Color equations could exist if the plate series were taken with large hour angles and/or with different lenses and/or different telescopes, etc. It can also be complicated by the possible existence of a cosmic stellar color–proper motion correlation. The way to define the non-cosmic color equation is very similar to that for the magnitude equation – find a group of stars of the same proper motion with a large color range. We found that the color equations of the plates we used to derive proper motions are usually very small or negligible. We believe this is because these plates were taken within a small hour angle range with the same camera on the same telescope.

The final relative proper motions, μ_x and μ_y, for 6667 objects were derived with an accuracy of about $\pm$1-2 mas/yr for V$\leq$20 and $>\pm$3 mas/yr for V$>$20. This accuracy indicates that image centering errors are the dominate error source, and that the other corrections applied beforehand are relatively accurate.

5. Absolute proper motions and tangential velocities

The zero-point of the absolute proper motions, $(\mu_x^g, \mu_y^g) = (-5.17 \pm 0.28, +5.25 \pm 0.25)$ mas/yr, was defined by 374 galaxies in the field of view. The 11 QSOs were excluded because their peculiar energy distribution or colors preclude them from defining the stellar proper motion zero-point.

The mean relative proper motion of the NGC 288, $(\mu_x^c, \mu_y^c) = (-0.34 \pm 0.10, +0.07 \pm 0.06)$ mas/yr, was derived from the proper motions of the 1401 candidate members which were selected, based on their space distribution on the plates and the proper motion vector point diagram (VPD). The absolute proper motion of the cluster is $(\mu_x^{c'}, \mu_y^{c'}) = (+4.8 \pm 0.3, -5.2 \pm 0.3)$ mas/yr.

With the mean radial velocity of the NGC 288 of $V_r = -48 \pm 1$ km/s (Peterson et al. 1986) and the mean distance modulus V-M_v =14.61 $\pm$ 0.14 (Lee et al. 1990; Bolte 1992), which corresponds to $r = 8.4 \pm 0.5$ kpc, its velocity components were derived as $(U, V, W) = (-56 \pm 17, -276 \pm 17, +47 \pm 1)$ km/s or $(u, v, w)_{LSR} = (+47 \pm 17, -264 \pm 17, +54 \pm 1)$ km/s, if $(u, v, w)_{LSR}^{\odot} = (-9, +12, +7)$ km/s (Mihalas & Binney 1981). The total transverse velocity is +268$\pm$17 km/s. Therefore the cluster is moving with a total speed of 274$\pm$16 km/s toward the direction of $(\ell_a, b_a) = (280° \pm 4°, 11° \pm 1°)$. If the rotational velocity of the LSR relative to the Galactic center is 220 km/s, the rotational velocity of the cluster will be -44$\pm$17 km/s, which means that the NGC 288 is moving on a retrograde orbit.

The absolute proper motions for the remaining field stars can also be obtained with respect to this derived zero-point. To avoid contamination of NGC 288 members, the field star sample was severely trimmed in both space and proper motion distributions, and 1555 field stars with minimal cluster star contamination were left. Since they are so close to the SGP, their (U, V) velocity components can be directly derived from their absolute proper motions and photometric parallaxes without being significantly affected by not knowing their radial velocities. The (U, V) compenents were also transformed into (u, v) using the solar motion adopted above.

As a preliminary stellar kinematics investigation, the mean values of (u, v) and their velocity dispersions, (σ_u, σ_v), for stars binned in vertical distances were derived. To avoid the influence of outliers on the mean velocities and dispersions as mentioned by Morrison et al. (1990), 5% stars at the lowest or highest velocity end in each bin were cut off. The results are for the combination of different stellar populations. The deconvolution of the thin, thick disk and halo components will be made in future work and their kinematics will be derived.

From the run of v with z, we can not see an obvious 'sharp break' at a height around $z = 5.5$ kpc as reported by Majewski (1992) from his measurement for field stars at the NGP. It is not clear if his break is a true feature or the result of small sample statistics. The flat zero run of mean u along z means that the Galaxy is neither expanding nor contracting. The mean v gradient of -35 ± 2 km/s/kpc below $z=1.5$ kpc is very consistent with the shear of -36 km/s/kpc obtained by Murray (1986). The mean v of the halo star dominated part ($z>5$ kpc) is less than 220 km/s, which is consistent with the results of most other observations (Morrison et al. 1990; Norris 1986; Ryan & Norris 1991; Soubiran 1993), but contrary to the value of Reid (1990) and Majewski (1992).

The distributions of σ_u and σ_v are roughly consistent with the results of Murray (1986) and Majewski (1992). But they are very noisy especially at high z. The final conclusions can only be made after more accurate estimations have been made using better techniques, such as the Maximum Likelihood Method.

6. Space distribution

The scale heights derived without deconvolving the different components are 305±17 pc for the Thin Disk, which is very close to 300 pc (Gilmore & Reid, 1983) or 325 pc (Reid & Majewski, 1993), and 1715±256 pc for the Thick Disk. The later is a little higher than most other measurements, which could be because the adopted metallicity gradient of −0.3 dex/kpc is too large, as the larger [Fe/H] change causes a flatter density law. Reid & Majewski (1993) claim that −0.1 dex/kpc may be more appropriate. Rose & Agostinho (1991) also reported that "the bulk of the thick disk is relatively homogeneous in age and metal abundance".

7. Final remarks and future work

In this presentation, we have discussed the procedures used to derive the photometric parallaxes from our CCD photometry, while taking into account the distortion effects on the photometric zero-point. We specifically stressed the need to pre-correct for various systematic errors before deriving proper motions. Considering the relatively short baseline (≤ 11 years) of the plate series, the final proper motion accuracy of about ± 2 mas/yr for most stars is good. This showed that our measuring procedures are adequate and that the various corrections applied beforehand are very accurate. From our preliminary kinematics, a sharp break at $z = 5.5$ kpc in the v velocity was not seen; the halo probably is not in retrograde rotation, but slightly prograde. The slightly higher scale height of the thick disk may indicate that the adopted metallicity gradient is too steep. In future work, we will reinvestigate the metallicity distribution based on new measurements, or, as is becoming very likely, we will derive this gradient from our UBV photometry and apply the metallicity deficiency correction to the absolute magnitudes for individual stars, since new observing time on the CTIO 4-m has been granted this year. We will study the stellar spatial, kinematic and possibly chemical abundance distributions in more detail, and apply this multi-dimensional data set to the new Galactic model being developed at Yale.

REFERENCES

Bolte, M. 1992, ApJS, 82, 145
Carney, B. W. 1979, ApJ, 233, 211
Chiu, L. -T. G. 1976, PASP, 88, 803
Chiu, L. -T. G. 1980, ApJS, 44, 31
Gilmore, G. F. & Reid, I. N. 1983, MNRAS, 202, 1025
Guo, X., Girard, T. M., van Altena, W. F., & López, C. E. 1993, AJ, 105, 2182
Hanson, R. B. 1979, MNRAS, 186, 875
Hanson, R. B. 1980, IAU Symp. 85, ed. J. E. Hesser (Dordrecht: Reidel), p.71
Lee, J. -F., & van Altena, W. F. 1983, AJ, 88, 1683
Lee, Y. W., Demarque, P., & Zinn, R. 1990, ApJ, 350, 155
Majewski, S. R. 1992, ApJS, 78, 87
Mermilliod, J. C. 1992, Bull. Inform. CDS, 40, 115
Mihalas, D. & Binney, J. 1981, Galactic Astronomy, edited by L. Olsen (Freeman, San Francisco), p. 400
Morrison, H. L., Flynn, C., & Freeman, K. C. 1990, AJ, 100, 1191
Murray, C. A. 1986, MNRAS, 223, 649
Norris, J. E. 1986, ApJS, 61, 667
Peterson, R. C., Olszwski, E. W., & Aaronson, M, 1986, ApJ, 307, 139
Reid, I. N. 1990, MNRAS, 247, 70
Reid, I. N., & Majewski, S. R. 1993, ApJ, 409, 635
Rose, J. A. & Agostinho, R. 1991, AJ, 101, 950
Ryan, S. G., & Norris, J. E. 1991, AJ, 101, 1835
Sandage, A., & Fouts, G. 1987, AJ, 92, 74
Soubiran, C. 1993, AA, (in press)
Taff, L. G. 1981, Computational Spherical Astronomy, (Wiley, New York), p. 79.

Expected phase space distribution of disc stars

By J. A. SELLWOOD

Department of Physics & Astronomy, Rutgers University, PO Box 849, Piscataway, NJ 08855, USA

Disc stars are scattered at the resonances of a spiral wave, moving onto much more eccentric orbits with slightly changed angular momenta. A resonance therefore creates a feature in the phase space distribution of disc stars which may persist for some time. If spirals are transient instabilities generated by a recurrent cycle, then it is possible that a resonance feature from at least one recent pattern lies within a few kpc of the Solar circle. In order to search for it, we must determine the angular momenta and epicycle energies of many stars. Inspection of the *Hipparcos* sample plotted in these variables is suggested as a test to discriminate between rival theories of spiral structure which differ fundamentally over the role of resonances.

1. Introduction

The spiral arm patterns of disc galaxies are now widely regarded as oscillations in the surface density of the underlying disc stars. Over the past 30 years, three rival theories have emerged amongst theoreticians working on the generation mechanism for such density waves, and observational tests are urgently needed to determine which, if any, is correct. One school, propounded most strongly by C C Lin and his co-workers (Lin & Shu 1964, 1966 through to Bertin et al. 1989), holds that the spiral patterns are slowly-evolving features with lifetimes of many orbital periods. In the latest version of their theory, the pattern is required not to have an inner Lindblad resonance (ILR), which would damp the mode on a short time scale. A second mechanism stems from the original papers by Goldreich & Lynden-Bell (1965) and Julian & Toomre (1966), and continues to be advocated by Toomre and Kalnajs. The spiral patterns, in this type of theory, are nothing more than the vigorous response of the disk either to tidal forcing (Toomre 1981) or to the molecular clouds orbiting within it (Toomre 1990; Toomre & Kalnajs 1991); accordingly, the spiral pattern changes shape and amplitude continuously on a timescale of less than an orbital period. These authors also do not expect strong Lindblad resonances to be present, essentially because the large-amplitude waves do not have a well defined pattern speeds. More recently, I have proposed a third alternative, in which a recurrent cycle of "groove modes" (Sellwood 1991; Sellwood & Kahn 1991) also causes the patterns to fluctuate rapidly. In this theory, however, the Lindblad resonances play a central role since each pattern seeds the growth of the next by changing the distribution function through resonant scattering.

While it would clearly be nice to know for its own sake how the spiral patterns are generated, a clear indication of resonances would also imply that the patterns are short-lived, which has much wider implications. Rapidly fluctuating spiral patterns are expected to alter the structure of the disc, changing the distribution angular momentum and causing the well-established rise in the velocity dispersion of stars with their ages (Carlberg & Sellwood 1985; Jenkins & Binney 1990). On the other hand, the mild long-lived spiral patterns favoured by Lin would not alter the disc structure much, since the integrals characterising stellar orbits are adiabatically invariant when the potential varies slowly on many orbital time-scales.

It is hard to devise an observational test which could discriminate between these theories. Since all regard the patterns as density waves, the existence of intensity variations in red photometric images and non-circular streaming velocities in the gas across spiral arms do not favour one over the other. The different lifetimes expected for the patterns give rise to slightly different expectations for the level of random motion amongst the disc stars, which is usually characterised by the dimensionless quantity

$$Q = \frac{\sigma_u \kappa}{3.36 G \Sigma}, \qquad (1)$$

where σ_u is the radial velocity dispersion, κ the epicycle frequency and Σ the surface density of the stars. Accurate estimates in the range $1 < Q < 1.5$ would be weak evidence in favour of the long-lived hypothesis while values in the range $1.5 < Q < 2$ are more what the proponents of the short-lived idea would expect. Since few would regard this very indirect test as decisive, there seems little point in encouraging observers to try to determine all three observables to a precision of perhaps 10% – an extremely tall order.

But very high quality astrometric observations of many stars in the Solar neighbourhood *may* afford a more decisive test of the recurrent "groove instabilities" theory. The recent existence of an inner Lindblad resonance (ILR, *e.g.* Binney & Tremaine 1987) in the Solar neighbourhood would be strong evidence in favour of it, while no resonance features should be seen if either of the other two theories is correct. Unfortunately, a non-detection, would not decisively rule out the groove mode theory, since we may be unlucky enough to have had no recent resonance in the observable range.

The remainder of this paper is structured as follows: in §2, I present a few details of my recent N-body simulations which show transient spiral waves and in §3, I illustrate the characteristic signature in phase space produced by an ILR. In §4, I describe the observational requirements for this test and in §5 I discuss some complications which may limit the conclusions from the test.

2. Transient spirals in N-body simulations

I have been led to propose the theory of groove modes from my recent work with N-body simulations, especially of models having exactly flat circular velocity curves, which linear analysis predicts to be stable. The simulations I have run do not verify this prediction, apparently for the reason that a finite number of particles, however large, behaves differently from a smooth fluid. This difference could always be due to a numerical inadequacy, but the number and quality of checks I have conducted (Sellwood 1983, Sellwood & Athanassoula 1986, Earn & Sellwood 1993) makes an artifact seem unlikely.

Figure 1 shows an N-body simulation using half a million particles of a $Q = 1.2$, half-mass disc model which appears to be linearly stable, and might therefore be expected to remain featureless. The linear theory prediction (Zang 1976; Toomre 1981; Toomre, private communication) was made for a slightly different model, having $Q = 1.5$ and a very gentle outer density taper which I had to sharpen in order to accommodate the disk on a finite grid; however, there is no indication that these differences are responsible for destabilising the model. Other details of the simulation are similar to those given in Sellwood (1991).

As may be seen, no features appear in the early evolution but Fourier analysis shows that fluctuating non-axisymmetric features grow from the noise level determined by the random distribution of this number of particles. As Toomre & Kalnajs (1991) have

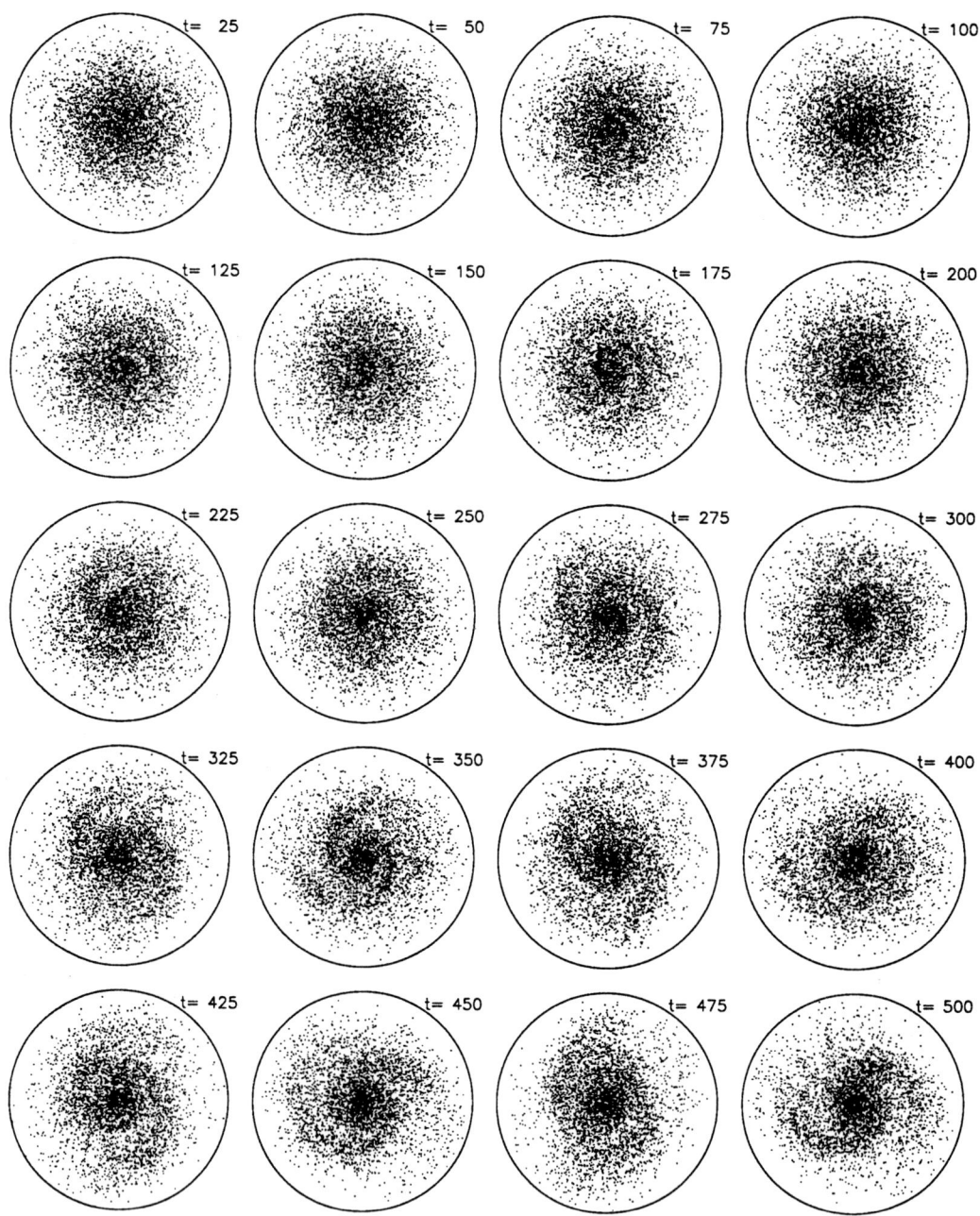

FIGURE 1. The long-term evolution of an N-body realisation of a linearly stable disc galaxy model. Only 1% of the particles is plotted. The circles represent the grid boundary and the orbital period at the half mass radius 20π of the indicated dynamical times.

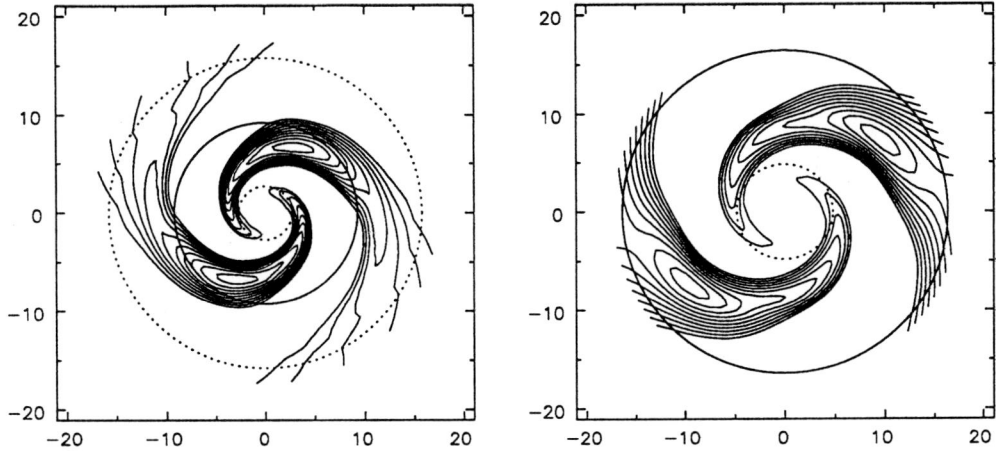

FIGURE 2. Two co-existing spiral wave patterns from the model which both grow and rotate at different rates. The circles mark co-rotation (full-drawn) and the two Lindblad resonances (dotted) for each pattern.

shown, the mass distribution in any grainy disc should become polarised by its own self-gravity giving rise to an excess amplitude in trailing spiral wakes – a phenomenon which can also be described as swing-amplified particle noise. However, the amplitude of tightly wrapped leading spiral components in this type of behaviour is expected to remain at the noise level, yet it rises significantly above that in the simulation. Thus some additional mechanism must be responsible for the vigorous spiral activity observed.

After a long time, individual large-amplitude spiral waves can be identified which maintain a phase coherence and grow exponentially over several orbital periods. Thus the vigorous patterns appear to behave as true linear instabilities of the system at the time each develops. All have the general form of a peak overdensity just inside the co-rotation resonance and extend inwards to the ILR and more than one pattern is present at most times in this model. The two co-existing patterns illustrated in Figure 2 have different angular rotation rates and saturate at different times; the estimated pattern speeds are: 0.108 ± 0.003 (left) and 0.061 ± 0.003 (right) in units if inverse dynamical times.

3. Resonance scattering

As expected, the vigorous disturbances raise the velocity dispersion of the disc stars. The usual way of showing this is to compute the second moments of the velocity distribution in radial bins; such measurements show that spiral heating takes place over a broad radial range. However, this is misleading, since the stars scattered by each pattern are quite localised in phase space, as shown in Figure 3. Here, I plot the energy of non-circular motion (or epicyclic energy) of one particle in 10 at the start and at time 400, soon after pattern on the left in Figure 2 has saturated. A comparison of these two plots reveals that most particles have not changed their initial energy or angular momentum by much, while there are a few narrow tongues of particles at low angular momentum which have been scattered up to high epicyclic energies.

It is straightforward to calculate the locus of points expected in this plot for particles

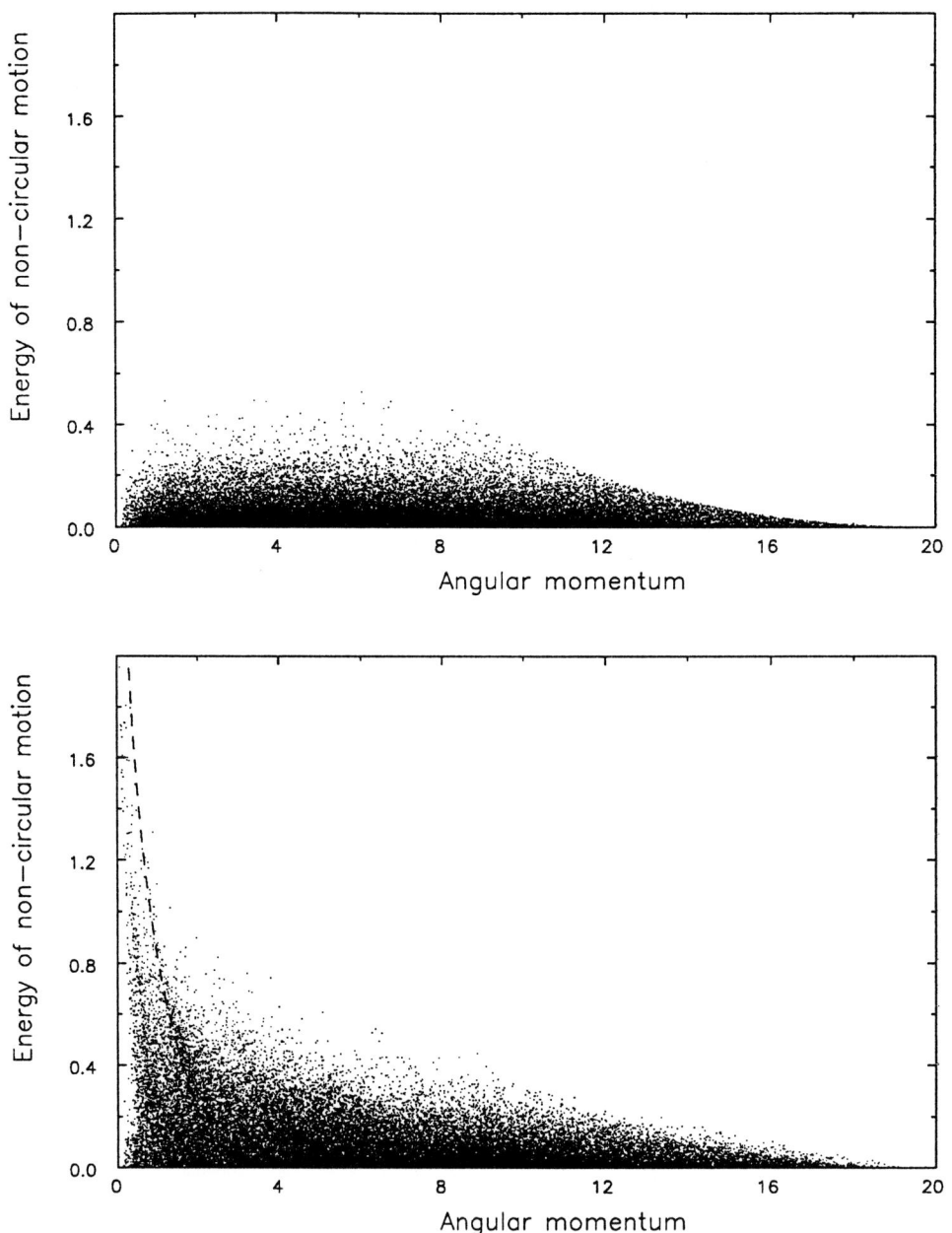

FIGURE 3. The distribution of particles in the $(J, E_{\rm rand})$-plane at the start (above) and at time 400 (below) in the simulation. Only 10% of the particles are plotted. The tongues rising upwards at low angular momentum result from resonant scattering by the stronger spiral waves. The dashed curve is drawn using equations (4) and (6) with $\Omega_p = 0.108$, $l = -1$ and $m = 2$.

scattered at a resonance with a rotating perturbation. Jacobi's integral, $E - \Omega_p J$ (e.g. Binney & Tremaine), is conserved for a particle of instantaneous energy E and angular momentum J moving in a potential rotating at the steady angular rate Ω_p; the change in energy in the inertial frame must therefore be related to the change in angular momentum by

$$\Delta E = \Omega_p \Delta J. \qquad (2)$$

The energy of non-circular motion may be defined as $E_{\text{rand}} \equiv E - E_{\text{circ}}$, where $E_{\text{circ}}(J)$ is the energy of a star having the same J but moving on a (minimum energy) circular orbit. This definition is easily shown to be equivalent to the epicyclic energy, defined e.g. by Lacey (1984).

In the Mestel (1963) disc,

$$E_{\text{circ}} = V_0^2 \left[\ln\left(\frac{J}{J_0}\right) + \tfrac{1}{2}\right]; \qquad (3)$$

therefore

$$\Delta E_{\text{rand}} = \Omega_p \Delta J - V_0^2 \ln\left(\frac{J_i + \Delta J}{J_i}\right), \qquad (4)$$

where J_i is the initial angular momentum. For small ΔJ this is

$$\Delta E_{\text{rand}} \simeq \Omega_p \Delta J - V_0^2 \frac{\Delta J}{J_i}. \qquad (5)$$

For resonant scattering from nearly circular orbits, J_i is related to Ω_p as

$$J_i = \frac{V_0^2}{\Omega_p}\left(1 + \sqrt{2}\frac{l}{m}\right), \qquad (6)$$

in the Mestel disc, where $l = 0$ at co-rotation, $l \pm 1$ at the Lindblad resonances, etc. Substituting into the expression (5) for small changes, we find

$$\Delta E_{\text{rand}} \simeq \Omega_p \Delta J \frac{\sqrt{2}l}{m + \sqrt{2}l}. \qquad (7)$$

Clearly $\Delta E_{\text{rand}} \equiv 0$ at co-rotation. At all other resonances, we must have $\Delta E_{\text{rand}} \geq 0$ for a particle starting on a circular orbit. Thus only one sign of scattering is possible at a Lindblad resonance; obviously, $\Delta J < 0$ at the ILR and $\Delta J > 0$ at the OLR.

The dashed line in Figure 3 shows the curve given by (4) for scattering from circular orbits at the ILR of the left hand wave shown in Figure 2. Since this line, which has no free parameters, is an almost perfect fit to the locus of scattered points in one tongue reaching up to large non-circular energies it shows conclusively that the tongue was produced by scattering at the ILR of the most recent strong spiral pattern in the simulation.

It does not seem too surprising that the dynamical behaviour of the simulation in which the DF has been sculpted in this manner by previous resonances is different from that predicted by a linear theory analysis of an assumed smooth DF. The difference between the N-body realisation and the smooth fluid assumed in semi-analytical work therefore stems from the finite capacity of resonances to absorb waves; in linear theory this capacity is assumed to infinite. It seems that the graininess of even this large number of particles is sufficient to start this non-linear cycle, in which the ashes of one disturbance seed the growth of a new, and apparently larger amplitude, instability. Exactly how this happens and what to expect for the properties of subsequent waves still needs to be understood, however.

As the idea developed so far is complicated and further progress is likely to make it more so, it would be desirable to have some observational evidence that resonance

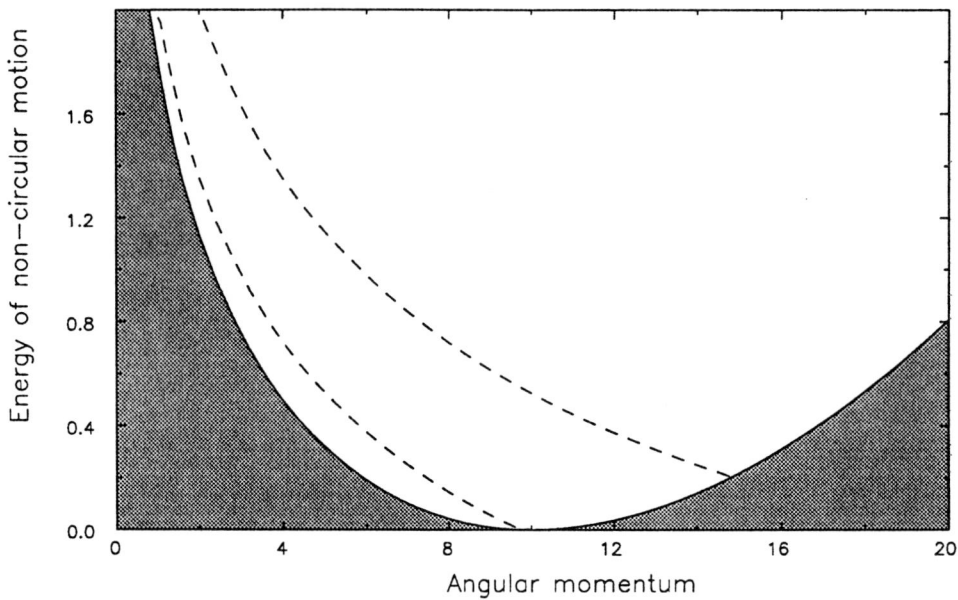

FIGURE 4. Boundaries, in the $(J, E_{\rm rand})$-plane of stars which visit the solar neighbourhood; stars in the shaded regions will be absent from a sample collected at the Solar radius. The dashed curves indicate possible resonant scattering loci.

scattering actually occurs in spiral galaxies. If it does not occur, the simulations will have been misleading us. The only place where there is any hope of finding evidence of resonances is in the Solar neighbourhood; if the spiral patterns in the disc of the Milky Way do behave as the simulations indicate then there should be several such resonance features over a Galactic radius range of a few kpc, and possibly one close enough to be detected.

4. Observational requirements

In order to search for resonance features, we need to determine the angular momenta and energies of non-circular motion of many stars. Inspection of a large sample plotted in these variables may reveal a narrow tongue of stars reaching up to high non-circular energies, demonstrating that it was caused by a resonance.

To create such a plot we need accurate distances and full space motions, both transverse and line-of-sight, for a large sample of stars. Fortunately, we do not need to know either the true distance to the Galactic centre or the circular velocity at the Sun, or even the precise shape of the rotation curve in the Solar neighbourhood. Errors in these poorly known parameters will cause complicated scale changes in plots such as Figure 3 but should not affect the topology of the plot – any tongues present should remain visible under the transformation caused by assuming an incorrect galactic model, provided both the assumed model and the Galaxy do not have any pathological properties.

Thus any convenient simple model potential, such as the isochrone (Eggen, Lynden-Bell & Sandage 1962) or logarithmic (Mestel 1963), can be used to estimate the home radius of a star – the radius at which a star of the same angular momentum would move on a circular orbit. Estimation of the non-circular (or epicyclic) energy in the assumed potential model is then straightforward.

Thousands of stars will be required for resonant tongues to stand out, but it will be important to use only stars with the best determined distances and motions. If the error in plotting each point is much larger than the width of the tongue, it will become very blurred and be hard to pick out. It is to be hoped that the *Hipparcos* data base will provide a large enough sample of stars with accurate distances and proper motions, and that the radial velocities of the same stars will be determined.

5. Possible complications

Probably the most severe difficulty in finding a resonance feature is that only a very limited range of possible pattern speeds are detectable. Stars with accurate distances will all be very close to the Sun, but they will have a wide range of angular momenta, since those with home radii quite far away can visit the Solar neighbourhood if their orbits are eccentric enough. The window in the (J, E_{rand})-plane created by observing stars in the Solar neighbourhood only will have the curved boundaries shown in Figure 4; the angular momentum scale is chosen so that the LSR has angular momentum of 10 in these arbitrary units, and a logarithmic potential has been assumed. A couple of possible loci of resonant scattering tongues are sketched by dashed curves in this figure. Obviously the selection boundaries make it impossible to identify a resonance which had existed inside the Solar radius, but ILR tongues created at larger radii could be detected.

(While no spiral wave can have an ILR very close to the outer edge of the disc, patterns with ILRs beyond the Solar radius are not ruled out. The reason the tongues in the N-body simulations all lie at low angular momenta is because the orbital period is shortest near the centre, causing spiral features to develop there first; it is likely that the Galaxy is dynamically old enough for low pattern speed features to have developed.)

A second cause of serious concern is that the tongues will become less distinct as they age, so that only those created recently will be sharp enough to be visible. The two principal sources of broadening are subsequent spiral activity and scattering by giant molecular clouds. While the largest part of spiral heating is the resonance scattering itself, stars in the wings of a resonance will be scattered slightly. As shown by Lacey (1984), molecular clouds both redirect the peculiar velocities of stars and also increase their epicyclic energies; the former process taking place on a shorter time-scale than the latter. The rate of scattering by GMCs is uncertain (it depends on the poorly-known number density and mass spectrum of the clouds) but it is slower than the heating rate, because the local stellar velocity ellipsoid does not have the axis ratios predicted for equilibrium under molecular cloud scattering (Lacey 1984). Thus the minimal requirement, that the tongues will not be erased by scattering faster than they are formed, is fulfilled – but an estimate of their survival time beyond this would be entirely dependent on the (unknown) contribution of GMCs to the overall heating rate.

Thirdly, the stars of the *Hipparcos* sample are not a randomly selected sample of Solar neighbourhood stars, and the selection criteria could prove troublesome. It may therefore be necessary, at a minimum, to discard large fractions of stars from specially targeted disc clusters which are likely to be moving as groups, in order not to create density clumps in the (J, E_{rand})-plane that are entirely due to sampling.

A further possible problem with this test is that the prediction comes from N-body simulations in which the motion of particles is confined to a plane. There have been no tests to date on the effects of 3-D motion on resonant scattering between stars and spiral waves. Simple arguments, such as the adiabatic invariance of motion normal to the plane when the stars in the plane drift slowly through potential perturbations, would seem to suggest that the vertical component of motion of the stars can simply be ignored. This should remain a valid approximation as long as bending instabilities (Sellwood & Merritt 1994; Merritt & Sellwood 1994) can be avoided.

6. Conclusions

The *Hipparcos* sample of stars offers a possible test of current ideas for the generation of spiral structure in galaxies. However, the difficulties discussed in the previous section are daunting and it seems likely that even this sample of stars will be inadequate for a conclusive result to emerge from such a test. Nevertheless, the distribution of the *Hipparcos* stars in the (J, E_rand)-plane may tell us more about the recent dynamical events in the Milky Way than we can learn from simply calculating moments.

REFERENCES

Bertin G, Lin C C, Lowe S A & Thurstans R P, 1989. ApJ **338**, 78
Binney J & Tremaine S, 1987. *Galactic Dynamics*, Princeton University Press
Carlberg R G & Sellwood J A, 1985. ApJ **292**, 79
Earn D J D & Sellwood J A, 1993. *in preparation*
Eggen O J, Lynden-Bell D & Sandage A, 1962. ApJ **136** 748
Goldreich P & Lynden-Bell D, 1965. MNRAS **130**, 125
Jenkins A & Binney J, 1990. MNRAS **245**, 305
Julian W H & Toomre A, 1966. ApJ **146**, 810
Lacey C G, 1984. MNRAS **208**, 687
Lin C C & Shu F H, 1964. ApJ **140**, 646
Lin C C & Shu F H, 1966. *Proc Nat Acad Sci (USA)*, **55**, 229
Merritt D & Sellwood J A, 1994. ApJ *to appear*
Mestel L, 1963. MNRAS **126**, 553
Sellwood J A, 1983. *J Comp Phys* **50**, 337
Sellwood J A, 1991. In *Dynamics of Disk Galaxies*, p 123, ed B Sundelius (Göteborgs University)
Sellwood J A & Athanassoula E, 1986. MNRAS **221**, 195
Sellwood J A & Kahn F D, 1991. MNRAS **250**, 278
Sellwood J A & Merritt D, 1994. ApJ *to appear*
Toomre A, 1981. In *Structure and Evolution of Normal Galaxies*, p. 111, eds S M Fall & D Lynden-Bell, Cambridge University Press
Toomre A, 1990. In *Dynamics & Interactions of Galaxies*, p. 292, ed R Wielen, Springer-Verlag:Berlin, Heidelberg
Toomre A & Kalnajs A J, 1991. In *Dynamics of Disc Galaxies*, p. 341, ed B Sundelius (Göteborgs University)
Zang T A, 1976. PhD thesis, MIT.

The Solar Neighborhood

By ARTHUR R. UPGREN

Van Vleck Observatory, Wesleyan University, Middletown, CT 06457, U.S.A.

The stellar composition of the solar neighborhood was not understood at the beginning of this century. Shortly thereafter, Schlesinger introduced photography and rigor to differential astrometry, and the subsequent large increase in the number of heliocentric parallaxes and proper motions allowed the Hertzsprung-Russell and Hess Diagrams to correctly shape the distribution of nearby stars. Since those times the set of nearby stars and the luminosity function of Kapteyn and van Rhijn have been continually enriched at its faint end, as proper motion studies and later objective-prism surveys have identified the low-mass red dwarf stars, but not otherwise altered in its major features. Completeness, at least to the maximum of the luminosity function, extends outwards from the Sun to about 15 parsecs, and to the 25 parsec limit of the new Third Edition of the Catalogue of Nearby Stars (CNS3) for stars brighter than absolute magnitude +9. In recent years photometry has been extended to over 90 per cent of its 3800 stars. HIPPARCOS, supplemented by ground-based parallax efforts will raise the number with reliable distances and increase the distance limit within which brighter stars will be abundant and complete. Magnitude-limited spectral surveys by Vyssotsky and Upgren and their colleagues and by Stephenson form a worthy counterbalance to the distance-limited CNS3 for the study of the Malmquist and Lutz-Kelker corrections and other systematic effects, inevitably present in the data.

---

1. Historical perspective

The Nineteenth Century – the century between William Herschel and J.C. Kapteyn – produced little in the understanding of the distances and luminosities of common stars and their distribution in space. To be sure, it was during this period that the tools were being perfected which would enable us to gain a comprehensive picture of just these properties of stars and the galaxy they form. The first distances to stars were derived from measures of the heliocentric parallax, ending a three-century effort to confirm the ultimate requirement of the Copernican Revolution. The differences between the spectra of stars was recognized and the first tentative steps were taken towards their identification and classification. But the application of these techniques to the stars around us remained in the future.

An examination of textbooks and books for the lay reader may be more revealing than the research papers of the times, for a consensus view of the nearest stars. Thus a surprisingly naïve view of stellar distances was offered by Loomis (1882) who summarized the Nineteenth Century viewpoint with the statement "varieties of magnitude are chiefly caused by difference of distance rather than by difference of intrinsic splendor among the objects themselves. Those stars which are placed immediately about our solar system appear bright in consequence of their proximity, and are called stars of the first magnitude; those which lie beyond are . . . called stars of the second magnitude; and thus as the distance of the stars increases, their apparent brightness diminishes. . .".

The situation had not materially improved by 1906, when Professor Todd's <u>New Astronomy</u> was published; we continue to find a garbled view of the solar neighborhood as reproduced in Figure 1, taken from his textbook.

Photography and rigor introduced into both astrometry and spectroscopy greatly increased the number of stars with reliable distances, luminosities and surface tempera-

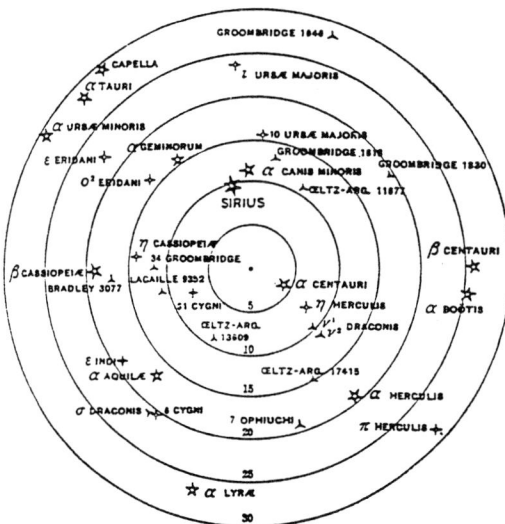

FIGURE 1. The locations of stars thought in 1906 to be within ten parsecs of the Solar System. The distances shown are in light years. Large errors in parallax and the absence of proper motion surveys to locate nearby faint stars make for an erroneous view of the region nearest the Sun.

tures and the Hertzsprung-Russell Diagram soon followed as a direct result (see reviews by Russell, 1914 and DeVorkin, 1977).

Further studies by Kapteyn, van Rhijn and others led to the Hess Diagram and the stellar luminosity function (van Rhijn 1936) as the stellar composition of the solar neighborhood finally took shape. In the half-century since those efforts, the number of nearby stars has been greatly expanded but they have not altered its major features. The picture took on its present form within two decades of Todd's (1906) flawed portrayal. However, the list of stars within five parsecs of the Solar System that now seems to form a de rigueur component of text appendices is only about forty years old; before that time the lists extended to four parsecs or less. Among the first to define five parsecs as the outer limit to these compilations were Bok and Bok, whose book The Milky Way went through four editions (Bok and Bok 1941, 1945, 1957, 1974).

2. Completeness of the five-parsec compilations

These editions by Bok and subsequent compilations show the changes in the stellar content of the five-parsec-limited sample up to the present time. They show a slow steady increase in the number of stars and stellar systems, which has occurred mostly among the faint dwarf M stars. This is in close agreement with Henry and McCarthy (1990) who listed the dwarfs north of Declination $-30°$. A number of five-parsec censuses are listed in Table 1.

The likelihood of further additions to this list is very small, except perhaps at the very faint end of the luminosity function. One indication of this has been provided by the recent photometric program completed by Weis (1988). Weis obtained photoelectric photometry of 2013 stars in the NLTT Catalogue (Luyten 1979, 1980) which fulfilled the following conditions; all were north of the Celestial Equator, with proper motions less

Author	Source	Year	stars	systems
Bok & Bok	The Milky Way, 2nd ed.	1945	48	37
Bok & Bok	The Milky Way, 3rd ed.	1957	56	42
Bok & Bok	The Milky Way, 4th ed.	1974	60	45
van de Kamp	Stellar Paths	1981	63	47
Batten	The Observer's Handbook	1987	65	50
Batten	The Observer's Handbook	1993	67	51

TABLE 1. Stars within five parsecs of the Sun.

than $0''\!.5$ per year and all were of color class m, the reddest color assigned to stars listed in the NLTT. Furthermore, none had a trigonometric parallax or were fainter than a red apparent magnitude of 13.5. The distances, determined photometrically, indicated that 295 of the stars lie within 25 parsecs of the Sun and that ten are closer than ten parsecs. Yet none were found to be within seven parsecs, indicating that future additions to the five-parsec list are likely to be very rare.

Although few stars have been added to the list in recent years, the precision of their distances has been greatly improved due to improved practices of present ground-based parallax programs over their predecessors. The data given in the Third Edition of the Yale Parallax Catalogue and its Supplement (Jenkins 1963) reveal that the distances of only ten stars were known with standard errors less than two percent, and that Barnard's Star alone had an error of less than one percent (Upgren 1977). Today, the Fourth Edition of the YPC (van Altena, Lee and Hoffleit 1993) contains 26 stars in 17 systems with distance errors less than one percent, the result of a remarkable improvement in the precision of ground-based parallax work. These stars are listed in Table 2. It is likely that the results from the program of HIPPARCOS will further increase this list.

Although a detailed analysis is beyond the scope of this paper, a simple test shows that the five-parsec sample appears representative of the general stellar population in the Galactic disk. For this purpose, six other points were adopted as centers of spheres of five parsecs in radius. They are all located at a distance of five parsecs from the Sun along each of the three orthogonal axes of the Equatorial coordinate system. In this frame, the x-axis connects the Equinoxes with the positive direction toward the Vernal Equinox. The y-axis also lies in the Equatorial plane with the positive direction oriented towards a Right Ascension of six hours and the positive z-axis is directed towards the North Celestial Pole.

Table 3 shows the cumulative star counts in each spherical volume including that centered at the Sun, or seven in all. It is apparent that the local stars are representative of a larger volume except perhaps for the stars fainter than absolute magnitude 13. This table cannot be taken as conclusive, but it shows no anomaly among the closest stars with respect to those lying somewhat farther away.

3. Distance- and magnitude-limited compilations

The volume of space enclosed within five parsecs is not sufficient for a definitive stellar sample. One way of defining the solar neighborhood is that it is the volume within which virtually all stars are known. It is obvious that under such a definition, no single distance limit can suffice. Incompleteness sets in for the faintest stars far closer than the distance

YPC	π(mas)±s.e.	μ''/yr	V	B−V	M_v	Name
49.00	282.7±2.3	2.91	8.07	1.56	10.33	Grb 34 A
			11.10	1.40	13.36	Grb 34 B
160.00	231.0±1.9	2.98	12.38	0.55	14.20	van Maanen 2
343.10	372.7±2.7	3.33	12.52	1.85	15.38	G 272-061 A
			12.56	1.88	15.42	G 276-061 B
1577.00	380.6±2.1	1.33	−1.46	0.00	1.44	Sirius A
			8.44	−0.03	11.34	Sirius B
1805.00	286.2±2.1	1.25	0.38	0.42	2.66	Procyon AB
2553.00	418.9±2.1	4.68	13.46	2.03	16.57	Wolf 359
2576.00	394.2±1.8	4.81	7.48	1.51	10.46	Lal 21185
2730.00	297.7±1.7	1.40	11.11	1.77	13.48	Ross 128
3278.00	770.2±6.1	3.85	11.22	1.90	15.65	Proxima
3309.00	750.0±5.4	3.67	−0.01	0.71	4.37	α Cen A
			1.33	0.88	5.71	α Cen B
3845.00	155.3±0.6	1.18	9.04	1.58	10.00	Wolf 630 A
			9.76	1.62	10.72	Wolf 630 B
4098.00	545.9±1.3	10.31	9.54	1.74	13.23	Barnard's Star
4330.00	285.4±2.5	2.23	8.91	1.54	11.19	+59° 1915 A
			9.69	1.59	11.97	+59° 1915 B
4494.00	172.2±0.7	1.47	9.13	1.50	10.31	+04° 4048 A
			17.53	2.20	18.71	+04° 4048 B
4706.01	220.2±1.0	0.68	13.41	1.90	15.12	G 208-044 A
			13.99	1.98	15.70	G 208-045 B
5077.00	286.9±1.1	5.22	5.21	1.18	7.50	61 Cygni A
			6.03	1.37	8.32	61 Cygni B
5736.00	315.6±1.1	1.83	12.27	1.92	14.77	Ross 248

TABLE 2. Stars with parallaxes known within one percent error.

Center	Number of stars with absolute magnitude brighter than				
	+6	+9	+11	+13	all magnitudes
(0, 0, 0=Sun)	7	14	22	34	61
(+5, 0, 0)	4	13	16	23	41
(−5, 0, 0)	8	12	23	31	50
(0,+5, 0)	6	11	16	29	50
(0, −5, 0)	6	17	27	39	51
(0, 0, +5)	5	12	22	34	46
(0, 0, −5)	10	16	24	30	41

TABLE 3. Numbers of stars in different volumes of space.

of the nearest of many types of bright stars, leading to different distances which depend on the types of stars. Data are mostly available for the establishment of limits within which specific types of stars are complete.

In order to accomplish this objective we need to be familiar with the primary sources of observational material. These can be divided into distance-limited and magnitude-limited catalogues of stars. Three in the first group are worth mention, all of which were made from objective prism surveys in order to avoid the high-velocity bias incorporated in proper motion surveys. Vyssotsky was the first to accomplish this goal and the lists published by him and his colleagues are still the benchmark against which all other lists of nearby stars must be measured.

Vyssotsky published five lists of stars and a summary (Vyssotsky 1943, 1956, 1958, 1963, Vyssotsky et al. 1946, Vyssotsky and Mateer 1952) containing 895 dwarf K and M stars and stellar systems detected as such on plates taken with the 65-cm refractor of the McCormick Observatory. His classification criteria were of such reliability that only one star in his lists is now known to be a giant. In the last few years, broad-band photometry in BVRI and precise positions and proper motions have been found for all of the stars (Gliese et al. 1993). The data indicate that a high velocity bias may still be present (Weis and Upgren 1993) and the author intends to examine the McCormick records in detail in order to detect possible reasons for it.

The other two magnitude-limited samples are those of Upgren et al. (1972) and Stephenson (1986a, 1986b). The first forms a southern extension to the Vyssotsky stars but does not quite cover all of the deep southern sky. It includes 624 stars to a limiting magnitude of between 10.5 and 11.0. The second is limited to regions more than ten degrees from the Galactic plane and contains 3989 stars in two lists brighter than the 12th magnitude and north of $-25°$. The spectral region examined by Stephenson was limited to a small portion of the visual region of the spectrum, and perhaps for this reason, about one quarter of its stars are in fact giants or subgiants as revealed by photometry (Weis 1991). Weis has identified the dwarfs among the later Stephenson stars in order to measure the extent of any bias in the Vyssotsky stars.

The distance-limited compilations of Gliese (1969), Gliese and Jahreiss (1979) and Woolley et al. (1970) are superseded by the revision of the former by Jahreiss and Gliese (1993) hereafter referred to as CNS3. The revision contains over 3800 stars within 25 parsecs, a great increase over its predecessors (Jahreiss 1993). Tests for completeness described and used by Upgren and Armandroff (1981) and Gliese, Jahreiss and Upgren (1986) indicate that the CNS3 is statistically complete to its limit for stars brighter than absolute visual magnitude +9 (or B−V of 1.40 or spectral class M0V) and closer to completeness for fainter stars than its predecessors. Figure 2 shows the degree of completeness for stars grouped by B−V color. Since more than 90 per cent of all stars in the CNS3 have photometry, estimates of B−V from spectral types needed to be made for only a few mostly very faint stars. It appears that the CNS3 provides samples of stars that are plentiful and essentially complete for the dwarf M stars to about 15 parsecs and for brighter stars up to about +3 or F2 on the main sequence to its limit of 25 parsecs. Stars brighter than F2 become too scarce for analysis of their properties unless the distance limit can be increased beyond 25 parsecs.

4. Definitions of the Solar neighborhood

The Solar neighborhood can be defined in different ways. One can be defined as that region limited by the point at which a trigonometric parallax is of "good" quality. Since the precision of a parallax is determined by the dimensionless ratio σ_π/π, where σ_π is

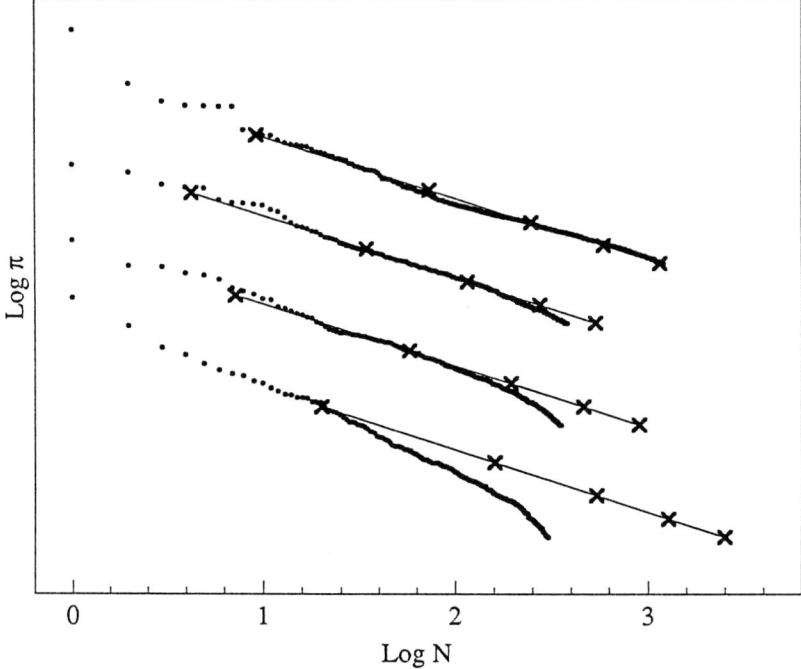

FIGURE 2. Completeness indications for CNS3 stars. The straight lines are of slope −3 and the stars are grouped by B−V color: From the top, the data represent stars of B−V ≤ 1.40; stars with 1.40 < B−V ≤ 1.50; stars with 1.50 < B−V ≤ 1.60 and stars redder than B−V = 1.60. Compare with Figure 7 of Gliese, Jahreiss and Upgren (1986) showing data from CNS2 and its Supplement. The principal difference between the two figures occurs for the stars with B−V between 1.40 and 1.50, the early M dwarfs. They now appear complete to a distance of about 20 parsecs. The reddest group here does not correspond to groups in their Figure 7.

the standard error in the parallax, we can set a distance beyond which this ratio cannot be maintained at a specific level. This horizon is adopted as the point at which the ratio exceeds +0.175 for two reasons. First, this is the limit at which the correction found by Lutz and Kelker (1973) can be evaluated, and second it is about at this point that secondary distance methods become more useful in the calibration of stellar physical properties. The dispersion in magnitude varies with $5 \log e$ times this ratio. Hence at the distance horizon, $\sigma(M_v) = \pm 0.38$ magnitudes.

In the days before rigorous measuring and computational methods; that is before about 1960, σ_π averaged about $\pm 0\rlap.{''}016$ (Hertzsprung 1952) setting the horizon at only about eleven parsecs. Stars brighter than the Sun are too scarce within this limit for reliable calibration.

Today, the precision of parallaxes from the ground-based programs has improved by almost an order of magnitude as indicated in Table 2. It is unfortunate that the number of these programs has decreased greatly over the last several decades. However, the promise of HIPPARCOS, the astrometric space satellite, in defining the solar neighborhood can hardly be overestimated. Prior to its launch in 1989, its error in parallax was estimated to be about $\pm 0\rlap.{''}002$ for stars brighter than its completeness limit of about visual magnitude eight, increasing to $\pm 0\rlap.{''}005$ at its detection limit at magnitude twelve. Preliminary results from it presented at this workshop indicate that these estimates are realistic. The impact

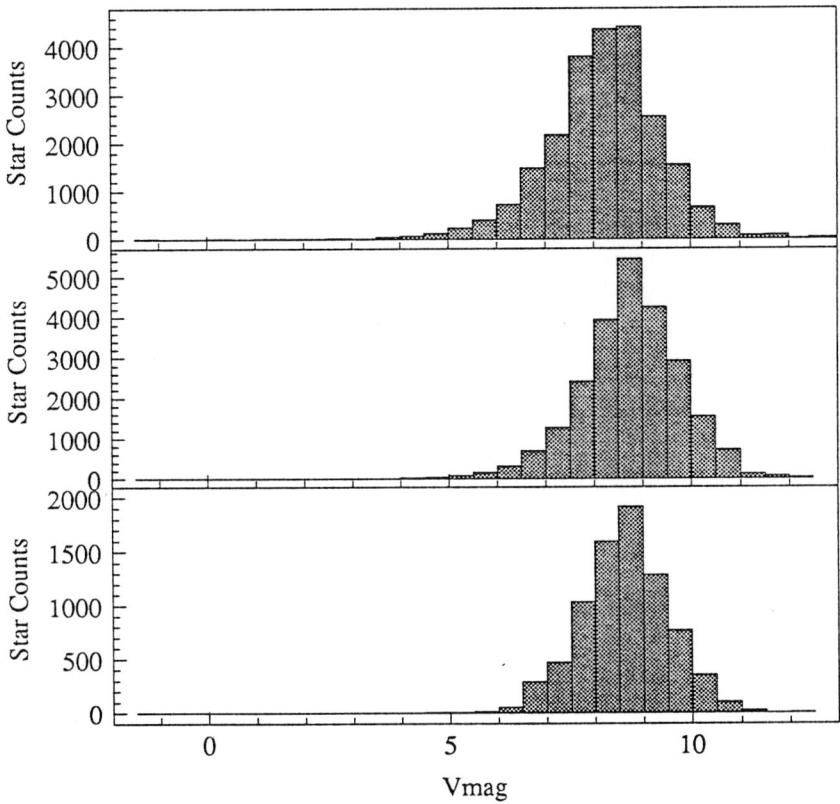

FIGURE 3. Stars included in the HIPPARCOS program. The distributions by magnitude are shown for the stars of luminosity classes IV, V and VI for spectral classes A0–F2, F3–G4 and G5–M, respectively, from top to bottom. The total numbers of stars included in the three figures are 23002, 24083 and 7917, respectively.

of its data complemented by those from the ground-based observatories, will be great indeed.

A second definition of the solar neighborhood is more strict. This would be defined as the volume over which the stellar sample is essentially complete in addition to its member stars being measured with precision. As discussed above, the limit for dwarf M stars is less than 15 parsecs and that for main sequence F, G and K stars is the 25 parsec limit of the CNS3. For brighter stars, HIPPARCOS will also provide a new definition of the solar neighborhood. Using the parallax error estimates available prior to its launch, Murray (1983) and Upgren (1983a, 1983b) both found that the nearby star census and the luminosity function would be most improved for stars between apparent magnitudes 0 and +6, or from about A0 through G0 on the main sequence as well as the giants and subgiants. These are the stars that are abundant at 50 to 100 parsecs but not at 25 parsecs. With the advent of the CNS3, the F and G dwarfs are reasonably plentiful at the smaller distance, but the A stars and the giants are not.

The HIPPARCOS Input Catalogue (INCA) contains information for about 118 000 stars for which parallaxes, proper motions and positions are to be obtained. The normal giant stars even brighter than the completeness limit of magnitude eight are not at all fully represented owing to their relative abundance, but the main sequence stars are all included as is shown in Figure 3. Completeness appears to hold for the main sequence at least to mag. 8.0, but for even the early A stars the 100 parsec limit occurs at a considerably brighter magnitude. The yet brighter stars such as the normal B stars, are found in too few numbers for good parallaxes. Even if they are not individually of much value, collectively they may be worth study as Pagel (1979) has demonstrated.

5. Conclusions

Two facts are now inescapably clear from this discussion; first the very nearby stellar neighbors are well cataloged and understood, except at the extreme faint end of the main sequence. Second, the program for HIPPARCOS, if successful, will give a whole new meaning to the solar neighborhood for brighter stars, both for increased precision in physical properties and for completeness. The census of nearest stars must clearly extend to different distances due to the great spread in their intrinsic luminosities. It is worth pointing out that some vital properties are not known even for some of the nearest stellar neighbors. Among them are radial velocities and age-related spectral measures such as calcium and hydrogen emission intensities and metal-to-hydrogen estimates. These data, if known, would provide an exceptionally detailed picture of the disk population of our Galaxy.

Acknowledgements

The author wishes to acknowledge funding from the National Science Foundation through Research Grant AST-9218605. He also wishes to acknowledge the personal friendship and support of Dr. Wilhelm Gliese extending over many years. Dr. Gliese died just before this workshop was convened. It is appropriate, therefore, that this paper is dedicated to his memory.

REFERENCES

Batten, A.H. (1987). Observer's Handbook, R.A.S. Canada, 164.

Batten, A.H. (1993). Observer's Handbook, R.A.S. Canada, 193.

Bok, B.J. & Bok, P.F. 1941, 1945, 1957, 1974, The Milky Way, 1st, 2nd, 3rd & 4th Editions, (Harvard U. Press, Cambridge).

DeVorkin, D.H. (1977). Dudley Observatory Report No. 13, in Memory of Henry Norris Russell, eds. A.G.D. Philip & D.H. DeVorkin, (Dudley Observatory, Albany).

Gliese, W. (1969). Veröff. Astron. Inst. Heidelberg, No. 22.

Gliese, W., & Jahreiss, H. 1979, A&AS, **38**, 423.

Gliese, W., Jahreiss, H. & Upgren, A.R. (1986). The Galaxy and the Solar System, eds. R. Smoluchowski, J.N. Bahcall & M.S. Matthews, (publ. University of Arizona Press, Tucson), 13.

Gliese, W., Jahreiss, H., Upgren, A.R. & Weis, E.W. (1993). in preparation.

Henry, T.J. & McCarthy, D.W. (1990). ApJ, **350**, 334.

Hertzsprung, E. (1952). Observatory, **72**, 242.

Jahreiss, H. (1994). These Proceedings.

Jahreiss, H. & Gliese, W. (1993). Catalogue of Nearby Stars, Third Edition.

Jenkins, L. F. (1963). General Catalogue of Trigonometric Stellar Parallaxes, (Yale University Observatory, New Haven).

Loomis, E. (1882). A Treatise on Astronomy, (Harper & Brothers, New York).

Lutz, T. E. & Kelker, D. H. (1973). PASP, **87**, 617.

Luyten, W.J. (1979). NLTT Catalogue, **I** & **II**, (University of Minnesota, Minneapolis).

Luyten, W.J. (1980). NLTT Catalogue, **III** & **IV**, (University of Minnesota, Minneapolis).

Murray, C. A. (1983). Proceedings of an International Colloquium on the Scientific Aspects of the HIPPARCOS Mission, Held at Strasbourg, France, 1982, p.115.

Pagel, B. E. J. (1979). European Satellite Astronomy, eds. C. Barbieri & P. L. Bernacca, (University of Padua, Padua, Italy), 211.

Russell, H.N. (1914). Popular Astronomy, **22**, 275 & 331.

Stephenson, C.B. (1986a). AJ, **91**, 144.

Stephenson, C.B. (1986b). AJ, **92**, 139.

Todd, D. (1906). New Astronomy, (American Book Company, New York).

Upgren, A. R. (1977). Vistas in Astronomy, **21**, 241.

Upgren, A. R. (1983a). Kinematics, Dynamics and Structure of the Milky Way, ed. W. L. H. Shuter, (Reidel: Dordrecht), 15.

Upgren, A. R. (1983b). IAU Colloquium No. 76, The Nearby Stars and the Stellar Luminosity Function, eds. A. G. D. Philip & A. R. Upgren, (L. Davis Press, Schenectady, N.Y.), 57

Upgren, A.R. & Armandroff, T.E. (1981). AJ, **86**, 1898.

Upgren, A.R., Grossenbacher, R., Penhallow, W.S., MacConnell, D.J. & Frye, R.L. (1972). AJ, **77**, 486.

Van Altena, W.F., Lee, J.T. & Hoffleit, E.D. (1993). General Catalogue of Trigonometric Stellar Parallaxes, (Yale University Observatory, New Haven).

Van de Kamp, (1981). Stellar Paths, (D. Reidel, Dordrecht).

Van Rhijn, P.J. (1936). Pub. Kapteyn Astr. Lab. Groningen, No. 47.

Vyssotsky, A.N. (1943). ApJ, **97**, 381.

Vyssotsky, A.N. (1956). AJ, **61**, 201.

Vyssotsky, A.N. (1958). AJ, **63**, 77.

Vyssotsky, A.N. (1963). Stars and Stellar Systems, Volume III, ed. K. Aa. Strand, (Publ. University of Chicago Press), 192.

Vyssotsky, A.N., Janssen, E.M., Miller, W.J. & Walther, M.E. (1946). ApJ, **104**, 234.

Vyssotsky, A.N. & Mateer, B.A. (1952). ApJ, **116**, 117.

Weis, E.W. (1988). AJ, **96**, 1710.

Weis, E.W. (1991). AJ, **102**, 1795.

Weis, E.W. & Upgren, A.R. (1994). Poster Paper at this Workshop.

Woolley, R.v.d.R., Epps, E.A., Penston, M.J., & Pocock, S.B. (1970). Roy. Observ. Ann., No. 5.

Disk density from a survey of F stars at the NGP

By J. KNUDE

Copenhagen University Observatory, Øster Voldgade 3, DK-1350 Copenhagen K, Denmark

Preliminary findings from a combination of proper motions and uvbyβ photometry of a magnitude-limited north galactic pole sample are reported. With sharp photometric definitions the density variations $\nu(z)$ seem to be identical for F5 and F8 stars. $\nu(z)$ may follow a gaussian. With $\sigma_W = 12$ km/s a value for $\rho_{total}(0) = 0.117\ \mathcal{M}_\odot pc^{-3}$ is deduced. The subgroups seem also to have similar U and V dispersions, contrary to what previously was found for brighter solar metallicity F stars. The data distribution is dense enough to compute $\partial^2 ln(\nu)/\partial z^2$, implying a total disk density of 0.109 $\mathcal{M}_\odot pc^{-3}$. The two estimates are almost identical to the density of identified local mass.

---

1. Programme

The programme which consists of the study of all A5-G0 stars brighter than B=11$\overset{m}{.}$5 with b $\geq$ 70° was undertaken for various reasons: e.g. the derivation of the density variations of groups perpendicular to the Galactic plane. Kinematical information – isothermality mainly – and a model of the Galactic potential together with the density law may result in an estimate of the total density in the disk.

2. Data

Candidate stars came from an objective prism survey of the NGP undertaken by Tarmo Oja. The list of stars was kindly put at my disposal and positions were determined from The Astrographic Catalog with assistance from Mogens Winther.

New uvbyβ photometry of $\sim$5500 stars were undertaken by the author. No previously published data are included in the photometric catalog, thus allowing an assesment of the external errors. About 1200 stars may be found in published uvbyβ catalogs, mainly Hill et al. (1982), Perry & Johnston (1982), and Olsen (1983). Particularly Hill et al. is useful since it is a catalog of A and F stars at b $\geq$75°. The main differences from the present catalog are that Hill et al. goes fainter and that their catalog does not have a single limiting magnitude, as does Tarmo Oja's spectral survey.

Since the Galactic latitude of these stars is large, proper motion and photometric parallax may be combined to give the velocity components U and V. The error introduced by neglecting the radial velocity is minor, only a few km/s at the distances of these stars. About 2000 of the programme stars are in the AGK3 catalog. The remaining 3000 stars have been on the CAMC programme for some years and their second epoch positions are now nearing completion.

3. Sample homogeneity

The requirements for a sample used to derive the density $\nu(Z)$ are homogeneity and that it must reflect the galaxy's gravitational potential. There has been little discussion of what a homogeneity requirement eventually should be in order to fit models of the potential. How sharply must color and absolute magnitude be defined? Bahcall (1984b,

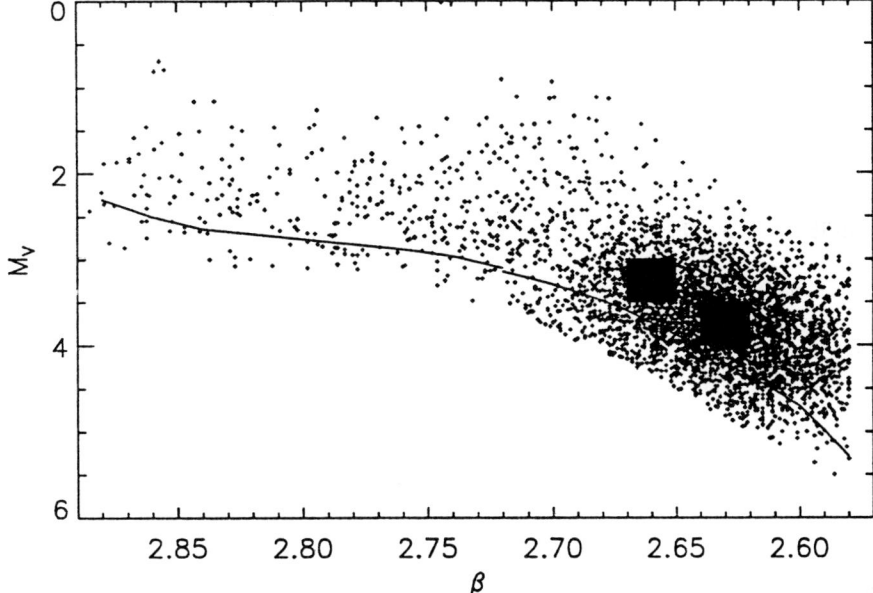

FIGURE 1. β vs. M_V diagram. Standard line and F5, F8 tracers indicated according to the definition in Table 2 (see also Fig. 6).

Table 1), summarizes two galaxy models for observed mass components. M_V of interest here ranges from 3.2 to 4.2. Compared to Figure 1 this corresponds to a β range 2.72–2.64, and, from Figure 6 below, to a range of 0.1 in $(b-y)_0$ for stars on the standard line. In Bahcall & Soneira's (1980) model this corresponds to one mass component. This must mean that the stellar groups defining the potential cannot be used as tracer populations since the quoted M_V range alone includes stars younger than 2×10^9 and older than 12×10^9 yr – stars that do not share a dynamical history. The tracers must ideally be identical within the observational errors, with the same mass and age. The same age is required since time-dependent heating changes the velocity dispersion and the stellar distribution in the Galactic potential. A luminosity-color selection may mix widely different age groups with distinct dynamical memory. This problem is probably most serious for slowly evolving low mass stars.

4. A density law from a volume-complete sample of F stars

Hill et al. (1979) used $uvby\beta$ photometry of their A and F star sample to deduce a well-determined run of the spatial density of F stars towards the NGP. They selected two groups of stars from the F star range, F5 and F8 [(B–V) = 0.42, 0.55 and M_V = 3.6, 4.2, respectively; σ_M = 0.5, σ_{B-V} not given]. The resulting density laws are not identical: the slope of the F5 law is much steeper. A more recent example is from Kuijken & Gilmore (1989a) using a faint sample of K dwarfs in the SGP area.

The Hill et al. sample of F stars has been used extensively to derive $\rho_{disk}(z=0)$ (Bahcall 1984b, Kuijken & Gilmore 1989b). Depending on the adopted law, F5, F8 or F5+F8, the disk density varies from about the identified local disk density to three times this value. Since Hill et al. do not state their complete sample definition it is hard to trace a reason for the difference of their F5 and F8 laws, but Fig. 6 below suggests that F8 is

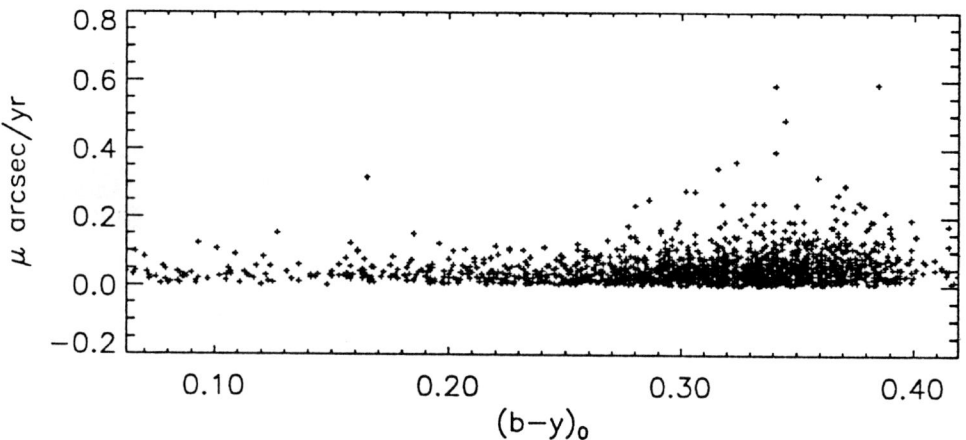

FIGURE 2. $(b-y)_0$ - proper motion. Note that ipII cuts in at $(b-y)_0 = 0.290$.

contaminated with ipII whereas F5 seems blue-ward of the cut in point. Contamination occurs because $\sigma_W(\text{ipII}) \geq 20$ km/s and $\sigma_W(\text{pI}) \approx 10$ km/s (Strömgren 1987).

Completeness of the photometric catalog is determined by the completeness of the objective prism survey. The incompleteness is probably not in excess of a few percent. Since there is some reddening towards the NGP, the uvbyβ photometry are reduced to intrinsic values. For this and for the computation of M_V the Crawford (1975) calibration has been used. The M_V calibration is particularly important so individual stellar distances are computed. No main sequence assumption is required. Figure 2 is a diagram of the proper motion vs. $(b-y)_0$, and since μ has a larger scatter for $(b-y)_0 \geq 0.290$, this gives a pre-warning for the tracer selection. If two groups were chosen to the red and to the blue of 0.290, we might introduce a difference in kinematics: the redder stars could, for the same distances, have a larger spead in velocities and therefore a less steep density variation than the bluer stars.

The photometry permits a selection by chemical composition, the δm_0 index, and by log g, the δc_0 index. One has to be extremely careful when introducing composition and gravity selection. The polar sample has a pronounced relationship between δm_0 and δc_0 caused by the sample's limiting magnitude, interstellar reddening, the age-metallicity relation and a probable local metal gradient in the polar direction (Knude 1991). An uncritical choice of e.g. δm_0 will bias both the stellar age range and the distance distribution, so the subgroup's dynamical information may be less relevant than anticipated. Since the velocity dispersion of homogeneous tracers scales with the disk density, some care must be taken choosing the color slot of the test population. Figure 3 shows clearly that the dispersion in U decreases with the β index. If a similar effect is present for W one would expect different density laws for F5 and F8 stars. The U and V dispersion variation with β (or equivalently with $(b-y)_0$) is discussed in more detail in §5.

The estimate of the chemical composition, δm_0, depends on the intrinsic calibration of the uvbyβ system. Crawford's calibration did not include very many metal-weak stars, which were not available . Olsen's (1988) more recent calibration includes the metal-poor stars. There seem to be at least two consequences of the newer calibration. First, the polar reddening becomes slightly larger, and secondly, a relatively larger fraction of ipII stars, $\delta m_0(\beta) \geq 0.050$, is found than with Crawford's calibration. The data quoted

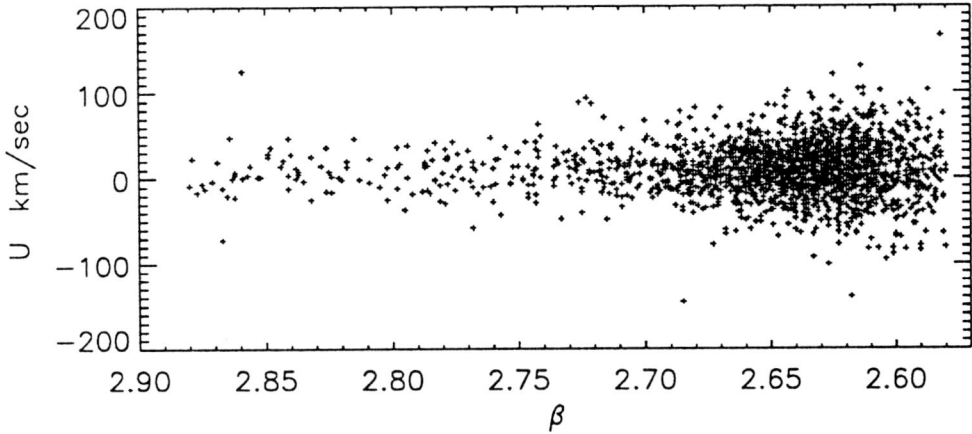

FIGURE 3. β - U. Apparently the dispersion increases when the stars become cooler. A stars are included to make any trend more pronounced.

Type	β	M_V	Number	Crawford's Standard	
				β	M_V
F5	2.67-2.65	3.0-3.5	258	2.665	3.67
F8	2.64-2.62	3.5-4.0	492	2.628	4.23

TABLE 1. Definition of the F5 and F8 tracers.

presently are exclusively based on the Crawford calibration. Our choice of M_V is about 0.5 brighter than the standard line corresponding to $\delta c_0 \approx 0\overset{m}{.}05$: the Hill et al. (1979) sample has $\delta c_0 \approx 0.00$. Table 2 below contains our F5 and F8 definitions, together with Crawford's values for these types on the standard line.

We have chosen slightly evolved tracers mainly for two reasons. According to Fig. 1 there are many more stars above the standard line than on it and they have less age mixing. This is particularly the case for the F8 stars, as Figure 6 below shows.

To achieve volume completeness a correction for effects of the limiting magnitude must be applied. This is done via a M_V – Z(pc) diagram. The upper distance limit is set by the less luminous stars. Without this correction a remote distance bin would be introduced, where the number of tracer stars would be underestimated, implying too steep a density relation, and thus, with a given W-dispersion, overestimating the total disk density. The M_V correction has, however, another effect which works differently on the two groups. The sharp M_V-distance confinement excludes **nearby** red stars which are more frequent in F5 than in F8. This probably means that we exclude the most metal rich stars in F5, stars that are not present in F8. Figure 4 shows the density variation for the F5 and F8 groups, respectively. There seems to be a slight difference, F8 being marginally steeper than F5. A reason for this could be that the two groups comprise tracers with different W dispersions. We do know of at least one difference between the two groups: F5 includes relatively fewer ipII stars than F8. We have therefore chosen to exclude ipII. Figure 5 shows what might be termed the F5 and F8 pI density variations, normalized to their nearest distance bin. The two laws now seem more alike. At least they do not show

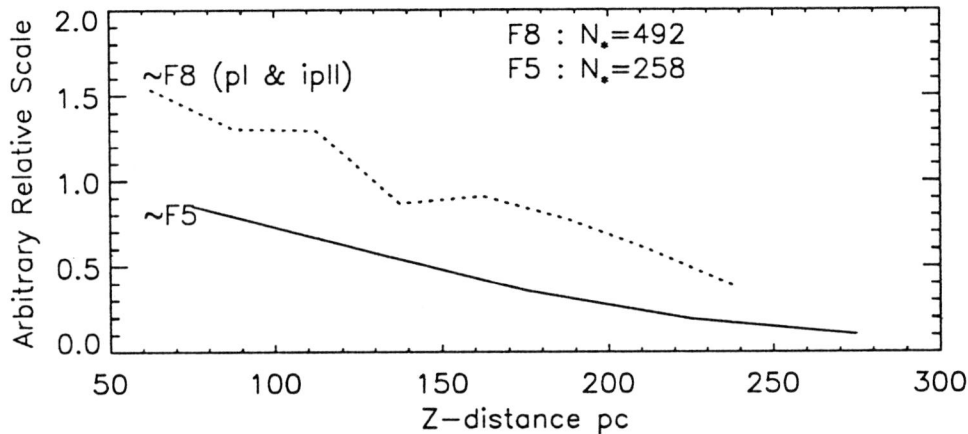

FIGURE 4. Density law data for F5 and F8 tracers prior to ipII correction.

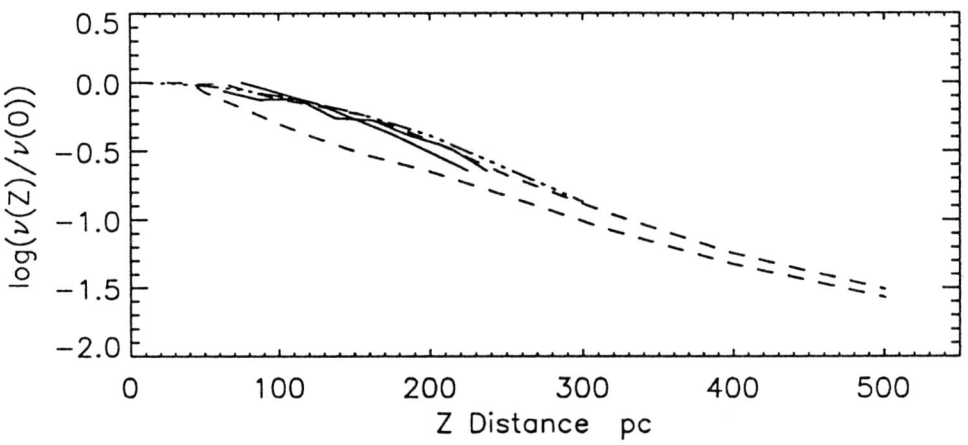

FIGURE 5. Normalized density data for F5 and F8 tracers after ipII correction (solid). Hill et al. (1979) F5, F8 laws also indicated (dashed). The "fit" $\exp(-\frac{Z^2}{2\times 150^2})$ is also shown (dash dots).

the Hill et al. (1979) distinction between F5 and F8. The Figure contains a comparison curve $\exp(-\frac{Z^2}{2\times 150^2})$. A more detailed study shows that the law close to the plane is too shallow for an exponential. The quoted "scale height" of 150 pc is not final but will probably not vary by more than about 10 pc.

5. Kinematics: are the tracer groups isothermal?

Unfortunately, there is no systematic radial velocity data available for the F5 and F8 groups. The situation has been remedied since Adamson et al. (1988) published an extensive catalog of radial velocities for a large fraction of the F stars in Hill et al. (1982). The W dispersion of the present groups may be estimated rather reliably from the stars

Tracer	N	$\overline{W}$ km/s	σ_W km/s	N_{JK}	% with W
F5	60	−7.7	12.0	258	23
F8	47	−9.1	11.9	492	10

TABLE 2. Present F5, F8 tracers common to Adamson et al. (1988).

Tracer	Completeness %	$\overline{U}$ km/s	σ_U km/s	$\overline{V}$ km/s	σ_V km/s
F5	~40	8.5	32.1	−16.7	21.2
F8	~40	10.1	31.9	−20.1	18.1

TABLE 3. Present F5, F8 tracers common to the AGK3.

in common. Bahcall (1984b) also quotes a W dispersion for the absolute magnitude range from 3.2 to 4.2 not deviating too much from the combined F5, F8 range, $\sigma_W = 11$ km/s. As Table 2 shows there are some stars in common with Adamson et al. (1988). It is odd that F5 and F8 have W measured for about the same number of stars, since there are twice as many stars in F8 as in F5.

In Table 3 U and V data are given for tracers also in the AGK3. The values may not be final since there is a systematic δm_0 difference between the stars in AGK3 and those not included. Stars without proper motion may increase the dispersions slightly, mostly for the F8 tracers.

It seems the two groups have the same velocity dispersions. $\sigma_U = 32$ km/s, $\sigma_V = 20$ km/s, $\sigma_W = 12$ km/s. Compared to Adamson et al. we find $\sigma_U > \sigma_U$(Adamson) and $\sigma_V > \sigma_V$(Adamson) indicating that Adamson's selection may be younger than the present one. The resulting identity of the U and V dispersions is a surprise since we previously found solar metallicity F stars with $V \leq 10^m05$ to have increasing dispersion for decreasing β (Knude et al. 1987). Figure 6 shows part of the NGP stars with different symbols between isochrones of age 6, 8, 10 and 12 Gyr (VandenBerg 1985). The standard $(b-y)_0 - M_V$ curve is also indicated. The F5 and F8 groups are indicated together with boxes showing the location of the solar metallicity middle and late unevolved and evolved groups respectively. In the solar programme the late groups thus have much larger age ranges than the middle groups. F5 and F8 tracers, on the other hand, have different ages but comparably small age ranges. If the heating of the disk is a power law, one would expect a stronger dependence of the range of the included velocity dispersions on age range than on age itself. With a power of $\frac{1}{2}$, $\sigma_U \sim A^{\frac{1}{2}} \Rightarrow \Delta\sigma_U \sim (\Delta A)^{+1} \cdot A^{-\frac{1}{2}}$. With coarse A and ΔA estimates from Fig. 6 the ratio of the included velocity dispersion ranges $(\Delta\sigma_U)_{late} / (\Delta\sigma_U)_{middle} \approx 2$, which is almost as observed. We may therefore conclude that with the sharp photometric definitions of the tracer groups F5 and F8 we can probably accept the identity of the velocity dispersions as a physical reality.

6. Total disk density

If we adopt Bahcall's (1984a, 1984b) model for the distribution of stars perpendicular to the Galactic plane, the total disk density may be estimated from any of the isothermal components:

$$\rho_{total}(0) = \frac{1}{2\pi G}\frac{1}{Z^2}(\sigma_W)_k ln[\frac{\rho_k(0)}{\rho_k(Z)}] \, , \, Z \ll 200 \text{ pc}.$$

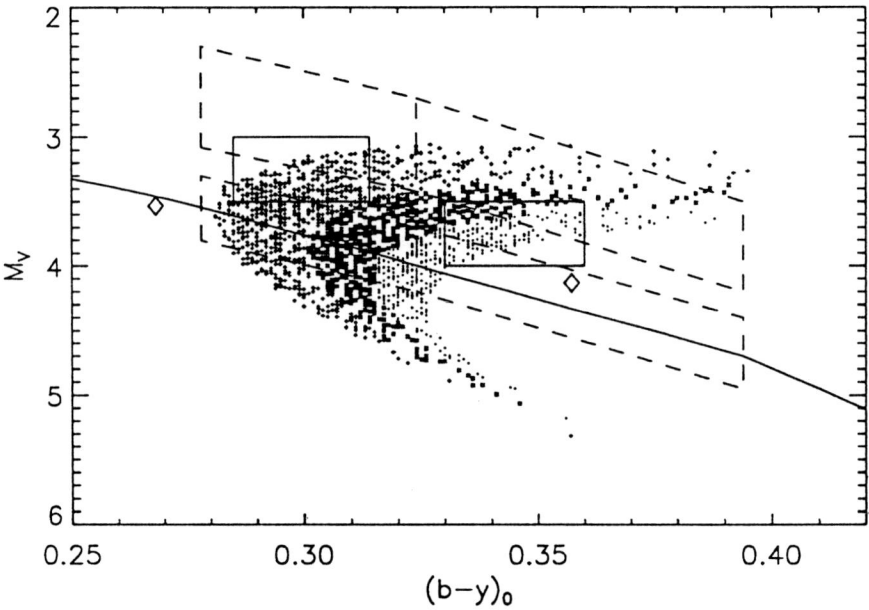

FIGURE 6. The boundaries between different plotting symbols are isochrones of ages 6, 8, 10 and 12 Gyr. ◇ indicates the location of Hill et al. (1979) F5 and F8 groups. Solid boxes are the present choice of F5 and F8 tracers. Dashed boxes indicate the middle and late F star divisions from Knude et al. (1987). ipII cuts in at (b-y)$_0$ = 0.290.

In our case k is either F5 or F8. All stars used to determine the density are almost fulfilling the Z demand, but with our gaussian adoption for the density variation, the Z dependence disappears anyway. The theoretically expected variation close to the plane is a gaussian (see Bahcall op. cit.). With the fit of 150 pc to the Z dispersion and σ_W = 12 km/s calculated from the stars common to Adamson et al., we find $\rho_{total}(0) = 0.117$ $\mathcal{M}_\odot$ pc^{-3}. If Poisson's equation is combined with the equation for K$_Z$, and the Galaxy's rotation curve is flat, $\rho_{total}(0)$ may be estimated from
$$4\pi G \rho_{total}(0) = \frac{\partial}{\partial Z}[\frac{1}{\rho}\frac{\partial}{\partial Z}(\rho \sigma_W)].$$
With the constant σ_W = 12 km/s we have computed the second derivative at Z = 125 pc using the F8 density distribution without smoothing. Neighbouring counts are thus independent and the result is $\rho_{total}(0) = 0.109$ $\mathcal{M}_\odot$ pc^{-3}. These two estimates of the disk density compare rather well with the local density of identified matter: $\rho_{total}(0) = 0.096$ $\mathcal{M}_\odot$ pc^{-3}, Bahcall (1984b) and $\rho_{total}(0) = 0.108$ $\mathcal{M}_\odot$ pc^{-3}, Hill et al. (1979). Not much room for any unidentified matter in the disk is thus left by the F star data. A conclusion also drawn by Kuijken & Gilmore (1989b) from a reanalysis of available F dwarf (and K giant) data.

7. Conclusions

With narrow ranges in β and M$_V$, 0.02 and 0.5 respectively, the F5 and F8 tracers follow similar density laws and have common dispersions for all three velocity components. These homogeneities indicate that the inferred total disk density is almost identical to the identified local mass density. The fundamental homogeneity may however be that the two tracing groups have similar age ranges.

REFERENCES

Adamson, A.J., Hill, G., Fisher, W., Hilditch, R. W. & Sinclair, C.D. 1988 MNRAS **288**, 273.

Bahcall, J.N. 1984a ApJ **276**, 156

Bahcall, J.N. 1984b ApJ **276**, 169

Bahcall, J.N., Soneira, R.M. 1980 ApJS **44**, 73

Crawford, D.L. 1975 AJ **80**, 955

Hill, G., Hilditch, R.W., Barnes, J.V. 1979 MNRAS **186**, 813

Hill, G., Barnes, J.V., Hilditch, R.W. 1982 Publ.dom.Astrophy.Obs. *XVI*, 111

Knude, J. 1991 A&A **249**, 88

Knude, J., Schnedler Nielsen, H., Winther, M. 1987 A&A **179**, 115

Kuijken, K., Gilmore, G. 1989a MNRAS **239**, 605

Kuijken, K., Gilmore, G. 1989b MNRAS **239**, 651

Olsen, E.H. 1988 A&A **189**, 173

Olsen, E.H. 1983 A&AS **54**, 55

Perry, C.L., Johnston, L. 1982 ApJS **50**, 451 Strömgren, B. 1987 *The Galaxy* p. 229, eds. Gilmore, G., Carswell, R., Reidel, Dordrecht

VandenBerg, D.A. 1985 ApJS **58**, 711.

Proper motions of Stephenson's spectroscopically selected Red Dwarfs

By D. H. P. JONES AND M. AZZARO

Royal Greenwich Observatory, Madingley Road, Cambridge CB3 0EZ, UK

In 1986 Stephenson published two substantial papers listing 4000 K and M dwarfs discovered from an objective prism survey. Proper motions for more than half these stars have been measured or garnered from the literature and used to infer their space density. We believe about twenty per cent. to be giants.

1. Introduction

There are three questions about the solar neighbourhood which still require further investigation.

(i) Are there significant numbers of stars in the solar neighbourhood which remain uncatalogued?

(ii) Is there a discrepancy between the density of gravitating matter in the solar neighbourhood demanded by dynamical studies and the mass density of catalogued stars?

(iii) Are the velocity dispersions of faint stars exaggerated by the fact that many of them have been discovered from their large proper motions?

Light on some of these problems can be found from the work of Stephenson (1986a, 1986b) who published two major papers listing 4200 K and M dwarfs discovered from an objective prism survey. The first paper listed stars of unknown proper motion and the second those with published proper motions, the NLTT (Luyten 1980) being the major contributor. Although the lists were published separately they were both based on the same original material; objective prism plates taken with a Schmidt telescope which were subsequently examined by eye. The survey covered the whole sky north of declination $-25°$ and more than $10°$ from the Galactic plane within the magnitude range $7.5 - 12.5$. Because of its lack of kinematic bias and well-defined flux limits this sample is an excellent one to study.

2. Proper motions

Stephenson's stars without known proper motion were put on the observing programme of the *Carlsberg Automatic Meridian Circle* in 1987 and the positions of the great majority have now been published in the *Carlsberg Meridian Catalogue, La Palma, No 6* (1992) and earlier volumes. Proper motions have been measured by comparing the Carlsberg positions with earlier meridian positions or with the Astrographic Catalogue (Fabricius *et al.* this conference). We now have positions for 1822 stars but proper motions for only 1268. Because of the heavy reliance on the Astrographic Catalogue for first epoch positions the flux limit of the Astrographic Catalogue at about $m_{pg} = 11.5$ is also the flux limit of the proper motions. This means that the stars lacking proper motions are chiefly the fainter and redder ones.

In order to fill this gap we measured first epoch positions of a further 138 stars from glass copies of the *Palomar Observatory Sky Survey I* relative to the AGK3 Catalogue. 112 of the stars were selected as those which (if dwarfs) would have the largest spectroscopic parallaxes.

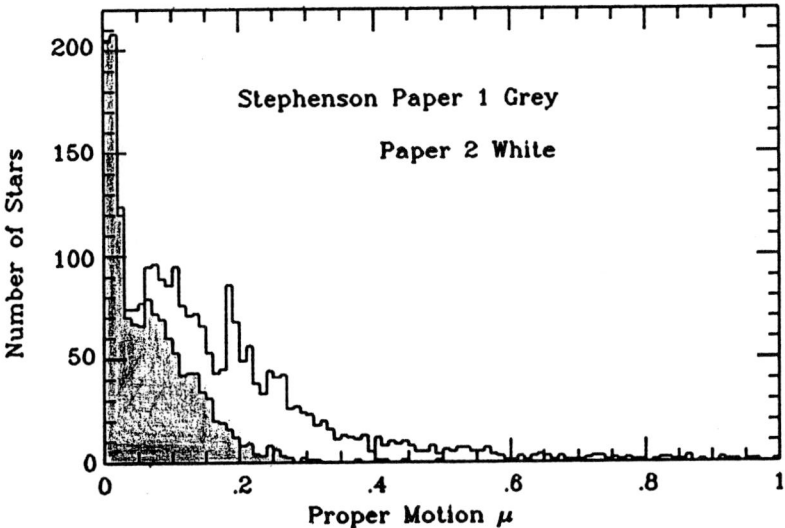

FIGURE 1. Histogram of proper motions from both Stephenson papers.

We have proper motions of 63 per cent. of the stars in Paper I and a quick search of the literature produced proper motions for 93 per cent. of the stars in Paper II. Figure 1 displays the histogram of the proper motions of all the Stephenson stars; the frequencies from Paper II have been scaled down to allow for their greater degree of completeness.

The most noticeable feature in Figure 1 is the large number of stars with proper motions < 0.03 "/yr and the small peak at 0.2. The small proper motions are probably distant giants and the second peak seems to be associated with an excess of stars at the lower bound of the NLTT Catalogue.

3. Modelling the proper motions

In order to model the proper motion histogram we adopted a conventional velocity ellipsoid for red dwarfs and generated artificial proper motion histograms by a Monte Carlo technique, assuming the stars to be uniformly scattered throughout a sphere. Random errors were added to the Monte Carlo histogram with the same standard deviation as our own measures. Stars outside the declination and galactic latitude limits of Stephenson's sample were rejected. The radius of the sphere was varied to achieve the best fit to the observed histogram. The best fit seems to be with a sphere of radius 60 parsec as shown in Figure 2 which requires that about 500 stars (20% of Figure 2 of the smallest proper motion must be distant giants.

The division into giants and dwarfs can be readily checked against the $UBVRI$ photometry of Weis (1991). There are 178 stars from Stephenson's Paper I which have both Weis photometry and known proper motions. Weis divided his sample into 'dwarfs' and 'non-dwarfs' from his $UBVRI$ photometry and these are identified in the proper motion histogram Figure 3.

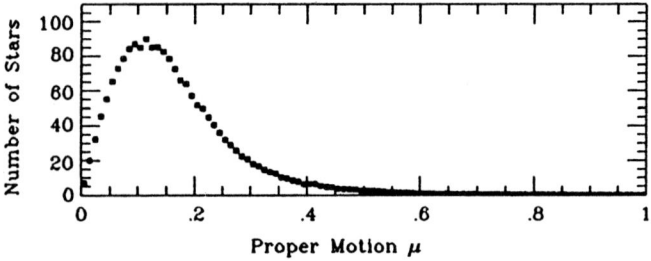

FIGURE 2. Histogram of a Monte Carlo simulation with radius 60 pc.

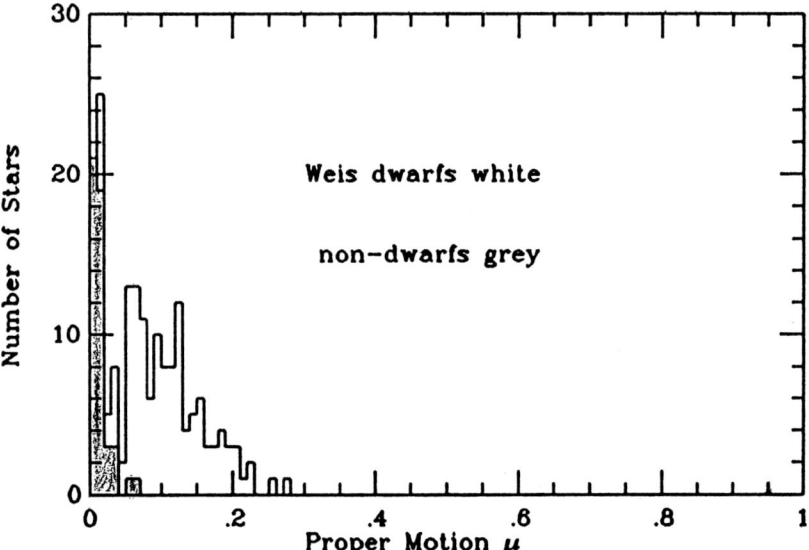

FIGURE 3. Histogram of proper motions for Weis 'dwarfs' and 'non-dwarfs'.

The corresponding means are:

For 48 non-dwarfs	$<\mu> = 0.014$	″/yr
For 130 dwarfs	$<\mu> = 0.106$	″/yr
	$<\pi> = 0.029$	″
Mean Tangential Velocity	$<\tau> = 17$	km/s

These results are in good accord with our discussion of Figure 1; a homogeneous sphere of radius 60 pc should yield a mean parallax of 0.″025. The density of stars in Stephenson's Paper I is 0.002 stars per cubic parsec or 0.001 solar masses per cubic parsec. This represents a one per cent. increase in the identified local mass density (Kuijken & Gilmore 1989).

REFERENCES

Fabricius, C., Morrison, L.V. & Helmer, L., 1994, Improving Proper Motions with the Carlsberg Automatic Meridian Circle. This workshop

Kuijken, K. & Gilmore, G., 1989, MNRAS, 239, 651
Luyten, W.J. (1980). NLTT Catalog (University of Minnesota, Minneaopolis)
Stephenson, C.B., 1986a, AJ, 91, 144
Stephenson, C.B., 1986b, AJ, 92, 139
Weis, E.W., AJ, 102, 1795.

Observations in the region of the Orion Association

By R. L. SMART

Dept of Astronomy, University of Florida, Gainesville, Fl., U.S.A.

This paper summarizes observations taken and reductions currently in progress to find the proper motions of stars in the Orion Association. In order to ensure consistency across the large region – over 30 square degrees – and to maximize precision, the frames are strongly overlapped. The final result will incorporate star positions from the San Fernando and Algiers sections of the Astrographic Catalogue, with observations taken at the McCormick Observatory in 1955 and 1992 to give three epochs spanning 90 years.

1. Introduction

A chart of the Orion Region is shown in Figure 1: this is the center of the I-Orion Association.

Many properties of this association make it an important region for observational tests of evolutionary theories:

(i) It is the second closest known association, containing over 50 O/B type stars.

(ii) It is of a moderate angular and physical size.

(iii) It is the nearest site of active star formation.

(iv) It contains at least four subgroups, I-Orion $a,b,c,$ and d believed to be of different evolutionary ages.

(v) It is covered extensively spectroscopically and in the infrared by the IRAS.

However, despite these properties there have only been a few kinematic studies in the Ia and Ib-Orion subgroups: Lesh (1968) using proper motions of 16 stars derived from meridian catalogues found an expansion age of 4.5 million years. The error in this age was very large due to imprecise proper motions and a small sample size. Warren (1978) determined membership of the whole association by combining previous photometric and kinematic data with new photometric data. Gieseking (1983) using newly determined radial velocities for 66 stars in the association found evidence of fragmentation within subgroups. **Both Warren and Gieseking lament the lack of precise proper motions, and because of this they were unable to give conclusive results.**

The reasons for the lack of kinematic studies are probably threefold: the groups subtend a large angular size making astrometric reductions difficult; any reduction would require computational ability that has only recently become available; and most of the work in the association has been limited to the very young regions in the c and d subgroups.

2. Observations

The region of Orion we observed is outlined in Figure 1. The observations consist of photographic plates from three epochs.

(i) 1900. We used 19 frames from the San Fernando Observatory and 12 frames from the Algiers Observatory. The frame orientation is shown in Figure 2A, and results in up to 500 per cent overlap. The image diameters, x,y coordinates (with an average position precision of four microns, or $0\rlap{.}''24$) of all stars found on the plates down to a limiting

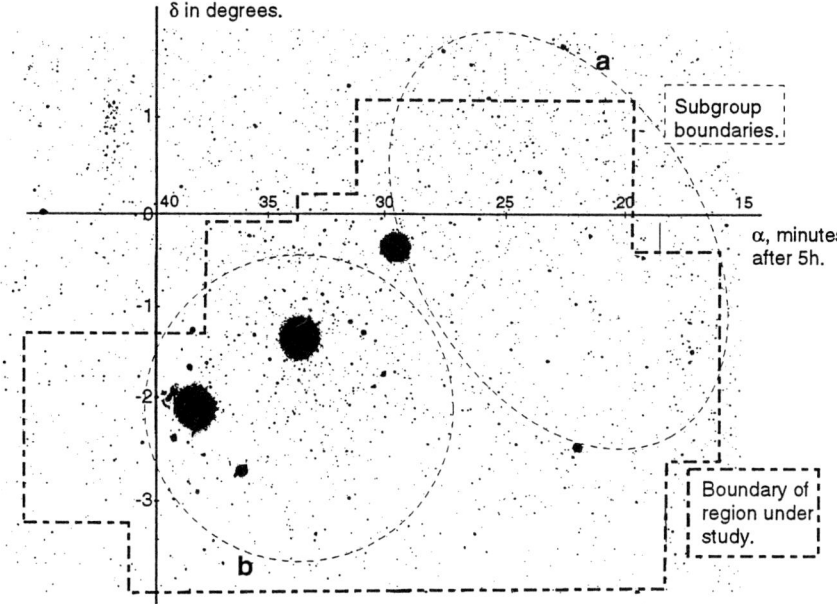

FIGURE 1. I-Orion Subgroups and Region under Study. (Reproduced from the Vehrenberg 1950.0 *Atlas Stellarum* and Blaauw 1988).

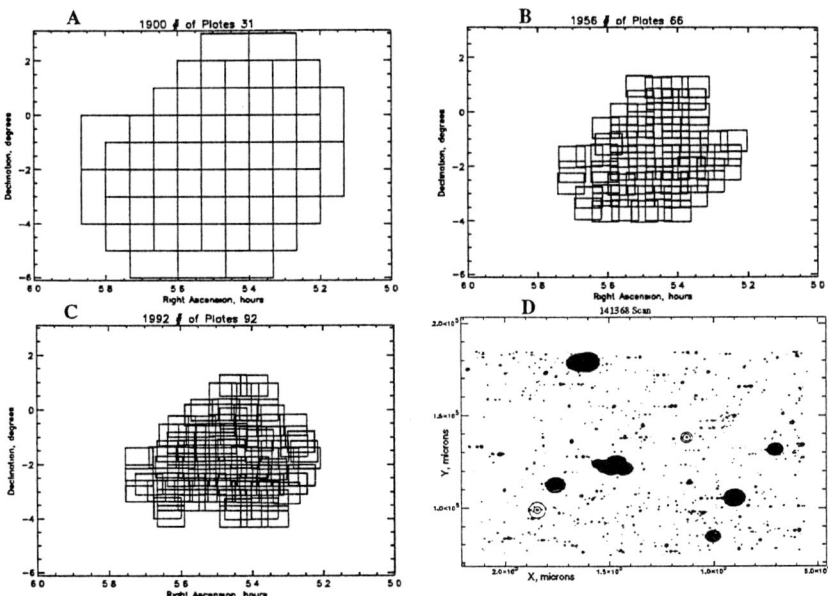

FIGURE 2. Plate orientations and a low threshold plate scan.

magnitude of 12-13 were published in the observatories' reports. These stars form the search database for the next two epochs.

(ii) 1956. Eichhorn photographed 67 regions, as shown in Figure 2B, on the McCormick Observatory Leander Refractor. Each region was exposed four times with a 180° rotation after the second exposure. Ignoring bad exposures, the final reduction consists of 240 frames. Each plate was exposed with a two-magnitude diffraction grating, giving up to two measurable grating images. Due to the two-thirds overlap and multiple exposures, a stars's image may appear on up to 10 frames.

(iii) 1992. Smart photographed 93 regions, as shown in Figure 2C using the same telescope and grating as for the 1956 epoch. Each region was exposed three times, and in the final reduction there were 260 frames. The region was covered twice and a star's image may appear on up to 14 frames.

The second and third epoch plates were measured on the University of Minneapolis Automated Plate Scanner (Pennington et al. 1993). This is an exceptionally fast machine of the flying spot type which gives a positional precision of 1-2 microns or $0\farcs044$. The speed of the machine enabled the plates to be scanned in both forward and reverse orientations. In the second scan, dual threshold density scanning was used.

Figure 2D is a typical low threshold density scan of plate 141368 exposed by the author in 1992. This photographic plate actually has three exposures of three regions, making a total of nine exposures. The large ellipses are grating images that were merged by the scanning process. In the higher threshold scan these bright images are resolved, although some of the fainter images are lost.

3. Reduction

We use a overlapping plate technique as described by Eichhorn (1985). Assume the standard coordinates $\xi(\alpha,\delta)$ and $\eta(\alpha,\delta)$ are related to the rectangular coordinates by the relation:

$$(x,y) = s(\xi,\eta) + \Xi a \qquad (3.1)$$

Where Ξ is a model matrix. By varying the number and dependence of the terms in the model matrix we can account for departures from a gnomic projection. In the simplest six-constant model we correct for origin shift, rotation and focal length errors by the model matrix:

$$\Xi = \begin{pmatrix} \xi & \eta & 1 & 0 & 0 & 0 \\ 0 & 0 & 0 & \xi & \eta & 1 \end{pmatrix} \qquad (3.2)$$

We also analyzed terms in $\xi^2, \xi\eta, \eta^2, m - \langle m \rangle, (m - \langle m \rangle)\xi$ and $\xi(\xi^2 + \eta^2)$. These terms correct for tangential point shift, plate tilt, guiding errors, radial terms, coma, and magnitude-dependent radial terms. Figure 3 shows the residuals *vs* magnitudes after a six-constant single plate fit.

The parabolic shaped curve results from two difficulties: bright stars produce large images in which it is difficult to find the centroid, and; faint images produce small, non-circular images, which also make finding the centroid difficult. Incorporating the grating images as described by Eichhorn (1970) allows the magnitude dependant terms to be accurately modelled.

Ultimately, as we are using least squares, we choose a model that minimizes the sum of the squares of the residuals, but, in addition to the graphical approach exemplified above, we also use a variety of techniques to aid in model selection. In particular ratios of a

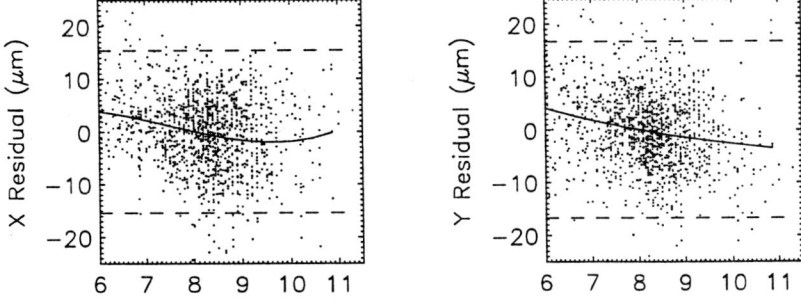

FIGURE 3. Residuals *vs* magnitudes for a six-constant fit.

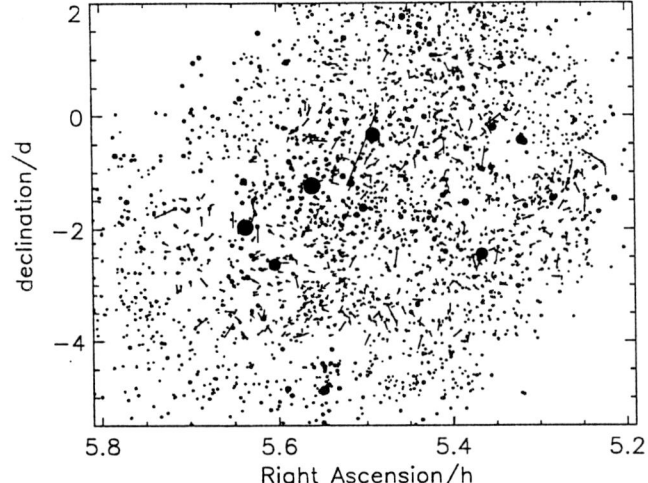

FIGURE 4. Proper motions found using a basic six-constant fit.

parameter value, formal error and variance highlight parameters that can be constrained or dropped completely.

4. Results

Figure 4 shows provisional results using a six-constant fit.

The vectors represent the proper motion of the star over a 10 000 year period. The final results will be published in the form of a catalogue.

Acknowledgements

This work was supported in part by the Graduate School, North East Regional Data Center through the Research Computing Initiative of International Business Machines, and the Division of Sponsored Research, at the University of Florida; the Sigma Xi Foundation; the American Astronomical Society; the Florida Space Grant Consortium; and the Leander McCormick Foundation.

REFERENCES

Blaauw, A. 1964. *Ann. Rev. Astr. and Ap.*, **2**, 213.

Blaauw, A. 1988. *Astrofizika.*, **29**, 23.

Eichhorn, H.K. 1970. *Conf. of Phot. Astr. Technique, Tampa, Florida.*, 241.

Eichhorn, H.K. 1985. *AA.*, **150**, 251.

Gieseking, F. 1983. *AA.*, **118**, 102.

Lesh, J.R. 1968. *ApJ.*, **152**, 905.

Pennington, R.L., Humphreys, R.M., Odewahn, S.C., Zumach, W., and Thurmes, P.M. 1993. *PSAP.*, **105**, 521.

Warren, W.H.,JR. and Hesser, J.E. 1978. *ApJ Supp.*, **36**, 497.

Stellar Clusters, Superclusters and Groups

By OLIN J. EGGEN †

Cerro Tololo Inter-American Observatory, National Optical Astronomy Observatories,
Casilla 603, La Serena Chile

One of the strongest growth areas in nineteenth century astronomy was the concept of star streaming, and the recognition of the three largest superclusters in the Solar Neighbourhood—the Hyades, Sirius and Pleiades. This paper reviews the more recent developments in our knowledge of the local membership and characteristics of these three groups and the older HR 1614 group.

---

1. Introduction

One of the strongest growth areas in nineteenth century astronomy was the concept of of star streaming. The recognition of the three largest superclusters in the solar neighbourhood can be directly traced to three star streams - Stream 1 (Hyades), Stream 2 (Sirius) and Stream 3 (Pleiades). The subsequent increase in accuracy of available astrometry has isolated other possible candidates but, with a single exception, only the major superclusters will be examined here, on the basis of those features which provide the most compelling evidence of the existence of superclusters. The following definitions and basic assumptions are adopted.

(a) Most stars are formed in clusters.

(b) Stars evaporate from clusters, beginning with those of lowest mass, but retain, within the error of the astrometry, the isoperiodicity of the galactic orbit characteristic of cluster members. Such evaporated stars form a cluster halo and then a supercluster.

(c) In time, the length of which is dependent upon environmental forces, the cluster is completely disrupted and the galactic orbit is replaced by a tube of orbits, whose dimensions are determined by the alteration in space motion suffered by the individual stars. However, when the horizon defined by the accuracy of our available astromety intersects this tube, only the original cluster members with a very small range of V-velocity (velocity along the tube) will be seen. Stars identified by this small range in V-velocity are called members of a 'stellar group'.

(d) The space motion of stars is characterised by $V^2{}_{tot} = U^2 + V^2 + W^2$, where U, V and W are velocities directed, respectively, away from the galactic centre, in the direction of galactic rotation and toward the North Galactic Pole.

(e) The apparent convergent point of the proper motions of members of a cluster or supercluster is designated by $(\alpha, \delta) = (A, D)$. The proper motion of a star directed toward (A, D) is labelled v and that perpendicular to this direction by τ. The peculiar velocity is $PV = 4.74\tau d$, where d is the distance in parsecs. The parallaxes of individual members of a moving cluster are obtained from

$$\pi = 4.74\, v/V_{tot} sin\lambda \qquad (1.1)$$

where λ is the angular distance between the star and (A, D).

† Operated by the Association of Universities for Research in Astronomy, Inc., under cooperative agreement with the National Science Foundation

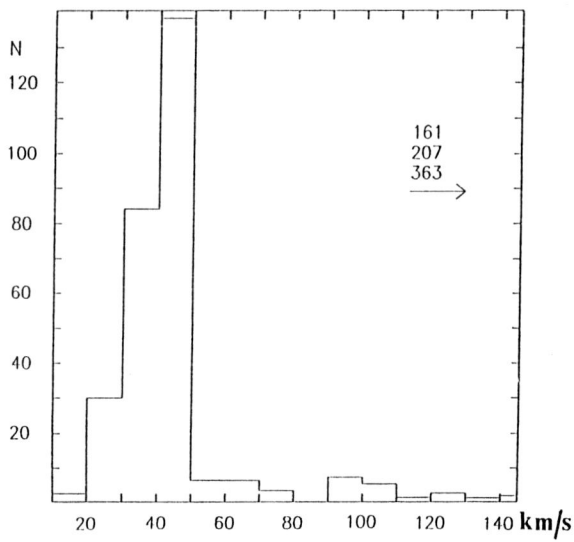

FIGURE 1. Distribution of V_{tot} for the stars in the region of the Hyades that pass the convergent-point test for membership.

2. The Hyades cluster and supercluster

Membership is discussed in Eggen (1992a, 1993a). Proper motions are available for several thousand stars between R.A. $3^h.75$ and 5^h, and between Dec. $+5°$ and $+25°$ (Luyten et al. 1981). Adopting $(A, D) = (6^h.4, +6°.5)$, derived from the bright, cluster stars (Eggen 1992a), 218 stars with $(R - I) > 0.85$ mag. were isolated, having 90% of their proper motion directed toward the convergent point. (R, I) photometry is available for all of the objects brighter than 17 mag. (see Eggen 1993 for references), and this, together with the relation

$$M(I) = 4.56\,(R - I) + 3.95 \tag{2.2}$$

leads to individual distances. Equation 1.1 can be written as

$$V_{tot} = 4.74\,\mu\,d/\sin\lambda, \tag{2.3}$$

and values of V_{tot} computed for all 218 objects. A histogram of the resulting values is shown in Figure 1.

The bright stars give $V_{tot} = 42.5 + 0.045X$, where X is the galactic radial distance in parsecs from the Sun. The peak in Figure 1 between 40 and 50 km/s is obviously formed by Hyades cluster members. The secondary peak from 30 to 40 km/s contains binary stars that are also members, but Equation 2.2 introduces an incorrect parallax because of the presence of companions. Equal components in a Hyades binary would lead to V_{tot}=30 km/s from Equations 2.2 and 2.3. The smaller peak between 20 and 30 km/s is known to contain a few multiple systems of member stars. Radial velocity results are available for only one half of the stars in the 30 to 50 km/s range of Figure 1, but they all confirm cluster membership by values consistent with those computed from the cluster motion, $\rho = V_{tot}\cos\lambda$.

The $(R - I, M_I)$ diagram for the stars in the 40 to 50 km/s bin of Figure 1, obtained from Equation 1.1 is shown in Figure 2(a), where Equation 2.2 with an assumed uncertainty of 0.02 mag. in $R - I$, is also represented. The stars in the 30 to 40 km/s bin are shown as crosses in Figure 2(b). The continuous line represents Equation 2.2 and

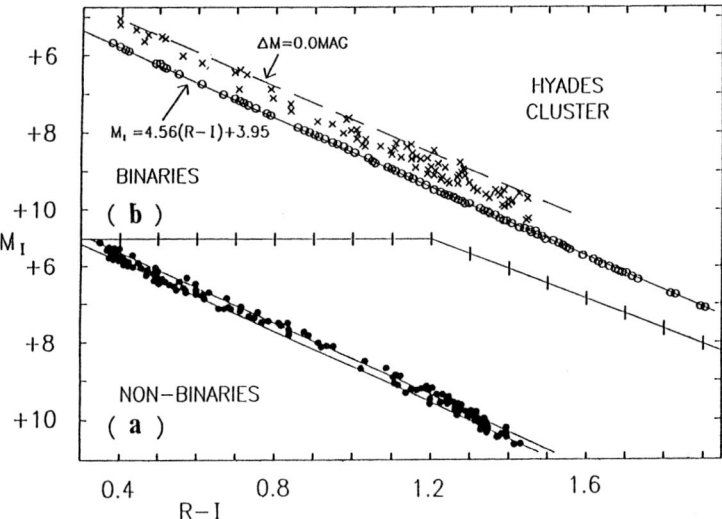

FIGURE 2. (a) The single Hyades cluster stars in the $(M_I, R-I)$ plane. The parallel lines represent Eq. (2.2) with ±0.02 mag spread in $(R-I)$. (b) The combined light of binary members of the Hyades cluster (crosses) and the individual components (open circles). The significance of the lines in (a) and (b) is explained in the text.

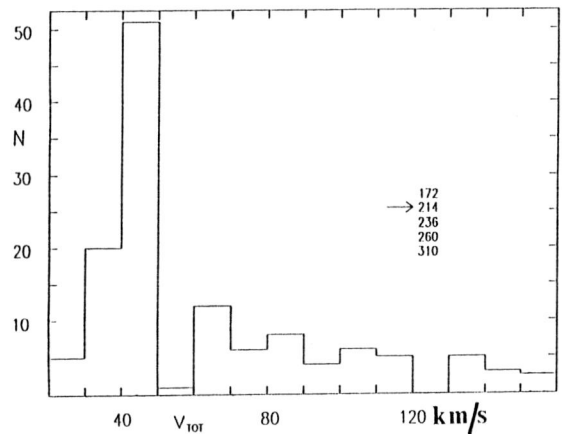

FIGURE 3. The distribution of V_{tot} for the candidate members of the Hyades supercluster within 20 pc of the Sun.

the broken line the locus of equal component binaries. The clear circles represent the individual stars obtained from the combined light and colour and Equation 2.2. Only binaries with a magnitude difference between zero and about 3.0 mag. will be found in this way because of the uncertainties in both photometry and astrometry.

The same procedure was applied to field stars thought to be within 20 pc of the Sun and the resulting distribution of V_{tot} and M_I are shown in Figures 3 and 4, respectively.

The 75 field stars, 43 single stars and 32 binary components, contain only the minimum number of supercluster stars within 20 pc of the Sun, mainly because of the lack of the necessary photometry. Photometry is available for all stars with an annual proper motion of 0.″5 and brighter than visual magnitude 15. The maximum distance for a Hyad with

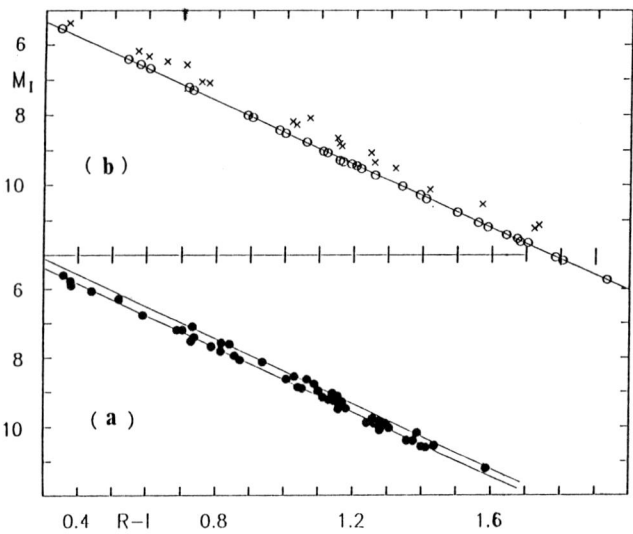

FIGURE 4. (a) The single Hyades cluster stars in the $(M_I, R - I)$ plane. The parallel lines represent Eq. (2.2) with ± 0.02 mag spread in $(R - I)$. (b) The combined light of binary members of the Hyades cluster (crosses) and the individual components (open circles). The significance of the lines in (a) and (b) is explained in the text.

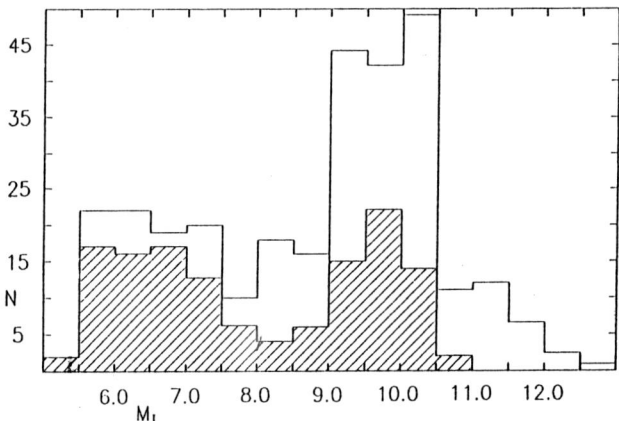

FIGURE 5. The distribution in bins of 0.5 mag. of the luminosities of Hyades cluster; single stars are (cross-hatched) and binary components (clear).

this motion is

$$d(pc) = V_{tot} \sin \lambda (4.74 \times 0\rlap{.}''5) \approx 20 \sin \lambda.$$

For stars such as the Hyades cluster members, only some 30° from (A, D), $\sin\lambda$ is 0.5 and the proper motion limit of $0\rlap{.}''5$ reaches only to about 10 pc for supercluster members.

The luminosity functions of the cluster and of the supercluster are shown in Figures 5 and 6, respectively. The single stars are represented by hatched areas and the components of the binaries by clear areas. Figure 6 also shows the results in visual luminosity (dots) compared with 2.5% of the general luminosity function (GLF) obtained from stars known to be within 20 pc (Wielen et al. 1983). An interesting feature of Figure 5 is the increasing ratio of binary to single stars with decreasing luminosity. There are fewer supercluster

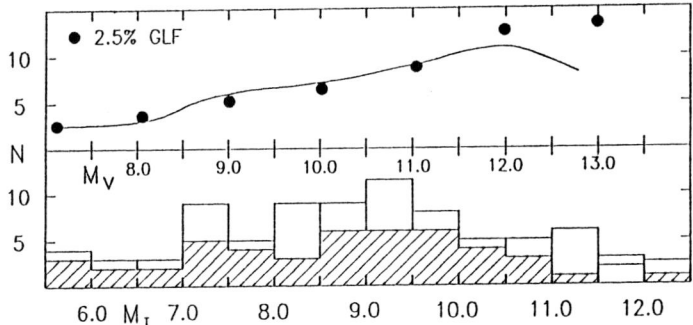

FIGURE 6. The lower panel shows the distribution of luminosities of single stars (cross-hatched) and binary components (clear) in the Hyades supercluster within 20 pc of the Sun. The continuous curve in the upper panel is the translation of the luminosity function to visual magnitudes and the dots represent 2.5% of the luminosity function of field stars within 20 pc of the Sun.

members, but this feature is not obvious in Figure 6. The shape of the GLF is matched well by the supercluster members, which appear to contain more than 2.5% of the total within 20 pc. On the other hand, the luminosity function of the cluster stars is markedly different. There are strong indications that the lowest mass stars have evaporated from the cluster and, together with those from other clusters, constitute the field.

3. The Pleiades supercluster

This supercluster contains several clusters; Pleiades, α Persei, NGC 2602, Sco-Cen (Eggen 1992b). The values of $(A, D) = (5^h.98, -35°.15)$ and $V_{tot} = 26.1 + 0.025X$ isolate the red and white dwarfs in Table 1. The red dwarfs are all within 18 pc of the Sun. The low space velocity leads to a maximum distance of $11 \sin\lambda$ pc for supercluster members with proper motion greater than $0''.5/yr$. So there are many members still to be isolated amongst stars of smaller proper motion. The white dwarf members are predominantly GD stars (Giclas et al. 1980), so the majority of candidate stars have not been located yet ($\mu < 0''.2/yr$, see Eggen 1993b), and the apparent motions of those available are poorly defined. The faintest white dwarf has M_V near +12 mag., compared with +14 mag. in the Hyades supercluster (Eggen 1993b).

The most interesting stars in Table 1 may be the pair 0148+6411 = G244-37 (red dwarf) and G244-36 (white dwarf), separated by 16″. The red star has a predicted radial velocity of -9.3 km/s from the supercluster motion, compared to an observed value of -15 ± 9 km/s (Wegner 1981). Star 0817+3741 is a magnetic white dwarf and 1334+4844 (GD 325) has a trigonometric parallax of $0''.0295$ (USNO), compared with the supercluster value of $0''.0305$. Star 0349+2447, in the Pleiades cluster, has been discussed extensively by Wegner et al. (1991).

The red dwarf 0449+0624 is Gl 179 (Gliese 1969), with a trigonometric parallax of $0''.078$, compared with the supercluster value of $0''.086$. Star 0530+4447 has a trigonometric parallax of $0''.0645$ (USNO), compared with $0''.0655$ from the supercluster motion. The pair 0537+5328 is Gl 211/2, with a trigonometric parallax of $0''.093$ compared with the identical cluster value of $0''.093$, and the observed radial velocity of -1.9 km/s compares well with the cluster value of $V_{tot}\cos\lambda = +0.6$ km/s. Star 0737+0218 is Gl 281, with an observed radial velocity of +18.5 km/s from two observations (Wilson 1967), compared with +19.0 km/s from the cluster motion. The pair 1255+3530 = G123-75, G164-31, is separated by 18″ and is also Gl 490 with $V_{tot}\cos\lambda = -12.9$ km/s, compared

Position/Number	Name	μ_α 0."001	μ_δ 0."001	ν 0."001	τ 0."001	PV km/s	X pc	V_{tot} km/s	Mod. m	V m	V−I m	Sp.T.
Red dwarfs												
0042+1221	G32−44	264	−199	330	−3	−0.2	+5	26.1	1.12	12.83	2.47	M
0148+6411	G244−37	225	−182	290	0	0.0	+11	26.4	1.27	11.37	2.15	dM2
0449+0624	G82−52	160	−300	337	44	+2.2	+9	26.3	0.33	11.95	2.51	dMYe
0530+4447	G96−45	50	−357	360	13	+1.0	+15	26.5	0.92	10.94	2.14	dM
0537+5328	+53.934	0	−520	518	−42	−2.1	+10	26.3	0.15	6.23	0.64	KIVe
0537+5328	+53.935	0	−520	518	−42	−2.1	+10	26.3	0.15	9.74	1.84	dM1
0737+0218	+2.1729	153	−258	300	−8	−0.5	+10	26.3	0.54	9.63	1.63	dM0
1233−3436	+34.8280A	−220	−169	258	12		−9	25.9	1.27	7.91	0.95	K4V
1233−3438	+34.8280B	−220	−169	258	12		−9	25.9	1.27	11.90	2.20	M
1255+3530	+36.2322	−234	−114	260	8	+0.7	+1	26.1	1.33	10.60	1.84	M0Ve
1255+3530	G164−31	−234	−114	260	8	+0.7	+1	26.1	1.33	13.16	2.69	dM4e
2145−7220	−72.1700	345	−310	464	−3		−5	26.0	0.16	9.84	1.92	dM
2330−1702	−17.6768	347	−219	410	24		−2	26.0	0.63	10.35	2.22	dM5
2330−1707	−17.6769	347	−219	410	24		−2	26.0	0.63	8.60	1.28	dK5
White dwarfs											$b-y$	
0031−2725	GD619	81	−57	99	−10	−2.6	−3	26.0	3.56	14.22	−0.125	DA1
0148+6411	G244−36	225	−182	290	0	0.0	+11	26.4	1.27	14.00	0.160	DA6
0317+3432	GD45	63	−142	154	−20	−3.5	+38	26.9	2.81	14.14	−0.028	DA3
0349+2447	LB1497			Pleiades Cluster					5.65	16.55	−0.088	DH2
0410+1145	HZ2	55	−95	110	4	+0.7	+37	27.0	3.10	13.89	−0.075	DA3
0637+4747	GD77	−10	−130	130	7	+1.5	+44	27.1	3.21	14.82	0.039	DH5
0871+3741	GD90	−50	−87	100	−6	−1.0	+48	27.3	3.78	15.70	0.090	DAH5
1334+4844	GD325	−122	−30	126	1	+0.2	+3	26.1	2.56	14.00	0.130	DA4
2256+2500	GD245	136	−63	149	18	+0.6	+2	26.1	2.59	13.70	−0.020	DA4
Variables										M_v	$B-V$	
369084	AB Dor	61	130	104	−34	−3.7	−2	26.1	1.82	+5.0	0.85	K1Vp
139084	V343 Nor	−53	−105	118	4	+0.8	−36	25.2	3.25	+4.9	0.83	K0V
174229	PZ Tel	10	−100	100	−8	−2.0	−48	24.9	3.60	+4.9	0.81	K0Vp

TABLE 1. Red and white dwarfs, and variables in the Pleiades supercluster.

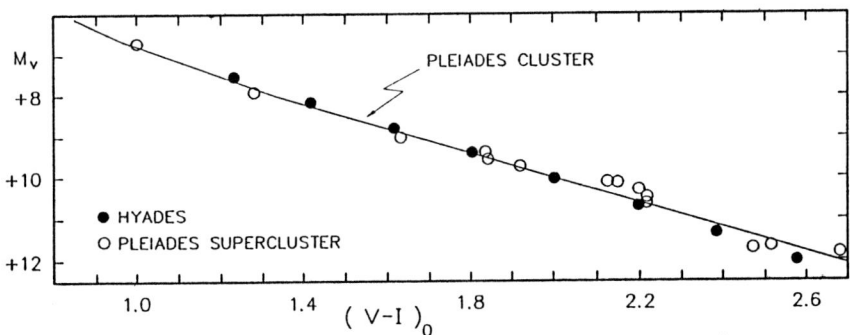

FIGURE 7. The Pleiades supercluster stars in the $(M_V, V-I)$ plane.

with the uncertain value of –9 km/s for the bright star. Star 2145–7220 is the variable AY Ind, which is probably a spotted star. The pair 2330–1702/7 is Gl 897/8, with a poorly determined trigonometric parallax of $0\rlap{.}''083$, compared with the cluster value of $0\rlap{.}''075$; and $V_{tot}\cos\lambda = +1.8$ km/s, compared with –1.5 km/s for the bright star and +1.3 km/s for the fainter. The bright component is itself a close binary with equal components and is separated from the faint star by 6'.

The Pleiades modulus of 5.65 mag. (Eggen 1986a) and a reddening of $E(V-I) = 0.10$ mag. has been applied to the observed colour-magnitude diagram of the cluster derived by Prosser et al. (1991), to give a ridge line of the Pleiades lower main sequence shown as a continuous curve in Figure 7. The mean Hyades relation, Equation 2.2, has been transformed to the $(M_V, V-I)$ plane with the relation $V-I = 1.925(R-I) + 0.26(V-I)$ between 1.0 and 2.5 mag., and the result is represented by filled circles in Figure 7. The filled circles represent Equation 2.2 and the continuous line the main sequence for the Pleiades. The Hyades and Pleiades main sequences are remarkably similar to $V-I = 2.5$ mag. The red dwarfs in Table 1 are represented by open circles in the Figure.

Several field, K-type stars with extremely active chromospheres, but slightly hotter than those in Figure 7, can be identified as supercluster members. Those with the best available astrometry are AB Dor (0.5 d), PZ Tel (0.9 d) and V343 Nor. The periods range from 0.5 day (AB Dor) to 0.9 day (PZ Vel), and all three show strong lithium. There are several objects of comparable colour and chromospheric activity in the Pleiades cluster. The radial velocity of HD 37605 (AB Dor) shows a range from +64 to –20 km/s, but this is believed to be due to the broad lines and the velocity is assumed to be constant, near +25 km/s, compared with the supercluster value of +22.5 km/s. HD 139084 (V343 Nor) has a velocity of +3.3 km/s, which is also the value predicted from the supercluster motion, and that of HD 174429 (PZ Tel) is uncertain, between +12 and –14 km/s, compared with $V_{tot}\cos\lambda = -1.7$ km/s.

4. UMa cluster and the Sirius supercluster

The sparse UMa cluster in the Sirius supercluster (Eggen 1992c) has $(A, D) = (20^h55, -38°1)$ and $V_{tot} = 18.5 + 0.005X$. The members of the cluster are in Table 2 and represented by open circles in the $(M_V, B-V)$ plane of Figure 8. The continuous curve in the Figure represents an isochrone for 6×10^8 year old solar models (Bertelli et al. 1990). Because of the low space motion, members of the supercluster with annual proper motion greater than $0\rlap{.}''5$ are only within $8 \sin\lambda$ pc of the Sun. So the majority of the supercluster members have yet to be found. However, the available data isolates the 20 stars in Table 3 as members within the same distance of the Sun (35 pc) as the

HD	Name	Mod. m	X	Y pc	Z	M_v m	$B-V$ m	Sp.T.	v 0"001	PV km/s
91480	37 UMa	1.84	12.8	6.8	18.3	3.31	0.33	F1V	78	−0.15
95418	β UMa	1.73	10.9	6.5	18.2	0.57	−0.02	Am	81	0.0
103287	γ UMa	1.96	9.2	7.5	21.6	0.30	0.00	A0V	96	+0.45
	+59.1428	2.10	9.6	8.4	22.2	7.94	1.28	M0V	95	+0.2
106591	δ UMa	1.93	8.4	9.1	20.9	1.31	0.06	A3V	104	+0.15
109011	+55.1536	1.88	6.9	8.6	20.6	6.21	0.93	K2V	111	+1.1
109647	+52.1638	1.80	5.9	7.3	20.9	6.69	0.95	K3V	117	+1.1
110463	+56.1618	1.69	6.1	8.4	19.1	6.58	0.95	K3V	124	+0.3
111456	HR4867	2.04	7.3	11.4	21.5	3.81	0.46	F6V	108	−0.45
112185	ε UMa	2.00	6.5	10.2	21.9	−0.24	−0.02	Hp	112	+0.3
113139	78 UMa	1.98	6.1	10.6	21.7	2.95	0.36	F2V	114	−0.2
115043	+57.1425	2.04	5.7	11.4	22.2	4.80	0.60	G2V	115	−0.15
	+58.1441	1.94	5.3	11.5	20.9	7.80	1.26	M0V	121	−0.3
116656	ζ UMa A	1.95	4.6	10.7	21.6	1.03	0.02	A2V	123	+0.5
116657	ζ UMa B					2.00	0.13	Hm		
116842	80 UMa	2.06	4.7	11.4	22.7	1.95	0.16	A5V	117	+0.7
									mean	+0.2
									σ	±0.45

TABLE 2. UMa cluster members

UMa cluster. These stars are represented by crosses in Figure 8. Observed and predicted values of the radial velocity of the cluster and supercluster members are in Table 4 and the individual objects are discussed in the notes to that

The cluster and supercluster members are represented by filled and open circles, respectively, in the (X,Y) and (X,W) planes of Figure 9: X is positive in the direction away from the galactic centre, Y in the direction of galactic rotation and W in the direction of the North Galactic Pole. Soderblom et al. (1993) have published equivalent widths, $W(Li)$, for several members of the cluster and supercluster and these stars are shown in $(B-V, \log W(Li))$ plane of Figure 10 as filled and open circles respectively. Also, lithium abundances for single, Hyades dwarfs (Soderblom et al. 1990) are represented by crosses in the Figure. Obviously more data are needed, but the possible separation of various stellar ages in Figure 10 is promising. The supercluster members with $(R-I) > 0.4$ mag. (see notes to Table 4) are represented in Figure 11 by filled circles. The straight line in the Figure represents Equation 2.2, adjusted by +0.20 mag. for the different main sequence level of stars with Hyades abundance ([Fe/H] = +0.1) and the supercluster members for which [Fe/H] = −0.08 (Eggen 1992c and 1986b).

5. The HR 1614 supercluster and group

Although several superclusters have been isolated in the old disk population, none of them is associated with clusters now in the solar vicinity. One of these, the HR 1614 supercluster (Eggen 1992d), is especially useful in the present context because of a characteristic feature that distinguishes most members from the majority of field stars. The nine supercluster members within 40 pc of the Sun and 29 near the main sequence

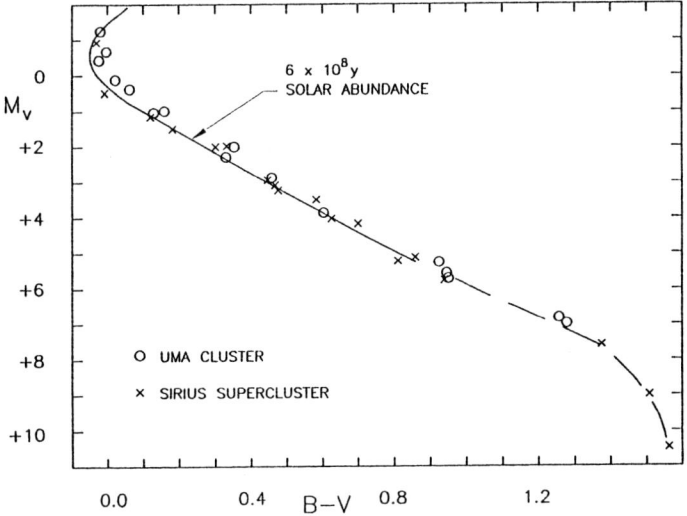

FIGURE 8. The $(M_V, B-V)$ diagram for the UMa cluster (open circles) and Sirius supercluster (crosses) members.

HD	Name	Mod. m	X	Y pc	Z	M_V m	$B-V$ m	Sp.T.	v 0."001	PV km/s
11131		−11.351	1.70	7.7	2.0	−20.4	5.09	0.63	G0V	173	−0.55
11171	χ Cet					2.97	0.33	F3III		258	−2.8
	G78−28	0.53	10.6	6.6	−2.2				258	−2.8
	G5−32	0.70	11.0	1.6	−8.2	11.49	1.56		273	−2.4
26913	+5.613	1.87	20.2	−2.4	12.0	5.15	0.70	G5V	154	+2.0
26923	V774 Tau					4.44	0.59	G0V		
	−11.916	0.36	7.5	−2.6	−8.7	9.98	1.52		318	−0.7
38392	HR1982	−0.50	5.0	−5.3	−3.3	6.75	0.94	K2V	471	−1.5
38393	γ Lep					4.10	0.47	F6V		
41593	+15.1065	0.93	14.9	−3.7	−0.7	6.30	0.81	K0V	165	−0.3
48915	α CMa	−2.94	1.7	−1.9	−0.4	1.48	−0.01	A0V	1326	+0.15
87696	211 Mi	2.06	14.7	−2.6	21.0	2.43	0.18	A7V	49	+1.2
97603	δ Leo	0.50	3.2	−3.1	10.6	2.04	0.12	A4V	191	−1.3
118972	−33.9242	0.82	−9.1	−9.2	6.7	6.11	0.86	K1V	263	−0.1
	+5.2767	1.47	−7.1	−4.1	17.9	8.50	1.38	M0V	192	−0.3
134083	45 Boo	0.97	−6.4	4.8	13.4	3.95	0.44	F5V	248	−1.25
139006	α CrB	2.07	−11.4	10.3	20.8	0.07	−0.03	A0V	150	+1.2
180777	59 Dra	2.17	7.6	23.4	12.0	2.95	0.30	F2V	131	+0.95
	LP452−10	0.94	−9.6	12.1	0.0	12.18			212	+0.2
184960	HR7451	1.58	−2.3	19.9	5.2	4.14	0.47	F7V	189	−0.75
									mean	−0.3
									σ	±1.3

TABLE 3. Members of the Sirius supercluster

HD/Name	Gliese	π 0″.001 C	π 0″.001 O	ρ km/s C	ρ km/s O	HD/Name	Gliese	π 0″.001 C	π 0″.001 O	ρ km/s C	ρ km/s O
11131		46	46	+5	0V	109647		44		−13	
11171						110463		46		−13	
G78–28		78		−10		111456		39		−13	−13V
G5–32		72		−5		112185		40	41	−13	−11V
26913		42		−7	− 8	113139		40	34	−13	−12
26923						115043	503.2	39	50	−12	−10V
−11.916	173	85	90	−5	−10V	+58.1441	509.1	41		−12	− 7
38392	216	126	123	−6	− 8V	116656		41	47	−12	−10V
38393						116657					
41593	227	65	65	−14	−13	116842		39	45	−12	−15V
48915	244	387	377	−9	− 8	118972		69		+4	
87696	378.3	39	45	−18	−18	+5.2767		51		−5	+ 7
91480		43	36	−16	−15V	134083	578	64	64	−3	− 6V
95418		45	53	−16	−13V	139006		39	45	−2	+ 1
97603	419	79	48	−15	−17V	180777		37	49	−8	− 5
103287		41	28	−15	−14V	LP452–10		65		+10	
+59.1428	457	38		−14	−10	184960		48	37	0	0V
106591	459	41	61	−14	−14						
109011		42		−14							

Notes

11131/71	Separation 184″.
G78–28	$(I, R - I) = (9.94, 1.13)$ mag.
G5–32	$(I, R - I) = (9.83, 1.09)$ mag.
26913/23	Separation 65″. The variable is a spotted star.
38392/3	Separation 100″.
48915	Visual binary: P=50.04 yr, a=7″.55. The companion is a white dwarf.
87696	δ Sct. Variable, P=0.1 d.
+59.1428	$(I, R - I) = (8.64, 0.555)$ mag.
109647	$(I, R - I) = (7.55, 0.39)$ mag.
112185	Spectrum variable, 5.09 d. Possibly sp. binary.
113139	ADS8739: ΔV=2.4 mag., P=115.7 yr, a=1″.256, with a mass sum of 2.3 solar masses.
+58.1441	$(I, R - I) = (8.33, 0.57)$ mag.
116656/7	Separation 14″. 116656 is a sp. binary, P=20.5 d and equal components of 2.45 solar masses.
+5.2767	$(I, R - I) = (8.38, 0.68)$ mag.
139006	Sp. binary (eclipsing), P=17.4 d. Masses and radii (in solar units) of A and B, respectively are (2.6, 0.9) and (3.0, 0.9). The luminosity of B is +5.1 mag.
LP452–10	$(I, R - I) = (10.54, 1.18)$ mag.

TABLE 4. Parallaxes and radial velocities of Sirius supercluster members.

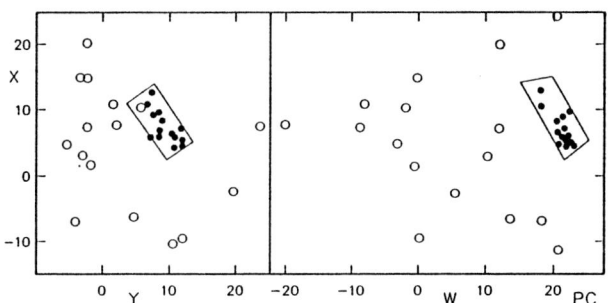

FIGURE 9. The space distribution of UMa cluster (filled circles) and Sirius supercluster (open circles) members.

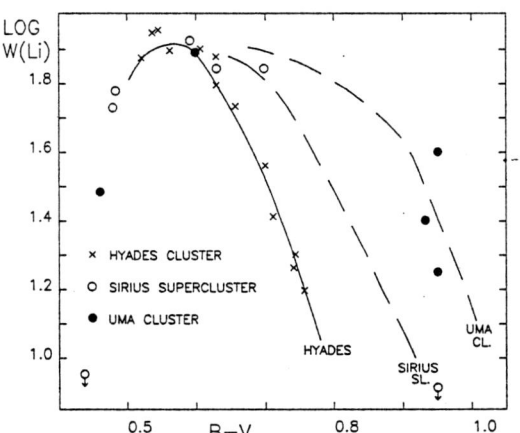

FIGURE 10. Hyades cluster, UMa cluster and Sirius supercluster members in the $(B-V, \log W(Li))$ plane.

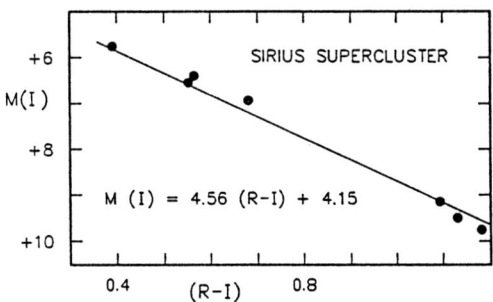

FIGURE 11. The reddest Sirius supercluster members in the $(M_I, R-I)$ plane.

group members are represented by filled and open circles, respectively, in Figure 12 (see Eggen 1992d). These stars are as old as those in M67 – near 5×10^9 yr. The upper panel of Figure 13 represents the $(m_2, B-V)$ diagram for the supercluster (filled circles) and group (open circles) members, compared with the Hyades main sequence stars. The index m_2 in the Geneva system (Rufener 1988) measures the abundance of heavy elements and despite their age, the group and supercluster members have the Hyades' abundance. The lower panel of Figure 13 shows the CN abundance of the group and supercluster members, compared with the Hyades Main Sequence stars. The stars of

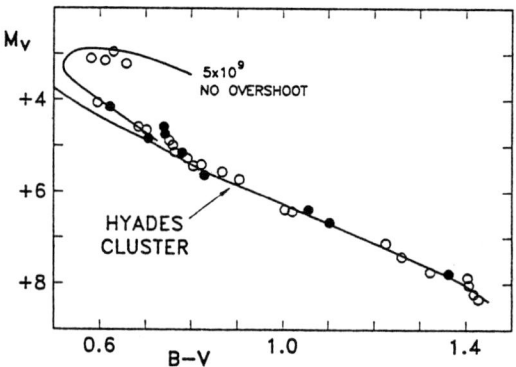

FIGURE 12. The HR 1614 supercluster (filled circles) and HR 1614 group (open circles) members in the $(M_V, B-V)$ plane.

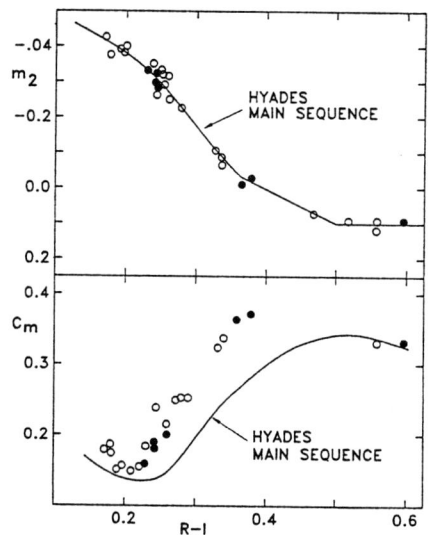

FIGURE 13. The HR 1614 supercluster (filled circles) and HR 1614 group (open circles) members in the $(m_2, R-I)$ and $(C_m, R-I)$ planes.

C_m are from DDO photometry (e.g. McClure 1970) and the CN strengths of the group and supercluster members are matched by less than 5% of the field stars near the Sun. This rare combination of strong heavy element abundance, age and nearly unique CN abundance gives some firm confirmation to the assumption that the supercluster stars evaporated from a cluster which finally disrupted, leaving the group stars as relics.

REFERENCES

Bertelli, G. et al., 1990, A&AS, 66, 91
Eggen, O.J., 1986a, PASP, 98, 775
Eggen, O.J., 1986b, AJ, 92, 910
Eggen, O.J., 1992a, AJ, 104, 1482
Eggen, O.J., 1992b, AJ, 104, 2141
Eggen, O.J., 1992c, AJ, 104, 1493

Eggen, O.J., 1992d, AJ, 104, 1906

Eggen, O.J., 1993a, AJ (in press)

Eggen, O.J., 1993b, AJ (in press)

Giclas, H., Burnham, R. & Thomas, N., 1980, Lowell Obs. Bull. No.166

Gliese, W., 1969, Veröff Astron. Inst. Heidelberg No.22

Luyten, W., Morris, S. & Hill, G., 1981, Proper Motion Survey with the 48-inch Schmidt Telescope No.LX

McClure, R., 1970, AJ, 75, 41

Prosser, C., Stauffer, J. & Kraft, R., 1991, AJ, 101, 1361

Rufener, P., 1988, Catalogue of Geneva Photometry (Geneva Obs.)

Soderblom, D. et al., 1990, AJ, 99, 595

Soderblom, D. et al., 1993, AJ (in press)

Weilen, R., Jahreiss, H. & Kroger, R., 1983, IAU Coll. 76, ed. A. Phillip and A. Upgren (Schenectady, Davis Press), p.163

Wegner, G., 1981, AJ, 86, 264

Wegner, G., Reid, I.N. & McMahan, R., 1991, ApJ, 376, 186

Wilson, O., 1967, AJ, 72, 905.

The Hyades Cluster

By NEILL REID

California Institute of Technology, 105-24, Pasadena, CA 91125, U.S.A.

This paper summarises the most recent results concerning the stellar content of the Hyades cluster, in particular, the luminosity function at faint magnitudes. The distance of the cluster centre is now well determined as 48 ± 1 parsecs ($(m - M) = 3.40$), but cluster members are distributed over a range of ± 10 parsecs. Mass segregation is clearly present, and some of the outlying M dwarfs may be in transition from the cluster proper to the surrounding supercluster.

---

1. Introduction

While the naming of the Hyades dates back to antiquity, it was only in 1908 that Louis Boss, using proper motions from the General Catalogue, actually demonstrated that there were stars forming a coherent, moving group in Taurus. He was the first to apply convergent point analysis to the cluster members, deriving a distance modulus of 2.98, or a distance of 39 parsecs, based on 39 of the brightest members (Boss, 1908). With the growing understanding of stellar evolution that accompanied the development of the Hertzsprung-Russell diagram, the Hyades acquired considerable importance as a nearby cluster whose distance could be determined directly, using geometrical arguments, and whose colour-magnitude diagram could therefore serve as a direct constraint on, first, stellar evolutionary theory and, latterly, the cosmic distance scale. As a result, the primary concern of most studies has centred on determining the mean distance of the cluster. Perversely, just as we have reached good agreement on the distance, the (metal-rich) Hyades have been supplanted as a preferred distance indicator by the more distant, but solar abundance, Pleiades.

In this review I shall summarise the most recent results on the Hyades distance, but I will concentrate more on the structure of the cluster itself. In particular, I will use the results derived from my own recent survey (Reid 1992, 1993), covering ~ 110 square degrees to V=19, to examine the Hyades luminosity function at faint magnitudes and the spatial distribution of stars of different masses. I shall not attempt to expand the scope of this talk to cover stellar evolutionary studies – such as lithium abundance or chromospheric activity and stellar magnetic fields - except insofar as these topics touch on cluster membership.

2. The distance of the Hyades

The evolution of the Hyades distance scale, and earlier studies of the cluster, has been discussed thoroughly by Griffin et al. (1988). To summarise, most estimates fell near van Buren's (1952) modulus of 3.03 until Hanson's (1975) survey, which was based on Lick plates and used galaxies to define the proper motion reference frame, arrived at a significantly higher modulus (3.42 initially, but revised to 3.18 by McAlister (1977)).

The convergent point method of distance determinations rests on the equation

$$r = \frac{V_s sin\lambda}{\kappa <\mu_u>} \qquad (2.1)$$

where $<\mu_u>$ is the mean absolute proper motion towards the convergent point, V_s is the space velocity, λ the angular distance between the convergent point and the cluster centre, and $\kappa = 4.74$. (The space velocity of the Hyades cluster is 46.7 km/s (Detweiler et al., 1983), while the velocity dispersion is negligible (~ 0.25 km/s, Gunn et al., 1988) - were the latter not the case, the Hyades would have dispersed spatially by now.) The two crucial steps are the determination of the position of the convergent point, and the conversion to absolute proper motions, and of those two the latter is the more difficult.

The two most recent complete analyses are the radial velocity survey by Gunn et al. and Schwan's (1991) analysis of the FK5 data. The convergent points are nearly identical. Gunn et al. derive

$$(\alpha_{CP}, \delta_{CP}) = (98°.2 \pm 1°.1, 6°.1 \pm 1°.1),$$

while Schwan gives

$$(\alpha_{CP}, \delta_{CP}) = (97°.7 \pm 0°.4, 6°.0 \pm 0°.2).$$

The difference in $\sin\lambda$ (and hence in r) amounts to less than 1%. However, Gunn et al., using Hanson's proper motions, derive a distance modulus of 3.28, while Schwan, using the meridian circle FK5 measurements, finds a modulus of 3.40. Of these two results, Schwan's is to be preferred, since the individual proper motions are both more accurate than are Hanson's data and are tied directly to an absolute reference frame. Moreover, the mean absolute proper motion derived from my own Schmidt survey, which is referred to the independent absolute system defined by ~ 1300 galaxies in each field, gives a distance modulus of 3.45. Direct trigonometric parallax measurements give a smaller modulus (3.18, Patterson & Ianna, 1991), but the major uncertainty in these observations is the conversion to absolute parallax and, allowing for that, the discrepancy is only $\sim 1\sigma$. Thus, the current best estimate of the mean distance of the Hyades cluster (giving Schwan's result double weight) is $(m - M) = 3.42 \pm 0.04$, or 48.3 ± 0.9 parsecs.

3. The Schmidt Survey

Most of the results discussed here are based on a proper motion survey covering four Palomar Schmidt fields towards the central and south-western parts of the Hyades (Reid, 1992). The first epoch plates were taken by the Oschin Schmidt as part of the original Palomar survey (POSS I - epochs 1950.77 to 1955.93), while the second epoch B, V plates were taken by the UK Schmidt between 1985.9 and 1987.8. These plates were scanned by COSMOS in 1989 and the total area covered is 112 square degrees, with ~ 0.5 degree gaps between the fields. The limiting magnitude for Hyades stars is set by insisting that all objects be detected on both POSS I plates - hence the POSS I O limits completeness to $B \sim 20.5$, or V=19 for Hyades M dwarfs. Thus the survey is neither as deep nor as extensive in areal coverage as the Luyten, Hill & Morris (1981) survey, also based on Palomar plates. However, the longer baseline in time gives proper motions of higher accuracy, allowing the use of more selective criteria in identifying candidate Hyads and hence the possibility of more complete follow-up observations - both accurate photometry and spectroscopy. The present survey extends approximately 3.5 magnitudes fainter than the Lick surveys.

Based on an internal comparison - between O/B and E/V plate pairs - I estimate the proper motions to be accurate to better than $0''.006$/yr (6 mas/yr) for $V < 16.5$, rising to ~ 10 mas/yr at V=19. An external comparison (with Lick astrographic astrometry) suggests that the uncertainties are as low as 4.5 mas/yr for $V < 15$, and a comparison with previous surveys shows that the Schmidt data are of comparable or superior accuracy (see Reid, 1992, for full details and extensive cross-referencing).

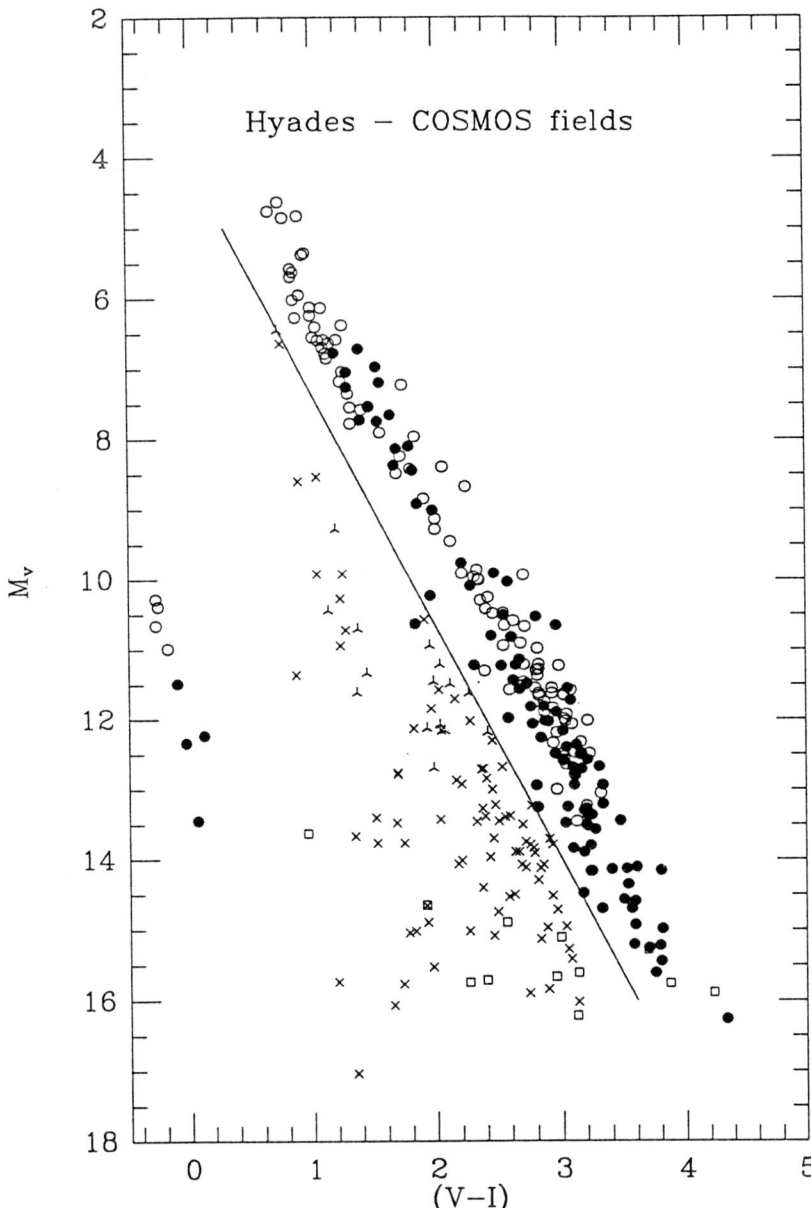

FIGURE 1. The $(M_v, (V-I))$ diagram for Hyades candidates. The photometric members are marked as open circles (data from literature) or solid points (data from P60); non-members are marked as crosses. The open squares mark the stars identified by Bryja et al. (1992)

4. Hyades membership

I have employed a three-stage approach in selecting stars likely to be cluster members. Ideally, one can determine membership probabilities based on the proper motions (μ_u, μ_t) (where μ_u is directed toward and μ_t perpendicular to the convergent point) and the uncertainties in μ. However, this approach requires both an accurate knowledge of the *distribution* of uncertainties, ϵ_μ and of the proper motion distribution of the background field stars. Both are a function of magnitude. Since neither distribution is defined unequivocally by data from the COSMOS scans alone, the initial criteria adopted err towards liberality (as, indeed, we all should) in selecting potential Hyads. In essence, all stars with

$$+0\overset{''}{.}075/yr \leq \mu_u \leq +0\overset{''}{.}17/yr,$$

and

$$-0\overset{''}{.}025/yr \leq \mu_t \leq +0\overset{''}{.}025/yr,$$

were selected in the first cut (the proper motion limits were varied slightly as a function of magnitude – see Reid (1992)). This produced a sample of 411 candidates, 66 of which have proved to be spurious in nature.

Proper motions consistent with those of the Hyades are necessary, but not sufficient, conditions for cluster membership. The proper motion defined sample is still contaminated by significant numbers of field stars, particularly at low μ_u and negative μ_t, close to the reflex solar motion vector. However, μ_u gives a direct distance estimate for each star from (2.1). Thus, the second step is to require photometry consistent with the Hyades colour-magnitude diagram. The photographic $(B-V)$ data are not adequate to this purpose, primarily since $(B-V)$ saturates at $M_v \sim +9$, or $V \sim 12.5$ – hence there is no means of distinguishing background early-type M-dwarfs. This is not the case for $(V-I)$, and I have collected together VRI photometry for the complete sample, either from the literature or directly, using the Palomar 60-inch telescope. Figure 1 shows the resultant $(M_v, (V-I))$ diagram. 210 stars have photometry consistent with their being Hyads (Reid, 1993). Figures 2a and 2b show the proper motion distributions of the stars in our survey. The average motion, $<\mu_u> = 0\overset{''}{.}109/\text{yr}$, implies a mean distance of 49.4 parsecs, in excellent agreement with Schwan (1991). Note that the non-photometric members are, as expected, concentrated towards the bottom left. However, there are also stars with photometry *consistent* with membership which lie well away from the cluster locus and which have, as a consequence, low membership probabilities. Rather than reject these stars outright as members, as step 3 of our process, we (Reid, Hawley, Mateo) have been carrying out follow-up spectroscopic observations, since the (relatively young) Hyades M-dwarfs are characterised by strong $H\alpha$ emission (Stauffer et al., 1991). Fewer than 10% of field stars have emission of comparable strength – and some of these field stars are probably members of the Hyades supercluster. All of the stars surveyed so far - we have data for 75% of those with $M_v > 11$ – have strong emission, even those with proper motion membership probabilities as low as 4%. (We shall return to the latter stars in the next section.) Finally, the low-mass Hyads are also strong X-ray emitters – we have ROSAT observations which detect stars as faint as $M_v = +13$ (Reid & Hawley, in prep.). Hence this is another criterion which can be used to identify cluster members.

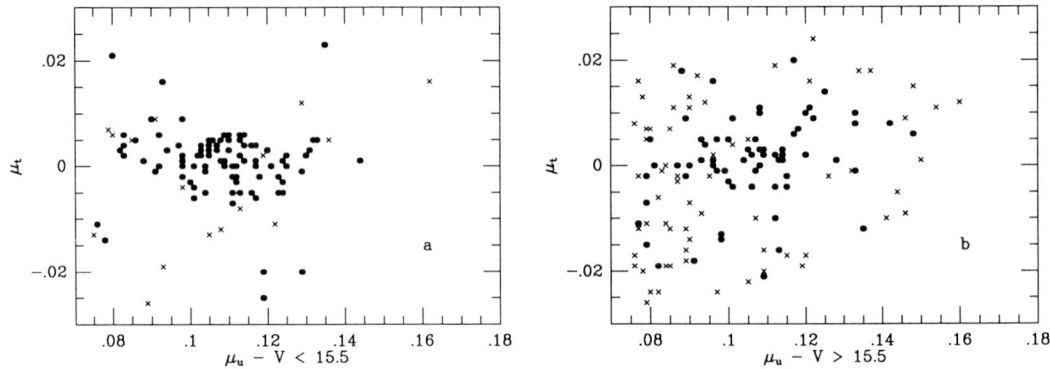

FIGURE 2. The proper motion distribution of photometric Hyades members (solid points) and non-members (crosses). We have divided the sample at $V = 15$.

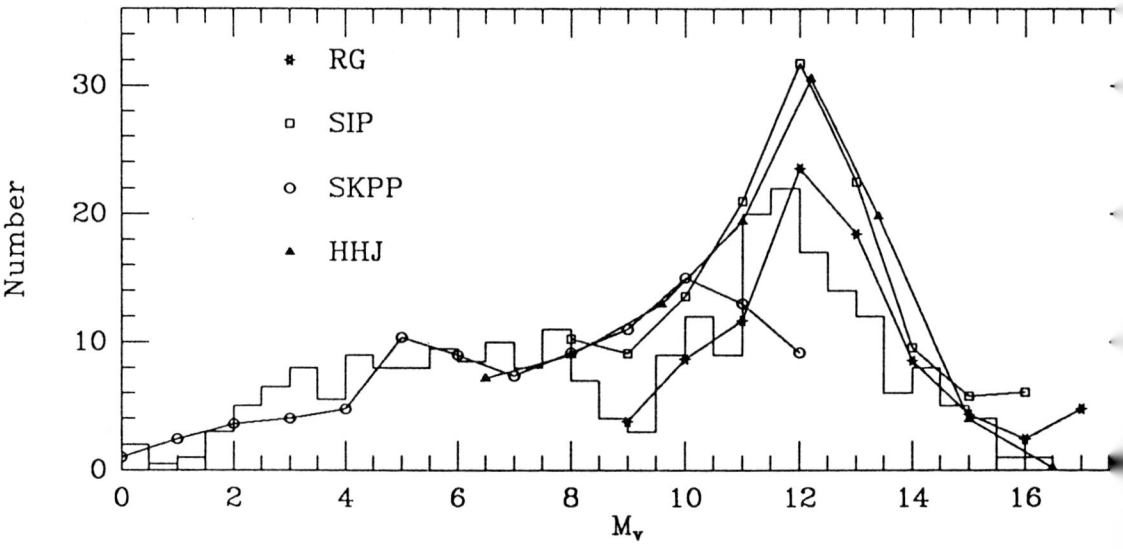

FIGURE 3. The Hyades luminosity function, $\Phi(M_v)$, compared with data for the general field and for the Pleiades cluster. See text for references.

5. The luminosity function and mass segregation

Figure 3 compares the luminosity function for the Hyades against data for the general Galactic field (Reid & Gilmore, 1982 – RG; Stobie, Ishida & Peacock, 1987 – SIP) and the Pleiades (Stauffer et al., 1992 - SKPP; Hambly, Hawkins and Jameson, 1992 – HHJ). We have included all photometric members in the hyades function and extended coverage to magnitudes brighter than $M_v = +7$ as described in Reid (1993). Two points are clear – first, the Hyades luminosity function peaks at $M_v \sim +12$, well above the completeness limit of the survey. There are few low-luminosity M-dwarfs in the Hyades.

Two recent Hyades surveys have produced very low luminosity (VLM) candidates. Bryja et al. (1992) used proper motions to identify 12 VLM Hyades candidates, but our VRI photometry (Fig. 1) shows that 9 are background stars. (One of the bona-fide Hyads was also identified in our survey.) Leggett & Hawkins (1989) have identified several VLM candidates on a purely photometric (I-K) basis, but the number of stars is consistent with predictions for the field (given the area surveyed) – and the one star observed spectroscopically has no $H\alpha$ emission. Future surveys will benefit from the availability of the POSS II IIIaF and IVN plates which, combined with POSS I E-plates, should extend the depth of coverage by ~ 2 magnitudes.

The second point is that the relative number of stars at $M_v = +12$ to that at $M_v = +7$ is lower for the Hyades than for both the Pleiades and the field. Over the lifetime of the Hyades ($\sim 3 \times 10^8$ yr) we expect stars to evaporate from the cluster. If mass segregation is present – and our observations show that the massive stars do indeed follow a more compact distribution both in the plane of the sky (Reid, 1992) and along the line of sight (Reid, 1993) – than the more widely-distributed low-mass stars will be lost preferentially. For the massive stars, Weidemann et al. (1992) estimate that there should be $\sim 20 \pm 4$ white dwarfs present, whereas only 7 are observed – a deficit of a factor of 2.5 to three. On this basis, the depletion of the low-mass stars is up by another factor of two – or, overall, a factor of between 5 and 6. (Weidemann et al. estimated a factor of 10, but that was on the basis of comparing a flat Hyades luminosity function with a van Rhijn initial luminosity function.) Most of the mass resides in the more massive stars, so with a present day total mass of $\sim 400 M_\odot$, this implies an initial cluster mass of $\sim 1000 M_\odot$.

Finally, returning to the matter of the Hyades M-dwarfs that are outliers in Figure 2b, while the position of these stars may simply reflect increased (non-Gaussian) astrometric uncertainties, it is also possible that evaporation could account for the discrepant proper motions. These stars may be making the transition between the cluster-proper and the surrounding supercluster (see Eggen's paper, this conference). We note that Hyades supercluster M dwarfs should show the same marked $H\alpha$ emission as do their contemporaries in the cluster itself.

6. Summary

The distance of the Hyades is now defined to an accuracy of $\sim 2 - 3\%$ as 48.3 ± 1 parsecs – a value supported by two independent calibrations of the absolute proper motion zeropoint. Based on this distance estimate, I have used the results from my Schmidt proper motion survey to derive the luminosity function of the cluster to $M_v \sim +15.5$ and find that there is a good qualitative match to both recent studies of the Pleiades and of the general field. Mass segregation is clearly present in the cluster, however, and we attribute a possible deficit of a factor of two in the number of M dwarfs to preferential evaporation of lower-mass stars. The initial cluster may have been as massive as $\sim 1000 M_\odot$.

The Hyades are no longer a prime distance indicator. However, the cluster will remain important for many years as a testbed for theories of stellar binarity, activity in low-mass stars and the dynamical evolution of clusters.

Acknowledgements

The photometry described in this paper was obtained using the Palomar 60-inch telescope which is operated jointly by the Observatories of the Carnegie Institution and by the California Institute of Technology. The spectroscopic observations were obtained using the 200-inch Hale telescope, which is operated by the California Institute of Technology. Part of this work was funded by a NASA ROSAT grant. I should also like to acknowledge the hospitality of Observatoire de Paris, Meudon, during the time that this paper was written.

REFERENCES

Boss, L., 1908, *Astr. J.*, **26**, 31
Bryja, C., Jones, T.J., Humphreys, R.M., Lawrence, O., Pennington, R.L., Zumach, W., 1992, *Astrophys. J.*, **388**, L23
Detweiler, H.L., Yoss, K.M., Radick, R.R., Becker, S.A., 1984, *Astr. J.*, **89**, 1038
Griffin, R.F., Gunn, J.E., Zimmerman, B.A., Griffin, R.E.M., 1988, *Astr. J.*, **96**, 172
Gunn, J.E., Griffin, R.F., Griffin, R.E.M., Zimmerman, B.A., 1988, *Astr. J.*, **96**, 198
Hambly, N.C., Hawkins, M.R.S., Jameson, R.F., 1991, *Mon. Not. R. astr. Soc.*, **253**, 1
Hanson, R.B., 1975, *Astr. J.*, **80**, 379
Leggett, S.K., Hawkins, M.R.S., 1989, *Mon. Not. R. astr. Soc.*, **238**, 145
Luyten, W.J., Hill, G., Morris, S., 1981, *Proper motion surveys with the 48-inch Schmidt*, **LIX**, Minnesota Obs., Univ. of Minnesota, Minneapolis, Minnesota
McAlister, H.A., 1977, *Astr. J.*, **82**, 487
Patterson, R.J., Ianna, P.A., 1991, Astr. J., **102**, 1091
Reid, I.N., Gilmore, G.F., 1982, *Mon. Not. R. astr. Soc.*, **201**, 73
Reid, I.N., 1992, *Mon. Not. R. astr. Soc.*, **257**, 257
Reid, I.N., 1993, *Mon. Not. R. astr. Soc.*, in press
Stobie, R.S., Ishida, K., Peacock, J.A., 1989, *Mon. Not. R. astr. Soc.*, **238**, 709
Schwan, H., 1991, *Astr. Astrophys.*, **243**, 386
Stauffer, J.S., Giampapa, M.S., Herbst, W., Vincent, J.M., Hartman, L.W., Stern, R.A., 1991, *Astrophys. J.*, **374**, 142
Stauffer, J.S., Klemola, A., Prosser, C., Probst, R., 1991, *Astr. J.*, **101**, 980
van Buren, H.G., 1952, *Bull. astr. inst. Neth.*, **11**, 385
Weidemann, V., Jordan, S., Iben, I., Casertano, S., 1992 *Astr. J.*, **104**, 1876.

Is the Stellar Luminosity Function universal?

By IMANTS PLATAIS

Yale University, Department of Astronomy, P.O. Box 6666, New Haven, CT 06511, U.S.A.

Observational evidence strongly indicates a complex nature for the luminosity function, at least for so-called poorly-populated open clusters. The most striking feature of luminosity functions for these clusters is a pronounced turnover at $M_v \approx +3^m$. This feature is well determined and cannot be explained by incompleteness of cluster members or statistical uncertainties. A short list of specific young open clusters, possibly having such luminosity functions, is given for the purpose of further study.

1. Introduction

The universality of the stellar luminosity function (LF) is a topic of great concern to astronomers interested in star formation and the structure of our Galaxy. Star clusters present an opportunity to study the luminosity function (along with the mass spectrum) for stars at the same distance, age and metallicity. Obviously, a considerable span in ages of open clusters may show eventual evolution of luminosity function with age. Unfortunately, time also works against the direct determination of the so-called Initial Mass Spectrum due to the cluster dynamical evolution. In addition, star clusters are contaminated by foreground and background field stars. Different techniques like star counts, proper motion and radial velocity memberships, color-color and color-magnitude diagrams in various photometric systems can be used to separate the true cluster stars from incidental field stars having at least something common with the cluster. Almost twenty years ago Taff (1974) concluded that there is no reason to reject the hypothesis of universality of the luminosity function and that the mass function is well represented by a power law. Ever since then, the dominant approach is to adopt a universal luminosity function for *all* open clusters despite a few discordant results. Among the latter the most prominent is a paper based on star counts for 20 open clusters mainly showing either a decrease or constancy of the luminosity functions below $M_{pg} = +5^m$ (van den Bergh and Sher, 1960). A comprehensive review of various aspects directly related to the problem of luminosity functions for open clusters can be found in Scalo (1986). Here, we address again the issue in light of results from recent astrometric studies of selected open clusters. It has been noticed (Platais, 1990) that some of the so-called poorly-populated open clusters (PPC) have peculiar luminosity functions. The rising part of luminosity functions discussed in that paper at the bright end of LF is quite short, just $2^m - 3^m$. Starting from $M_{pg} \approx +3^m$ the luminosity functions have a pronounced turnover and then remain essentially flat down to the limiting magnitude of completeness. This observational evidence strongly suggests a more complex nature for the luminosity function. Since those clusters may constitute a considerable portion of the Galaxy disk population, it is important to continue studies of younger clusters, as crucial tools in understanding the possible diversity of the stellar luminosity function. This should give a more decisive answer concerning the source of peculiarity of the luminosity function - dynamical depletion of low mass stars with age or a different mass spectrum at birth.

2. Luminosity function

2.1. Cluster membership

Given the fact that our observations of star clusters are contaminated by field stars, the construction of the luminosity function is in essence a problem of cluster membership. Among different membership techniques star counts are the simplest but also the most uncertain especially at low cluster member *vs.* field star ratio. Perhaps, for the poor clusters this technique is not applicable at all. Two kinematical techniques, namely, proper motions and radial velocities are most useful and commonly used for cluster membership. As a matter of fact our knowledge of luminosity functions is almost completely based on results of proper motion membership analysis. In a few cases radial velocities have been successfully used in the outer parts of clusters (e.g. Mermilliod et al., 1990), although an application of radial velocity techniques is hampered by the large telescope time required, relatively high percentage of spectroscopic binaries and inability insofar to derive radial velocities for a *complete* unbiased sample of cluster and field stars, which is an essential condition for deriving membership probabilities. In contrast, proper motions are generally free of these constraints but, of course, have others. For instance, since proper motions are derived from the measurements of photographic plates, the sky area under investigation is strictly limited by the size of the plate. Usually it is 2° by 2° for normal astrographs and $\approx$ 1° by 1° or even less for the long focus telescopes. Certainly, clusters which have angular diameters larger than the plate size will be affected by clipping their outer parts. An attempt to model the spatial distribution of cluster stars and thus estimate the number of missing stars (see Francic, 1989) is a step forward but still may be well off due to poor knowledge of the distribution of stars in the outer parts of clusters. Unfortunately, there is no single cluster for which we know exactly the full angular size. Another limiting factor of applying the proper motions is a possible presence of a star stream which is somehow associated with the cluster. The best known is the UMa cluster stream. Recently, we discovered probably a similar stream in the field of NGC 7092 which had previously been assumed to be a part of the cluster. For the time being it seems hopeless to study such streams in detail because they may cover large areas of the sky at extremely low densities. The color-color and color-magnitude diagrams seems to be marginally useful for luminosity function studies because the field star contamination in these diagrams is drastically increasing towards faint magnitudes. Moreover, the logarithmic nature of star magnitudes inevitably leads to a larger spatial volume at fainter magnitudes if the same dispersion for the Main Sequence of a cluster is maintained.

2.2. Luminosity functions from recent studies

A few years ago we started a program of determining the luminosity functions for poorly-populated open clusters (PPC). It is somewhat unclear which clusters fall into that category but usually PPC contains $\approx 10 - 15$ main sequence stars per magnitude bin down to the plate limit. Intuitively one might think that these clusters are quite common in our Galaxy. Indeed, among $\approx$1200 open clusters (Lyngå, 1987) maybe a half can be counted as poorly-populated. Unfortunately, they are also most poorly studied, with exception of some nearby clusters. Even those which have been studied using proper motions may still require better spatial and magnitude coverage. It should be noted that PPCs really require the most rigorous membership criterion - proper motions - since the ratio "cluster *vs.* field stars" can be as low as $\approx$ 1 : 100. In Figure 1 we present newly derived luminosity functions based on the most reliable proper motion membership probabilities available.

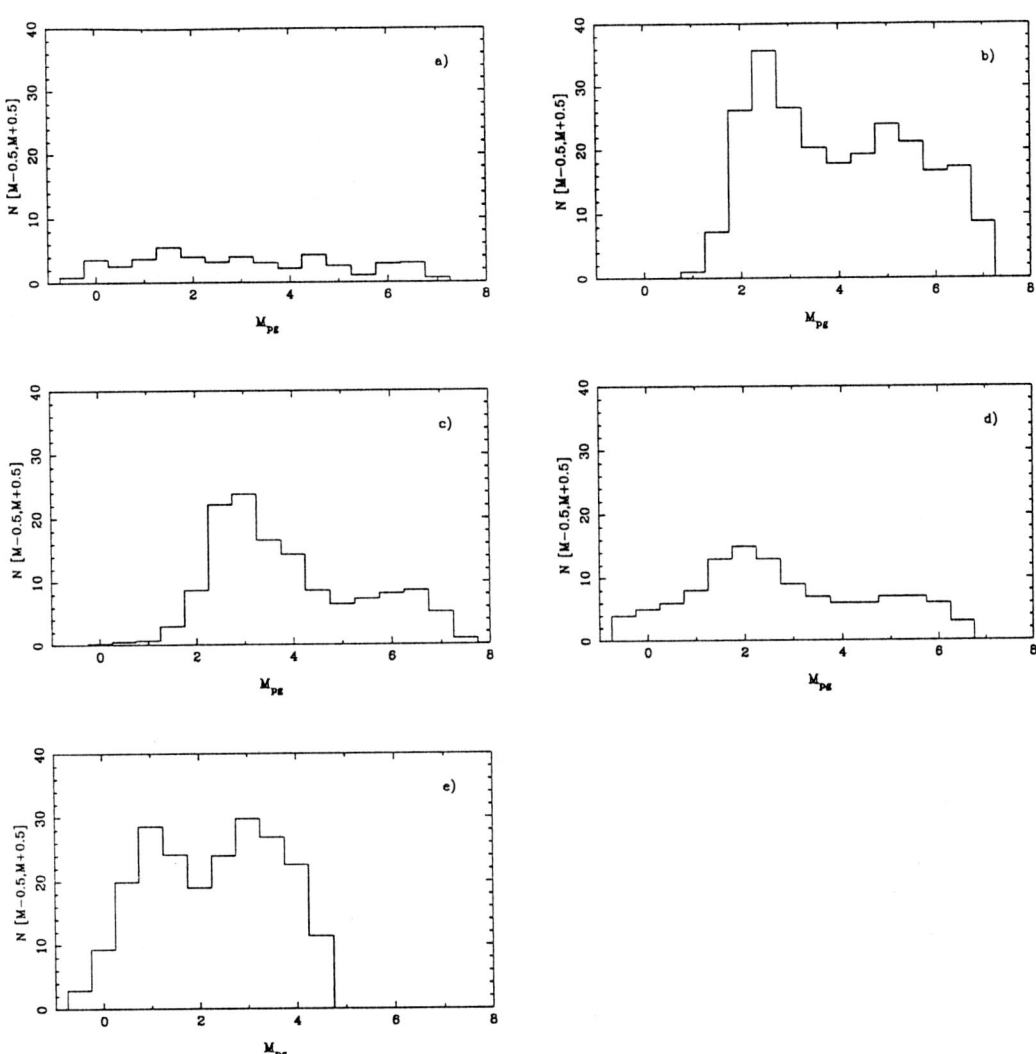

FIGURE 1. Luminosity functions of selected open clusters: (a) NGC 225: data from Lattanzi et al., 1991, *AJ.*, **102**, 177. (b) NGC 752. (c) NGC 3680: data from Girard et al., 1993, unpublished. (d) NGC 7092. (e) NGC 7209.

The luminosity function is formed as a sum of probabilities into bins of unit absolute magnitude. The known giant-branch stars have been excluded from the cluster member list. As can be seen, all luminosity functions illustrated differ considerably from that for the field stars and other clusters (Scalo, 1986). The bright, rising portion of the LF is very short: $2^m - 3^m$. At $M_{pg} \approx +3^m$ the luminosity functions have a pronounced turnover and then remain essentially flat down to the catalog limit. The presence of the turnover and the flat part of the luminosity function are quite distinct and cannot be explained by incompleteness. Although the plates used in the measurements cover at least three core radii, there still might be cluster members beyond the area studied. An analysis of the apparent radial density distributions for the bright and faint members strongly suggests that the number of missing faint cluster members is not significant. That means that the plotted luminosity functions really reflect true features of the complete luminosity function.

Discussion

The peculiar shape of the luminosity function for the poorly-populated clusters isn't a total surprise. Scalo (1986) hinted that UMa, NGC 3293, M67 and NGC 2506 are clusters having turnovers at lower masses. According to Scalo "...variations among clusters and between clusters and the field star IMF probably do exist in some cases". However, there might be another reason causing preferential depletion of low mass stars while not requiring us to abandon the hypothesis of a universal Initial Mass Function (IMF). Since all clusters in Figure 1 are older than log t=8.0, dynamical depletion in principle could carve out such a peculiar luminosity function. Existing dynamical evolution models (Terlevich, 1987), however, do not provide certain proof that such a scenario is the most likely. We may avoid the complication of dynamical evolution by studying young clusters. A detailed literature survey for all clusters with distances $D < 1$ kpc revealed a few young clusters which could be crucial showing whether dynamical processes or a diversity of the IMF is responsible for the peculiarity of the luminosity function. The clusters IC 2602 (log t=7.0), Coll 140 (7.4), IC 2391 (7.6), Blanco 1 (7.7) are of highest priority and should be studied by all techniques because they show some indirect features of peculiar luminosity functions, however, they lack comprehensive kinematical membership analyses. Another important field of research which can directly test the universality of the IMF is infrared imaging of very young clusters embedded in molecular clouds (Lada, 1993). The first results imply a turnover of the infrared K-luminosity function, though, at a lower corresponding visual absolute magnitude, namely, $M_v \approx +6^m$.

REFERENCES

Francic, S.P. (1989) AJ, **98**, 888

Lada, C.J. (1993) Harvard-Smithsonian Center for Astrophysics, Preprint Series, No. 3655.

Lyngå, G. (1987) Lund Catalogue of Open Clusters. Fifth edition, CDS: Strasbourg.

Mermilliod, J.-C., Weis, E.W., Duquennoy, A., Mayor M. (1990) AA, **235**, 114

Platais, I. (1990). In Nordic-Baltic Astronomy Meeting, eds. C.-I. Lagerkvist, D. Kiselman and M. Lindgren, 245. Uppsala, Sweden.

Scalo, J. M. (1986). Fundamentals of Cosmic Physics, **11**, 1

Taff, L.G. (1974). AJ, **79**, 1280

Terlevich, E. (1987) MNRAS, **224**, 193

van den Bergh, S. & Sher, D. (1960) Publ. David Dunlap Obs., **2**, 203.

Studies of nearby OB associations

By P. T. DE ZEEUW[1], A. G. A. BROWN[1]
AND W. VERSCHUEREN[1,2]

[1] Sterrewacht Leiden, P.O. Box 9513, 2300 RA Leiden, The Netherlands
[2] Astrophysics Research Group, Univ. of Antwerp (RUCA), Belgium

A description is given of a long–term project to study the formation, structure, and evolution of nearby OB associations and related star forming regions. The status of work on proper motions, radial velocities, multicolour photometry, and on the properties of the interstellar medium in and around the nearby assocations is discussed, with particular attention to the associations Sco OB2 and Ori OB1.

1. Project SPECTER

The study of OB associations is motivated primarily by the fact that they form the fossil record of star formation processes in giant molecular clouds. The stellar content and the internal kinematics of associations provide valuable information on the initial mass function, the local star formation rate and efficiency, the velocity distribution of young stars as a function of mass and position in the association, differential age effects between subgroups in an association, the characteristics of the binary population, the role of run–away stars, and the interaction between stars and the interstellar medium (Blaauw 1964, 1991).

The study of associations has long been hampered by a lack of accurate knowledge of the stellar content and the internal kinematics. For this reason, project SPECTER was started at Leiden Observatory in 1982, prompted by the HIPPARCOS mission. SPECTER is a collaboration between the observatories of Leiden, Brussels, and Antwerp, and has as main aim a study of the OB associations within 1 kpc from the Sun. The HIPPARCOS proper motions and parallaxes will become available by 1996, and in anticipation thereof a number of ground–based studies are being carried out. These include a photometric study of all the southern associations in the Walraven system, determination of accurate radial and rotational velocities for the early–type stars in some of the nearby associations, and a systematic study of the dust and gas in and around the associations. Here we give a brief progress report on the status of the SPECTER project.

2. Proper motions

The nearby OB associations generally cover many tens to hundreds of square degrees on the sky. Membership determination based on colour–magnitude diagrams suffers from the large distance spread within an association, especially for the nearest associations. The one–dimensional velocity dispersions in OB associations are of the order of 2 km/s, so that proper motions and radial velocities can be used to detect the common space motion of the stars, and hence establish membership. Measurement of the individual internal motions requires very high precision.

Much of the proper motion work on OB associations was summarized by Blaauw (1991). A number of studies based on fundamental and meridian catalogs were done for three subgroups of the nearby association Sco OB2, at 160 pc, which are usually referred to as Upper–Scorpius (US), Upper–Centaurus Lupus (UCL) and Lower–Centaurus Crux

Name	D(pc)	ℓ_-	ℓ_+	b_-	b_+	N
Run–Away		all		sky		162
Cas–Tau	140	all		sky		49
RCr A	150	355	5	−23	−13	165
Sco OB2 1	170	328	4	−20	8	1989
Sco OB2 2 (US)	170	336	4	6	33	1198
Sco OB2 3 (UCL)	170	312	338	3	32	1469
Sco OB2 4 (LCC)	160	292	314	−12	17	1489
Sco OB2 5	170	273	293	−21	6	1212
Per OB3	170	140	155	−11	−3	420
Per OB2	360	154	164	−22	−12	155
Ori OB1	380	196	217	−27	−12	810
Vel OB2	450	255	270	−20	5	641
Lac OB1	530	94	107	−20	−7	410
Mon OB1	720	201	206	−4	4	62
Cep OB3	720	108	113	1	7	105
Sct OB2	730	18	27	−4	3	55
CMa OB1	760	222	246	−16	−3	452
Cep OB2	790	96	108	−1	12	378
Cyg OB7	790	84	96	−5	9	204
Cep OB4	850	116	121	2	7	28
Cam OB1	1000	130	153	−3	8	424
Cyg OB4	1000	81	85	−9	−6	11

TABLE 1. The nearby OB associations: candidate member stars in the HIPPARCOS Input Catalogue

(LCC). This resulted in firm membership assignment for 77 stars with spectral type earlier than ∼B5, and 58 probable members down to B9 (Blaauw 1946, 1964; Bertiau 1958). Blaauw (1978) reanalyzed the motions of the 24 brightest members of Upper-Scorpius, and derived a kinematic age (defined as the time elapsed since the stars occupied the minimum area on the sky) of about 4 million years. The remaining two subgroups of Sco OB2 have received remarkably little attention.

Lesh (1968) investigated subgroup 1a of Ori OB1, at ∼380 pc. Smart (1993) employed an overlapping plate technique to establish membership and to study possible expansion of subgroups 1a and 1b. Numerous studies have been done of subgroup 1d, the very young Trapezium cluster. Its extent on the sky is sufficiently small to allow classical plate studies (e.g., McNamara & Huels 1983; van Altena et al. 1988; Jones & Walker 1988).

A third example is the nearby Cassiopeia–Taurus group identified and studied by Blaauw (1956). The 49 proposed members lie in an area of 140 by 100 degrees on the sky, and may well form an old, nearly dissolved, association at ∼140 pc. Small global systematic errors in fundamental catalogs become important when establishing membership via common motion of stars spread over such a large area of the sky.

The above examples illustrate that ground-based proper motion studies almost invariably have been confined to modest samples of bright stars in fundamental catalogs, or to small areas covered by a single photographic plate. As a result, membership for many associations has been determined unambiguously only for spectral types earlier than B5. Although this covers the important main–sequence turnoff region, it is generally not known what the lower mass limit is of the stars that belong to the association. This unfortunate and unsatisfactory state of affairs should improve rapidly in the next few years. Much can be learned from the PPM Catalog (Röser & Bastian 1989; Bastian et al. 1991), which provides proper motions with an accuracy of ~ 3 mas/yr.

The SPECTER proposal to the HIPPARCOS mission resulted in the inclusion in the final Input Catalog of 11888 candidate member stars of spectral type earlier than F8 in the OB associations within 1 kpc from the Sun. Table 1 lists the associations, gives their distance D, their extent on the sky in galactic coordinates (ℓ, b), and the number N of candidate members per association or subgroup. The HIPPARCOS proper motions will have an accuracy of $\lesssim 2$ mas/yr, and will provide vastly improved membership lists for all associations out to at least 600 pc. Measurement of parallaxes and internal motions will be significant only in the nearest associations (<200 pc). HIPPARCOS should also settle whether the Cas–Tau stars indeed form a physical group. If so, it should be possible to identify more members, and to investigate the possible relation with the α Persei cluster.

3. Radial velocities

SPECTER will supplement the proper motions of stars in the nearby OB associations with ground–based high precision radial velocities. This is being done in the context of an ESO Key Program (Hensberge et al. 1990). To date, multiple high resolution (0.25 Å) high S/N (between 30 and 200) echelle spectra (3700 - 5000 Å) were obtained with the CASPEC and ECHELEC spectrographs at La Silla for ~ 120 established and probable B–type members of Sco OB2. Also included in the radial velocity program are two more distant very young star clusters (NGC 2244 in Mon OB2, and Tr 14) for which the radial velocities will be the only accurate kinematical information available. The spectra also provide accurate rotational velocities, while repeated observations allow the detection of small and intermediate period binaries. Future plans include Ori OB1 and CMa OB1.

To study internal motions, or at least to derive the internal velocity dispersion, requires a precision better than 2 km/s in the radial velocity (Mathieu 1986). While this is routine for late type stars, the determination of such high precision radial velocities for early type stars poses a number of specific problems which are discussed in detail by Verschueren (1991). Here we briefly mention some of these.

Optimal observing (centering, guiding, slit angle, calibration) and reduction procedures are needed to reach the limit of the telescope-instrument-detector capabilities. By adding various improvements and new algorithms to the existing MIDAS echelle reduction package we are, e.g., able to obtain an rms wavelength calibration accuracy better than 0.5 km/s (Hensberge & Verschueren 1989; Verschueren & Hensberge 1990; David, Hensberge & Verschueren 1992).

The line density in early type spectra is 10–100 times smaller than in late type stars, while typical rotational velocities are 100–200 km/s. The line shapes and strengths depend sensitively on spectral type. Some optical absorption lines are formed at the base of the stellar wind and are influenced by atmospheric motions at the level of a few km/s. Therefore, i) close object–template matching, (ii) specific cross correlation techniques, and (iii) a rigorous line selection are crucial for radial velocity measurements (Verschueren, David & Hensberge 1992)

No early type radial velocity standard system has been established, i.e., there is not yet a set of early type stars with accurately known radial velocities on a consistent zero point throughout the spectral sequence. Several attempts to improve this situation are

Association	Prty 1	Prty 2	Age [Myr]
Sco OB2 1	30	739	
Sco OB2 2 (US)	130	764	6±1
Sco OB2 3 (UCL)	150	524	13±1
Sco OB2 4 (LCC)	41	634	11±1
Sco OB2 5	20	649	
Ori OB1 a	111	277	11.4±1.9
Ori OB1 b	95	51	1.7±1.1
Ori OB1 c	153	285	4.6±2.0
Ori OB1 d	14	–	<1.0
Mon OB1	9	51	
Sct OB2	46	–	
CMa OB1	109	378	

TABLE 2. Walraven photometry

being carried out (Fekel 1985; Latham & Stefanik 1991). We are frequently observing a set of 54 candidate early type radial velocity standards, selected because of their reported velocity constancy and because of their apparent recognition as candidate standards; we will determine their relative velocities by cross correlating stars of neighbouring spectral type in selected spectral regions—in order to avoid zero point sliding along the spectral sequence–and will use them as a template grid.

A preliminary analysis of the projected rotational velocities $v \sin i$ of 26 O4–B8 stars in NGC 2244 revealed an unexpected large fraction of slow rotators: 20% of the stars have $v \sin i < 30$ km/s (Verschueren 1991). Extension of the sample together with repeated radial velocity measurements will allow us to investigate a possible correlation with binarity and/or run–away character (Blaauw 1993). We found 13 stars with Hα emission in this cluster by means of low resolution spectroscopy. The $v \sin i$ values obtained for five of these stars lie between 100 and 300 km/s. Three of the thirteen stars are probably pre–main–sequence Herbig Be objects.

Measurement of radial velocities for spectral types later than A is possible with available cross–correlation techniques such as CORAVEL, but is practical for the nearby OB associations only after member selection by analysis of the PPM Catalog, or the HIPPARCOS results. A kinematically unbiased radial velocity program for these associations requires a dedicated ground–based effort, due to the large number of candidate members (Mayor et al. 1989).

4. Photometry

Even though the colour–magnitude diagram for the nearby associations is a poor guide to membership, multicolour photometry in combination with theoretical stellar models provides much information on the stellar content of an OB association.

Photometric observations were carried out between 1982 and 1989 with the 91 cm Dutch telescope at ESO, in the VBLUW Walraven system (Lub & Pel 1977). All O and B stars inside the association boundaries of Table 1 were selected from the CSI (Jung & Bischoff 1971), while A and F stars were selected within certain magnitude limits, so as to exclude foreground stars. The resulting stars were divided into two categories: priority

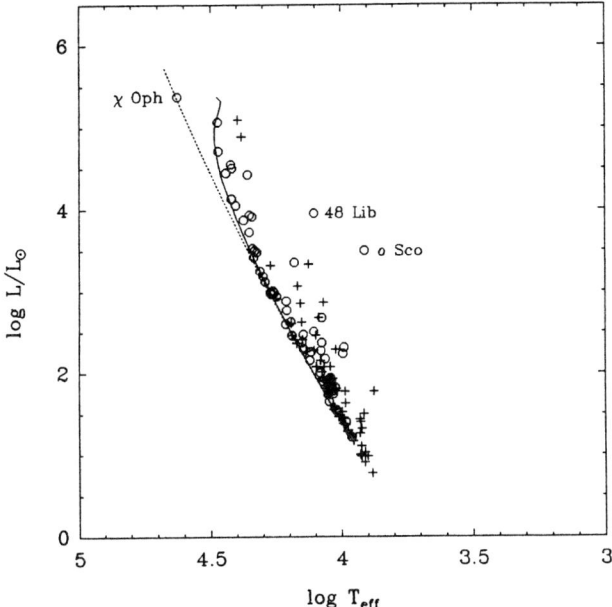

FIGURE 1. Hertzsprung–Russell Diagram for the subgroup Upper–Scorpius of Sco OB2. Circles denote the established members, and plusses indicate probable members. The three members with an anomalous location are stars with a peculiar spectrum (χ Oph and 48 Lib), and the giant o Sco. The solid line is the isochrone of 5.5 Myr (de Geus et al. 1989).

1 stars which are a) stars of spectral type O and B and/or b) stars that are established or probable members of the associations based on earlier proper motion, radial velocity and photometric work by a variety of authors. The remaining stars were given priority 2. Table 2 summarizes the number of VBLUW observations in the various southern associations obtained for a total of 5260 priority 1 and 2 stars (de Geus, Lub & van der Grift 1990).

De Geus, de Zeeuw & Lub (1989) analyzed the VBLUW data as well as published Strömgren uvbyβ photometry of the priority 1 stars in Upper–Scorpius, Upper–Centaurus Lupus, and Lower–Centaurus Crux. Model atmospheres and empirical calibrations were used to determine effective temperature T_{eff}, surface gravity $\log g$, and absolute bolometric magnitude M_{bol} of the stars, and to derive ages for the subgroups by isochrone fitting in the Hertzsprung–Russell diagram (Figure 1) and Table 2. It was found that UCL is the oldest subgroup, and US the youngest (cf. de Zeeuw & Brand 1985). The derived extinctions were compared with the IRAS 100μm emission, and it was shown that the Ophiuchus dark clouds, where star formation presently occurs, are on the near side of Upper–Scorpius, at ~ 125 pc.

Brown, de Geus & de Zeeuw (1993) carried out a similar study for the four subgroups of Orion OB 1, for both priority 1 and 2 stars. The distance to Orion is sufficient for photometric membership determination to be useful (see the classic study using UBV and uvbyβ photometry by Warren & Hesser 1977, 1978), and a number of new probable members were found. Brown et al. derived ages for the subgroups by isochrone fitting in the ($\log T_{\text{eff}}, \log g$) diagram, and they used Monte Carlo simulations of synthetic associations to obtain realistic error estimates on the ages (Table 2). The resulting sequence of increasing age in Ori OB1 is: 1d, 1c, 1b, 1a. These authors also derived initial mass functions, and showed that a single power law of the form $\xi(\log M) = AM^{-B}$ with $B = 1.7 \pm 0.2$ is consistent with the observed present day mass function in subgroups 1a, 1b and 1c. Analysis of the extinctions and the $100\mu m$ emission showed that the near edge of the Orion A and B clouds is located at ~ 320 pc while the far edge is at ~ 500 pc.

Similar studies of the remaining associations are planned. The northern associations will be investigated by use of published uvbyβ and Geneva photometry. Results on NGC 2244 can be found in Verschueren (1991).

5. Run–away stars

The importance of run–away stars was discussed at length by Blaauw (1991). Accurate proper motions, radial velocities, and in some cases distances, will help define the path of the known run–aways, so their paths can be traced backwards in time to find the place of origin. Analysis of the HIPPARCOS results will also provide new run–away stars. A sample of run-away candidates, selected on basis of isolated position and/or large radial velocity, is included in the SPECTER project (Table 1).

6. Interstellar medium

The effects of an OB association on its surrounding interstellar medium have been studied by numerous authors (eg. Tenorio–Tagle & Bodenheimer 1988; Elmegreen 1992). Most of these studies have concentrated on associations located at large distances, or on regions of active star formation, such as the Orion Molecular Clouds and the Ophiuchus clouds. Relatively little work has been done on the connection between the stellar content of young nearby associations and the surrounding interstellar medium.

De Geus (1992) derived initial mass functions for the three subgroups of Sco OB2 mentioned earlier, and he estimated the number of (massive) stars that have exploded already (the evaporation function, Blaauw 1984). He then calculated the energy and momentum input of the various stellar wind phases, and of the supernova explosions, and showed that these are consistent with the observed morphology and kinematics of the system of HI and CO shells that surround Sco OB2 (Cappa de Nicolau & Pöppel 1986; de Geus, Bronfman & Thaddeus 1990; de Geus & Burton 1991). Brown et al. (1993) carried out a similar analysis for Ori OB1, and showed that the hot cavity inside the large HI shell around the association (Reynolds & Ogden 1979) may blow out of the galactic HI layer in the near future.

7. Concluding remarks

The PPM Catalog, and especially the HIPPARCOS proper motions and parallaxes, will allow a much improved census of the stellar content of the nearby OB associations. Measurement of the internal velocity dispersions and expanding motions will require high precision, and is likely to be possible only for the associations closer than 200 pc.

Radial velocity measurements for B and A type stars can be done to a precision of ~ 2 km/s, but this requires a lot of hard work to eliminate systematic errors. An unbiased member selection based only on radial velocities is not practical at present. However, measurement of the radial velocity for the proper motion members is extremely useful.

Analysis of available photometry (and spectra) will provide accurate ages, initial mass functions, and the energy and momentum input into the interstellar medium, and will allow a detailed investigation of the influence of the young stellar group on the surrounding distribution of gas and dust. The Leiden–Dwingeloo HI survey (Hartmann & Burton 1993) will delineate the HI distribution and kinematics with unprecedented accuracy over the entire sky north of declination –30°. It will be interesting to use it to study the known shells around Sco OB2 and Ori OB1, and to look for similar structures around the other nearby associations.

On the theoretical side there is a strong need for careful N–body models of expanding associations. The total number of stars involved is not very large, so direct summation methods can be employed. However, a realistic initial mass function must be used, and the role of the gas must be included. Such models then need to be "observed" to investigate how well one can determine intrinsic properties of OB associations from observations.

Acknowledgements

This work was supported in part by ESA, NWO and the LKBF. The authors have benefitted from stimulating discussions with A. Blaauw, M. David, E. de Geus and H. Hensberge.

REFERENCES

Bertiau, F.C., 1958. ApJ, **128**, 533.

Bastian, U., Röser, S., Nesterov, V.V., Polozhentsev, D.D., Potter, Kh.I., Wielen, R., Yagudin, L.I., Yatskiv. YA.S., 1991. A&AS, **87**, 159.

Blaauw, A., 1946. Publ. Kapteyn. Obs., **52**, 85.

Blaauw, A., 1956. ApJ, **123**, 408.

Blaauw, A., 1964. ARA&A, **2**, 236.

Blaauw, A., 1978. In *Problems in Physics & Evolution of the Universe*, ed. L. Mirzoyan (Yerevan, USSR), p. 103.

Blaauw, A., 1984. In *Birth and Evolution of Massive Stars & Stellar Groups*, eds H. van Woerden & W. Boland (Dordrecht: Reidel), p. 211.

Blaauw, A., 1991. In *The Physics of Star Formation and Early Stellar Evolution*, eds C.J. Lada & N.D. Kylafis, NATO ASI Series C, **342**, 125.

Blaauw, A., 1993. In *Massive Stars: Their Lives in the Interstellar Medium*, eds J.P. Cassinelli & E.B. Churchwell, ASP Conf. Series, **35**, 207.

Brown, A.G.A., de Geus, E.J., & de Zeeuw, P.T., 1993. Submitted to A&A.

Cappa de Nicolau, C.E., & Pöppel, W.G.L., 1986. A&A, **164**, 274.

David, M., Hensberge, H., & Verschueren, W., 1992. In Proc. of 4th ESO/ST–ECF Data Analysis Workshop, (Garching: ESO), p. 163.

de Geus, E.J., 1992. A&A, **262**, 258.

de Geus, E.J., Bronfman, L., & Thaddeus, P., 1990. A&A, **231**, 137.

de Geus, E.J., & Burton, W.B., 1991. A&A, **245**, 559.

de Geus, E.J., de Zeeuw, P.T., & Lub, J., 1989. A&A, **216**, 44.

de Geus, E.J., Lub, J., & van der Grift, E., 1990. A&AS, **85**, 915.

de Zeeuw, P.T., & Brand, J., 1985. In *Birth and Evolution of Massive Stars & Stellar Groups*, eds H. van Woerden & W. Boland (Dordrecht: Reidel), p. 95.

Elmegreen, B.G., 1992. In *Star Formation in Stellar Systems*, eds G. Tenorio-Tagle, M. Prieto, & F. Sanchez (Cambridge Univ. Press), p. 383.

Fekel, F.C., 1985. In *IAU Coll. 88, Stellar Radial Velocities*, eds A.G. Davis Phillip & D.W. Latham (Schenectady: L. Davis Press), p. 335.

Hartmann, L.D., & Burton, W.B., 1993. In preparation.

Hensberge, H., & Verschueren, W., 1989. The Messenger, **58**, 51.

Hensberge, H., van Dessel, E., Burger, M., de Zeeuw, P.T., Lub, J., Le Poole, R., Verschueren, W., David, M., Theuns, T., de Loore, C., de Geus, E., Mathieu, R.D., & Blaauw, A., 1990. The Messenger, **61**, 20.

Jones, B.F., & Walker, M.F., 1988. AJ, **95**, 1755.

Jung, J., & Bischoff, M., 1971. Bull. d. Inf. Centre Donn. Stell., **2**, 8.

Latham, D.W., & Stefanik, R.P., 1991. Trans. IAU, Vol. XXIB, p. 269.

Lesh, J.R., 1968. ApJ, **152**, 905.

Lub, J., & Pel, J.W., 1977. A&A, **54**, 137.

Mathieu, R.D., 1986. Highlights of Astronomy, **7**, 481.

Mayor, M., Duquennoy, A., Grenon, M., Turon, C., Crifo, F., Imbert, M., Maurice, E., Prevot, L., Andersen, J., Nordstrom, B., & Lindgren, H., 1989. The Messenger, **56**, 12.

McNamara, B.J., & Huels, S., 1983. A&AS, **54**, 221.

Reynolds, R.J., & Ogden, P.M., 1979. ApJ, **229**, 942.

Röser, S., Bastian, U., 1989. PPM–positions and proper motions of 181731 stars north of −2.5 degrees declination. Astron. Rech. Inst. Heidelberg.

Smart, R., 1993. PhD Thesis, Univ. of Florida, Gainesville.

Tenorio–Tagle, G., & Bodenheimer, P., 1988. ARA&A, **26**, 145.

van Altena, W., Lee, J.T., Lee, J.F., Lu, P., Upgren, A., 1988. AJ, **95**, 1744.

Verschueren, W., 1991. PhD Thesis, Free University of Brussels.

Verschueren, W., Hensberge, H., 1990. A&A, **240**, 216.

Verschueren, W., David, M., & Hensberge, H., 1992. Mem. Soc. Astr. It., **62**, 953.

Warren, W., & Hesser, J.E., 1977. ApJS, **34**, 115 & 207.

Warren, W., & Hesser, J.E., 1978. ApJS, **36**, 497.

Measuring velocity dispersions in Open Clusters

By F. VAN LEEUWEN

Royal Greenwich Observatory, Madingley Road, Cambridge CB3 0EZ, U.K.

A review is presented of various aspects associated with the study of open cluster kinematics using proper motions of sub milli-arcsecond accuracy. The derivation of the proper motions and the definition of the system of proper motions using photographic plates is described as well as the effect of the relative motion of the cluster with respect to the Sun and the related problem of the distribution of stars in the cluster. The internal proper motion dispersions are derived for one particular example, the Pleiades cluster, and compared with numerical simulations.

1. Proper motion studies of open clusters

The only means currently available to obtain proper motions with relative accuracies at the level of at least 0.2 mas/yr is photographic plates, and in particular sequences of plates taken with the same instrument over a period of 70 to 100 years. Such sequences exist for some of the nearby open clusters: the Pleiades and Praesepe, where the projected cluster centre covers the area of a typical astrographic plate (16×16 cm, $2° \times 2°$ at $60''$/mm, and $1°\!.3 \times 1°\!.3$ at $40''$/mm). In particular astrographs for which the optics were provided by the brothers Henry in Paris around the turn of the century can produce plate material of relatively high quality (e.g. Paris, St.Petersburg, Shanghai, Leiden astrographs).

Photographic plates can be measured on a measuring device producing a smaller measuring error than the intrinsic inaccuracy of the positions of the stellar images on the plate (e.g. MAMA at Meudon, Paris, ASTROSCAN at Leiden Observatory). The measured star positions on a plate then only reflect the following information:

 (i) the position and orientation of the plate in the measuring machine,
 (ii) the pointing position of the telescope,
 (iii) the hour-angle and zenith distance of the exposure (differential refraction),
 (iv) the exposure time and guiding accuracy,
 (v) the position of the optical axis of the telescope and the tangential point of the plate
 (vi) and finally the positions of the stars on the sky at the epoch and time of the year of the exposure.

In order to recover the last item, the first five need to be carefully removed. On a photographic plate, however, items (i) to (iv) and the proper motions of stars represent the same effects on positions:

 ((a)) In comparing plates of different epochs, items (i), (ii) and the proper motion zero point can not be distinguished,
 ((b)) item (iii) and the rotation and expansion of the proper motion system can not be distinguished to sufficient accuracy,
 ((c)) item (iv) and relations between magnitudes and proper motions are correlated.

In all cases the problem is in the correlations between proper motions and the linear deformation of the image of the sky on the photographic plate (or CCD-frame). Non-linear deformations, such as due to variations in the tangential points of plates are much less of a problem, as these can not easily be correlated with proper motions.

Items (i) to (v) are removed from the measured positions on a plate using plate transformations determined by means of a least squares solution, fitting the measured positions to a reference frame. In the case of single field open cluster studies the reference field can be defined by the actual measurements, which will provide in general an internally more consistent and much larger number of reference points for transformations than a star catalogue. The determination of the plate transformation coefficients is done in four steps, where each step defines a new well defined reference frame, the last of which is the final catalogue of relative proper motions and positions.

To avoid comparing noisy positions with other noisy positions, a first reference frame is built up from a set of plates of the same epoch, preferably taken with the same plateholder for consistency of tangential points. They are corrected for the effects of differential refraction and hour angle which is given by (1.1), describing the transformation of a measurement of (x, y) under a parallactic angle ϕ and zenith distance z and with an atmospheric refraction index $n = n' + 1$ to (x', y') coordinates for a measurement at zenith (Van Leeuwen, 1983).

$$\begin{pmatrix} x' \\ y' \end{pmatrix} = \begin{pmatrix} 1 + n'(\cos^2 \phi + \frac{\sin^2 \phi}{\cos^2 z}) & -n' \tan^2 z \sin \phi \cos \phi \\ -n' \tan^2 z \sin \phi \cos \phi & 1 + n'(\sin^2 \phi + \frac{\cos^2 \phi}{\cos^2 z}) \end{pmatrix} \times \begin{pmatrix} x \\ y \end{pmatrix} \quad (1.1)$$

where ϕ and z are related to the siderial time of the exposure (T_S), the Right Ascension and Declination of the plate centre (α_p, δ_p), and the geographic latitude (b) of the telescope used through (1.2) and (1.3):

$$\cos z = \sin b \sin \delta_p + \cos b \cos \delta_p \cos(T_S - \alpha_p) \quad (1.2)$$

$$\sin \phi = \sin(T_S - \alpha_p) \cos b / \sin z \quad (1.3)$$

Subsequently, the new (x, y) positions for stars in common to all selected plates are averaged in a simple linear way. Stars not in common to all plates are not included in this definition of the first reference frame, which is now used to reduce all of the plates of the same epoch, solving for linear coefficients and tilt terms, as well as magnitude terms. The distribution of coefficients of the magnitude terms has to be investigated, and it should in general be avoided to include outliers in this distribution to form part of the the definition of the first reference frame. Using all positions now, this set of plates defines the second reference frame.

In the next step additional plates of different epoch are selected, and on those plates a set of stars in common to all these plates and the second reference frame. In the least squares reductions of the different epoch plates, without the knowledge of proper motions, the proper motion displacements take the role of the residuals, and it is the sum of squares of these replacements which is forced to a minimum. Including or omitting one or more stars at this level will thus create a significantly different transformation, and (x, y) coordinates defined in a different reference system. Transformations based on the selected common stars can, however, be applied to all objects on a plate. Proper motion solutions for the selected stars are all checked for outliers, and stars with outliers or plates causing many outliers should be removed from this reduction step.

The result at this stage is a set of preliminary proper motions, defining reference system three, in which for the group of stars used in the transformation determination equations (1.4) hold:

$$\begin{array}{lll} \sum_{i=1}^{N} \mu_{x,i} = 0 & \sum_{i=1}^{N} \mu_{x,i} x_i = 0 & \sum_{i=1}^{N} \mu_{x,i} y_i = 0 \\ \sum_{i=1}^{N} \mu_{y,i} = 0 & \sum_{i=1}^{N} \mu_{y,i} x_i = 0 & \sum_{i=1}^{N} \mu_{y,i} y_i = 0 \end{array} \quad (1.4)$$

Similarly, it is possible to re-define the proper motion system by applying a linear transformation as in (1.5) to all proper motions such that for a selected group of stars the conditions in (1.4) hold.

$$\begin{aligned} \mu_{x,i}' &= \mu_{x,i} + \mathrm{a} + \mathrm{b}x_i + \mathrm{c}y_i \\ \mu_{y,i}' &= \mu_{y,i} + \mathrm{d} + \mathrm{e}x_i + \mathrm{f}y_i \end{aligned} \qquad (1.5)$$

In particular, one can force condition (1.4) on a group of stars selected as probable cluster members. Proper motion systems defined in this way are related through transformations (1.5), and can only be properly compared after solving for and applying these transformations. Residual magnitude dependence of the proper motions can be removed in a similar way.

At this stage the reference frame is ready for the final reduction of all plate material, using positions and proper motions from reference frame three, to create a well defined positional and proper motion system. In a study of open cluster kinematics, the selection of cluster members at this stage should be obvious, allowing non-members only a very small chance to penetrate the sample. This implies that only clusters with a proper motion separating it well from the background can normally be considered for such studies. The same condition, however, also complicates the interpretation of the data, as is described in the next section. From here on it will be assumed that cluster members have been selected with high probability (but see final remarks), and that for the proper motions of these selected stars conditions (1.4) will apply.

2. The relative secular parallax

The influence of the relative secular parallax on the proper motions in an open cluster was described by (Binnendijk, 1946) in his study of the Pleiades proper motions. Due to the relative motion of the cluster with respect to the sun, every star in the cluster will acquire a proper motion relative to the mean cluster proper motion, depending on the position of that star within the cluster. Figure 1 shows the projections of the actual position and motion of a cluster star to observed proper motion and radial velocity and projected position. Defining the cluster proper motion as $(\mu_{x,c}, \mu_{y,c})$ and using the definitions in Figure 1, we find for the additional proper motion $(\mu_{x,p}, \mu_{y,p})$ of a cluster member:

$$\begin{aligned} \mu_{x,p} &= \mu_{x,c}(R\cos\theta)/D \\ \mu_{y,p} &= \mu_{y,c}(R\cos\theta)/D \end{aligned} \qquad (2.6)$$

where D is the distance between the sun and the cluster, in the same units as R. The effect of (2.6) on the dispersion of the proper motions can be measured from the proper motions and estimated from the a distribution function of the stars in the cluster. Both suffer, however, from low number statistics, with the additional complication that given the low number of stars in an open cluster, the position of the cluster centre is not well defined.

The relative secular parallax thus causes an extra proper motion dispersion in one direction, the direction of the proper motion of the cluster. Figure 2 shows this effect for Pleiades cluster stars. Around the projected centre of the Pleiades, which itself has an uncertainty in position of 0.5 pc, the dispersion in proper motion due to the relative secular parallax is 0.6 mas/yr as predicted from the spatial distribution of its members. Proper motions data show a dispersion of 0.5 mas/yr (Van Leeuwen, 1983). Further away from the centre the predicted dispersion increases to 1.5 mas/yr at a projected distance of $2°.5$.

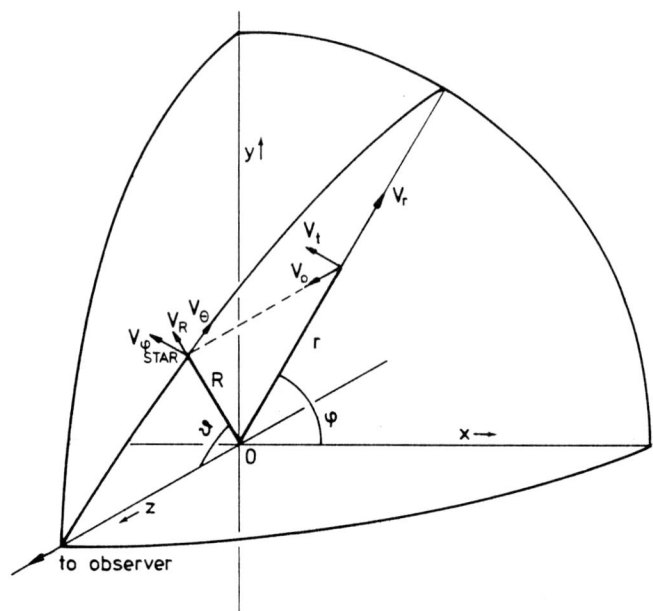

FIGURE 1. Projections of position and velocity components for a cluster star.

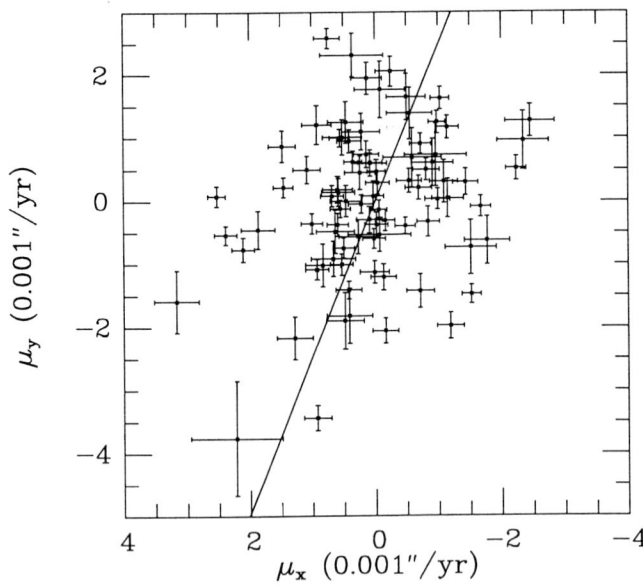

FIGURE 2. Proper motions for Pleiades cluster members and the direction of the cluster proper motion.

3. The convergent point

The radial velocity of the cluster will also reflect in the proper motions due to projection effects. This is the well known convergent point problem used to determine the distance of the Hyades. Here, the cluster proper motion and radial velocity are split, and the radial velocity is reflected only as a contraction (when moving away) or expansion (when moving towards us) of the cluster. The expansion rate is given by (3.7):

$$E = -V_{rad}/D \tag{3.7}$$

The average radial velocity of the Pleiades stars is 5.85 km/s (Rosvick et al, 1992), the distance is 130 pc (Van Leeuwen, 1983), from which is derived : $E = -4.6 \times 10^{-8}$ parts per year. As we made the proper motions of the cluster fulfil (1.4), this effect is no longer present in these proper motions. However, this elimination will be reflected in the proper motions of the field stars, which will now show an expansion of the same amount. The problem is, that the dispersion in proper motions of field stars makes it very difficult to observe any such effect. In the Pleiades field, (Van Leeuwen, 1983) showed that the error on the expansion rate for a uniform density of background stars ρ_b, a dispersion in the proper motions of these stars of $\sigma\mu_b$, and a field with radius r is given by:

$$\delta E = \sigma\mu_b \sqrt{\frac{2}{\pi\rho_b}} r^{-2} \tag{3.8}$$

In the Pleiades this is not a viable way of determining the cluster distance, contrary to the Hyades, where E is 20 times larger, and the field covered by the cluster 16 times.

4. The proper motion dispersions

The dynamics of an open cluster is largely determined by the distribution of the radial and transverse velocity dispersions as a function of distance to the cluster centre. Together they reflect the energy and angular momentum distribution of stars in the cluster. Figure 3 shows the three velocity components within the cluster: the tangential components V_ϕ and V_θ, and the radial component V_R. They are projected and observed as in (4.9) to (4.11) :

$$\mu_t = V_t/D = V_\phi/D \tag{4.9}$$

$$\mu_r = V_r/D = (V_R \sin\theta + V_\theta \cos\theta)/D \tag{4.10}$$

$$V_{rad} = V_o = V_R \cos\theta - V_\theta \sin\theta \tag{4.11}$$

Individual three dimensional positions of stars in a cluster are not yet available, and therefore the only observable quantity is the spread in proper motions and radial velocities as a function of projected distance to the cluster centre. If the cluster is assumed to be spherically symmetric, V_ϕ and V_θ will represent the same distribution, allowing in principle to derive all kinematic information on the cluster from the proper motions.

The observed transverse and radial components of the proper motions are related to μ_t and μ_r and the components of the relative secular parallax through (4.12) and (4.13) :

$$\mu_{t,obs} = \mu_t + \mu_p \sin(\phi - \phi_0) \tag{4.12}$$

$$\mu_{r,obs} = \mu_r + \mu_p \cos(\phi - \phi_0) \tag{4.13}$$

where μ_p is the relative secular parallax in the direction (ϕ_0) of the proper motion of the cluster. When we assume that the cluster is spherically symmetric, then the distribution function of μ_p is a function of the projected distance from the cluster centre only. Thus,

ring	stars	$\sigma\mu_r$	$\sigma\mu_t$	$\epsilon\mu$
0.0–0.5 pc	11	0.98 ± 0.26	1.03 ± 0.26	0.21
0.5–1.0 pc	22	0.98 ± 0.20	0.86 ± 0.20	0.19
1.0–1.5 pc	34	1.04 ± 0.15	0.69 ± 0.19	0.23

TABLE 1. Internal proper motion dispersions in the Pleiades in mas/yr

by collecting $\mu_{t,obs}$ and $\mu_{r,obs}$ in rings around the cluster centre, the effect of μ_p can be largely eliminated. The observed dispersions then give the following relations:

$$\sigma\mu_{t,obs} = \sqrt{(\sigma\mu_t)^2 + 0.5(\sigma\mu_{p,r})^2 + (\epsilon\mu)^2} \qquad (4.14)$$

$$\sigma\mu_{r,obs} = \sqrt{(\sigma\mu_r)^2 + 0.5(\sigma\mu_{p,r})^2 + (\epsilon\mu)^2} \qquad (4.15)$$

where $\epsilon\mu$ is the estimated measuring error of the proper motions. Using the data presented by (Vasilevskis et al. , 1979), the data presented in Table 1 were derived by (Van Leeuwen, 1980) for $\sigma\mu_r$, $\sigma\mu_t$ and $\epsilon\mu$. The main problem in interpreting these data in terms of V_R, V_ϕ and V_θ is in the distribution of stars in the cluster. Data like those presented in Table 1 can be obtained for the cluster centre as described above, but similar data for the halo stars is difficult to obtain. Without that information, interpretation of the proper motion dispersions depends on models, as in (Van Leeuwen, 1980), where an internal velocity dispersion in the cluster centre of 0.7 km/s was derived, for an average stellar mass of $2M_\odot$.

Another possibility relating these observations with actual velocity dispersions is provided by numerical simulations. The most realistic of simulations in the past 10 years were done by (Elena Terlevich, 1987), who showed a good agreement between the values of Table 1 and her results.

5. Final remarks

There are a few small complications disturbing the rather idealized picture described above. Numerical simulations (Terlevich, 1987) show that halos of clusters are flattened towards the galactic plane, which to some extent invalidates the density profile derivations and implementations. Stars are predicted and found to be escaping from open clusters mainly through three body interactions, two of the objects usually forming a binary system. This weakens the selection of "true" cluster members. In the Pleiades there are also a few of these stars, for which the proper motion is too large to be a member by around 5 times the observed dispersion for cluster members, but where the photometry puts the star very close to the Pleiades main sequence (see also Van Leeuwen et al, 1986).

Due to mass segregation the value of $\sigma\mu_{sec.par.}$ is not only a function of projected distance to the cluster centre, but also of the magnitudes of stars used, causing the effect to be smaller for the brighter stars, and larger for the fainter stars, with a distribution of values that is not gaussian, but reflecting the density profile of a group of stars in the cluster.

There are usually also complications in the available photographic material: most likely not all plates are taken with the same instrument. In that case it is advisable to reduce those plates taken with one and the same instrument first, define the proper motion and positional reference frame four, and then attempt to find a proper and simple transformation for plates taken with other instruments to this system. Other complications occur, when plates in non–central overlap are used. To make this work, the description of the imaging of the telescope on the photographic plate needs to be determined to a high level of accuracy (significantly better than the intrinsic positional noise of the stellar images on a plate) before data are combined.

The most promising prospect for open cluster research is the proposed ROEMER mission (see Høg and Lindegren, present volume), which will be able to measure accurately distances of many open clusters, detect their halo members, and thus improve the density distribution functions and the cluster centre determinations. The very high accuracy absolute proper motions and parallaxes will enable the reconstruction of three dimensional images of the nearest open clusters and allow for a far more detailed study of open cluster kinematics than is currently possible using photographic plates.

REFERENCES

Binnendijk, L., (1946). A study of stars in the Pleiades region. Annalen van de Sterrewacht te Leiden, **19**, part 2

Høg, E., & Lindegren, L. (1993). The ROEMER mission, present volume

Van Leeuwen, F., (1980) Mass and luminosity function of the Pleiades, in Star Clusters, ed. J.E.Hesser, 157.

Van Leeuwen, F., (1983). The Pleiades., PhD thesis Leiden University.

Van Leeuwen, F., Alphenaar, P. & Brand, J. (1986). A $VBLUW$ survey of the Pleiades cluster, A&AS **65**, 309.

Rosvick, J.M., Mermilliod, J.-C. & Mayor, M., (1992). Investigation of the Pleiades Cluster I. Radial velocities of corona stars. A&A, **255**, 130.

Terlevich, E., (1987). Evolution of N–body open clusters. MNRAS **224**, 224.

Vasilevskis, S., Van Leeuwen, F., Nicholson, W. & Murray, C.A. (1979). Internal motions in the central field of the Pleiades. A&AS **37**, 333.

The mass-luminosity relation from proper motions in Galactic open clusters

By I. V. PETROVSKAYA

Astronomical Institute of St. Petersburg State University, Bibliotechnaya pl. 2, St. Petersburg 198904, Russia

New, extremely precise proper motions of open clusters give the possibility of obtaining the velocities of stars of different luminosities. The interpretation of these velocities in the Hyades cluster is done using the stationary velocity distribution functions which have been found for the stars of different masses as a result of a purely discontinuous evolution scheme.

1. Introduction

In 1969 we first considered the evolution of stellar velocities in open clusters as a purely discontinuous random process (Petrovskaya 1969a,b). Such an approach takes into account not only faint distant interactions between stars but also strong close encounters, and gives the possibility of calculating more accurately the escape rate from the cluster. Then we found the velocity distribution functions which the stars of different masses reach under the action of stellar encounters after the relaxation time (Kaliberda and Petrovskaya 1970, 1971, 1972). The rms velocities q are found from these quasi-stationary distribution functions for the groups of stars with mean stellar masses $2m_0, m_0, 0.5m_0$ and 0, and equal to q equal to $0.947v_0, 1.072v_0, 1.10v_0$ and $1.41v_0$ correspondingly, where m_0 is the mean mass of the stars and v_0 is the rms velocity of the stars in the whole cluster.

The less massive stars have the higher velocities on average. Such a result is obvious because of the tendency towards energy equipartition. But the stellar system has the finite escape velocity, $v_e = 2v_0$, which corresponds to the absorbing screen in the theory of random processes. Therefore the rms velocity of each group is always less than v_e, even for zero mass stars.

2. The Hyades cluster

In the previous paper (Petrovskaya 1993) we used the results of Pels et al. (1975) who obtained the rms velocities 0.357 km/s and 0.403 km/s for two groups of stars in the Hyades cluster with $3^m < V < 10^m$ (group I) and with $V > 9^m$ (group II) repectively, and a rms velocity of v_0=0.386 km/s. Comparing these results with our theoretical results we found $<m_I> = 2.1m_0 = 1.36M_{sun}$ solar masses and $<m_{II}> = 1.1m_0 = 0.61-0.7M_{sun}$. Then we have $m_0 = 0.63M_{sun}$.

The highly accurate proper motions determined by Schwan (1990) for the 44 bright stars of the cluster with $B - V < 0.5^m (V < 7.5^m)$ give for these stars q=0.36 km/s and $m > 2.1 <m>$. Now we deal with the next paper (Schwan 1991) where he gives the proper motions of an additional 101 stars of the Hyades. Because of the presence of less luminous stars, the accuracy of the proper motions is poorer, and we find the new rms velocity for the whole cluster. For the calculation of v_0 we took the stars with $B - V > 1$, $<V> = 10^m.34$. The last value corresponds to the average mass in the cluster. So we find v_0=0.491 km/s. This value as well as the rms velocities, for the groups of stars contains

Group	1(n=45)	2(n=44)	3(n=23)	4(n=20)	5(n=4)
	$B-V < 0.5$	$0.5 \leq B-V \leq 0.7$	$B-V > 0.7$		
q (km/s)	0.472	0.547	0.553	0.201	0.554
q/v_0	0.961	1.11	1.13	0.409	0.554
$<m>/m_0$	1.8	0.5	0.4	2.8	2.7
$<m>/M_{sun}$	1.1	0.3	0.2	1.7	1.7

TABLE 1. The *rms* velocities and mean masses $<m>$ for the different groups of stars in the Hyades denoted in Figure 1. For the whole cluster we found $v_0 = 0.491 \, \mathrm{km/s}$, $m_0 = 0.63 M_{sun}$.

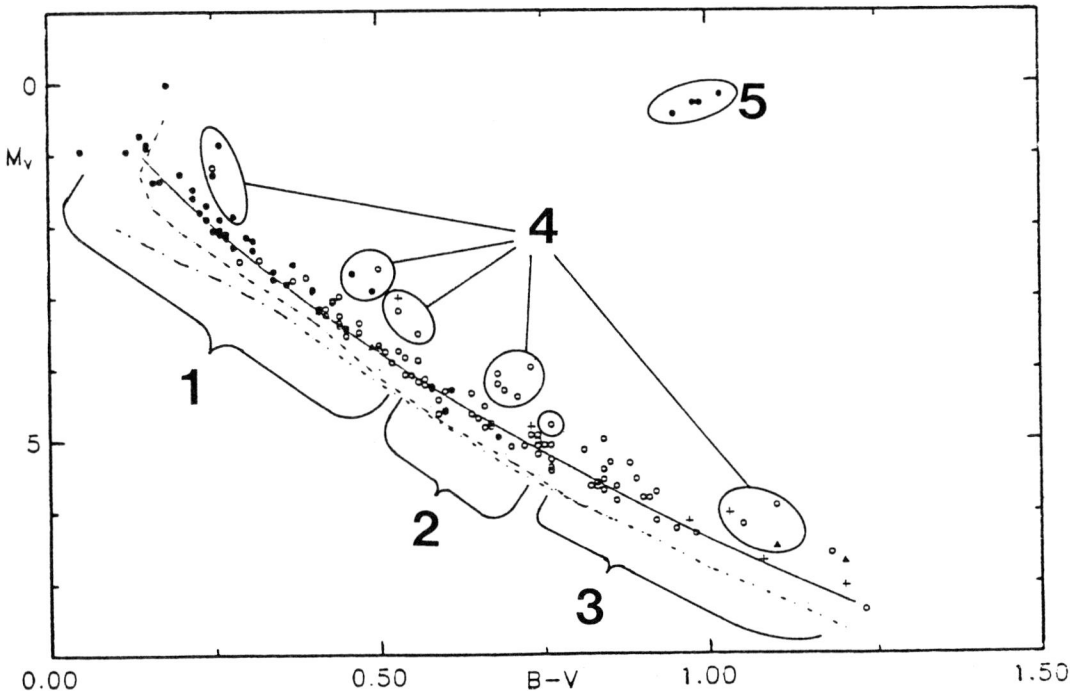

FIGURE 1. The colour-magnitude diagram for 145 Hyades members from the paper of Schwan(1991). Our grouping of the stars is as follows. Group 1, main sequence stars with $B - V < 0.5$ mag; Group 2, main sequence stars with 0.5 mag$\leq B - V \leq 0.7$; Group 3, main sequence stars with $B - V > 0.7$; Group 4, probable double stars above the main sequence; Group 5, stars in the top right hand corner of the diagram.

the errors of the proper motions. The ratio of q to v_0 is more accurate than either of these values.

We separated all stars of the Hyades main sequence (Schwan 1991) into three groups: 1, 2 and 3. We also considered 33 stars above the main sequence, 20 of which are probably double (group 4), and four in an extremely high part of the colour-magnitude diagram (group 5). These groups are presented in Figure 1. We found the relative *rms* velocity q/v_0 for each group, and then the mean mass. The results are presented in Table 1.

Using the observed proper motions we can find only an upper limit to the *rms* velocity, and therefore a lower limit of the mean stellar mass. We must also note that, following Schwan (1990), we used only the proper motion component μ_α as this is more accurate.

REFERENCES

Kaliberda, V.S. & Petrovskaya, I.V. 1970 Astrofizika, **6**, 135.

Kaliberda, V.S. & Petrovskaya, I.V. 1971 Astrofizika, **7**, 663.

Kaliberda, V.S. & Petrovskaya, I.V. 1972 Astrofizika, **8**, 305.

Pels, G., Oort, J.H. & Pels-Kluvers, N.A. 1975 A&A, **43**, 423.

Petrovskaya, I.V. 1969a Sov. AJ, **46**, 824.

Petrovskaya, I.V. 1969b Sov. AJ, **46**, 1220.

Schwan, H. 1990, A&A, **228**, 69.

Schwan, H. 1991, A&A, **243**, 386.

Proper motions of Galactic halo globular clusters and nearby Galactic dwarf spheroidal satellites

By R.-D. SCHOLZ[1] AND M. J. IRWIN[2]

[1]WIP-Astronomie, Universität Potsdam, Sternwarte Babelsberg, An der Sternwarte 16, 1590 Potsdam, Germany

[2]Royal Greenwich Observatory, Madingley Road, Cambridge CB3 0EZ, U.K.

The combination of Schmidt plates and an automatic measuring machine provides a powerful method for investigating the proper motions of both the Galactic halo globular clusters and the nearby Galactic dwarf spheroidal satellites (dSphs). We describe the first measurement of the mean tangential motion of two globular clusters, M3 and M92, directly with respect to a large number of background galaxies by the use of automated scans of Tautenburg Schmidt plates with the APM facility in Cambridge (UK). In both fields five pairs of plates centred on the cluster with epoch differences from 20 to 27 years were available. The proper motion determination was based on a stepwise regression method with 3rd order polynomials and used between 1200 and 2300 galaxies for each pair of plates to define an inertial reference frame. Encouraged by the accuracy level achieved for the mean absolute proper motion of the globular clusters (0.5 mas/yr) we have extended our work to other globulars and to the dSphs in Draco and Ursa Minor, where we combine Palomar and Tautenburg Schmidt plates with a base line of about 35 years. The dSphs pose a more challenging problem since with 0.5 mas/yr baseline accuracy the velocity uncertainty is still ~100 km/s at the ~70 kpc distance of the dSphs. The more accurate relative proper motions in both dSphs obtained by Stetson (1980) and by Cudworth et al. (1986) provide both an external comparison and a further chance to estimate the mean absolute proper motions of the dSphs.

---

1. Introduction

The kinematics of the galactic globular clusters and of local galaxies is of great interest in a discussion of dark matter and of the dynamics of the Milky Way. Webbink (1988) reviewed the efforts in the determination of radial velocities and proper motions of galactic globulars and encouraged new attempts to obtain their absolute proper motions directly with respect to background galaxies and quasars.

Proper motions of numerous globular clusters have been investigated by Cudworth and coworkers by measuring the relative motion of cluster stars with respect to field stars. This required various assumptions concerning the direction of the solar apex and the secular parallaxes of the chosen field stars around each cluster. Not only were the results affected by the choice of solar apex and motion but they were also afflicted by large errors due to the small number of field stars used (~ 50) and their inherent proper motion dispersion. Although a very high accuracy (± 0.2 mas/yr) in the relative proper motion of the cluster stars was achieved, the uncertainty in the absolute proper motion of the whole cluster remained of the order of 2 to 4 mas/yr (eg. Cudworth 1976, 1979). Recently, Cudworth & Hanson (1993) converted the relative proper motions of 14 globular clusters to absolute proper motions using new assumptions concerning the motion of the sun and of the field stars derived from the Lick absolute proper motion program. The accuracy of these re-reduced absolute proper motions lies between 0.5 and 2.0 mas/yr with the exception of M3 for which it remained about 3.0 mas/yr. In an attempt to circumvent the problem of the assumptions about the field stars and the solar motion Brosche et al. (1983, 1985, 1991) used Lick stars with known absolute proper motions as reference stars to connect globular cluster motions to an extragalactic reference frame. Tucholke (1992a,b) derived the proper motion of globular clusters relative to the background of the SMC using whole plate scans with a PDS measuring machine.

On Schmidt plates taken in higher galactic latitudes there are large numbers of background galaxies which define the local inertial reference frame for the determination of absolute proper motions very accurately. Small systematic errors in the absolute proper motion are vital for obtaining meaningful proper motions of very distant groups of stars and star clusters. Here we describe the determination of the absolute proper motion of the galactic globular clusters M3 and M92 based on complete scans of Tautenburg Schmidt plates with the APM facility in Cambridge. These new results have allowed us to obtain the orbits of these clusters in a galactocentric inertial system (Scholz, Odenkirchen & Irwin 1993a,b).

For the nearby dwarf spheroidal galaxies accurate relative proper motions have already been determined by Stetson (1980) for the dSph in Draco and by Cudworth, Olszewski & Schommer (1986) in the Ursa Minor dSph. In the latter paper the authors encouragl spectral observations of blue objects in order to find some QSOs for the definition of an absolute zero proper motion. We propose an alternative way to obtain the absolute proper motions of these dSphs by measuring large numbers of stars and background galaxies on Palomar and Tautenburg Schmidt plates. However, the distances to the dSphs in Draco and UMi are $\sim$70 kpc, seven times larger than that of the globular clusters considered in our former investigation, making it much harder to measure meaningful tangential motions of the dSphs.

2. Observations, measurements and plate matching

For the work on the globular clusters M3 and M92 we selected five pairs of plates in each field, all taken with the Tautenburg Schmidt telescope (main mirror = 2m, correction plate = 1.34m, focal length = 4m) and all with the globular cluster at the plate centre. Each plate covers some $3°\!\!.3 \times 3°\!\!.3$ of sky with a plate scale of $51''\!\!.4$/mm. For the dSphs in Draco and Ursa Minor we have used glass copies of the Palomar Schmidt sky survey plates taken in 1954 and 1956 for the first epoch. On the Sky Survey plates the dSphs are situated near the plate corners, whereas on the second epoch Palomar plates also included in this project the dSphs are close to the plate centre. Third epoch plates centred on the dSphs were obtained with the Tautenburg Schmidt telescope in 1991. Table 1 lists the plate material used for both the globular cluster and dSphs study.

All measurements were done using the Automated Photographic Measuring (APM) facility in Cambridge (UK). The APM has already been extensively used for astrometric work with UK Schmidt, Palomar Schmidt and Tautenburg Schmidt plates (Kibblewhite et al. 1982; Kibblewhite et al. 1984; Evans 1988, Scholz, Odenkirchen & Irwin 1993a,b).

The measured objects were classified into stars, nonstellar objects, noise images and merged objects using the standard APM software. Objects classified as nonstellar on the two deepest plates in each field were considered for the extragalactic reference frame.

The plate matching consists in pairing up objects between a reference plate and the comparison plates using several passes of ever decreasing search radius, starting with bright objects and large search radii (up to several mm), finally iterating down to faint objects within a target search radius of 30μm on the Tautenburg plates and 25μm on the Palomar plates. For one globular cluster plate (T7050) and all Palomar-Tautenburg plate matchings we had to include up to 3rd order terms in the coordinate transformation.

Field & Plate Number	Epoch	Pass-band	Field & Plate Number	Epoch	Pass-band
M3			Draco		
T2176	1966.2	B	E1148	1954.6	R
T7000	1989.2	B	O1148	1954.6	B+U
T2175	1966.2	B	PS29671	1982.5	Bv
T6999	1989.2	B	PS29667	1982.5	R
T2873	1969.3	B	T7564	1991.5	B
T7002	1989.2	B	T7664	1991.7	no filter
T2167	1966.2	B	T7717	1991.8	V
T6226	1986.4	B			
T2174	1966.2	B			
T6232	1986.4	B			
M92			UMi		
T2311	1966.7	B	E1575	1956.2	R
T7031	1989.4	B	O1575	1956.2	B+U
T1023	1963.6	B	PS29610	1982.4	Bv
T7055	1989.5	B	PS29666	1982.5	R
T2225	1966.5	B	T7585	1991.6	B
T7050	1989.5	B	T7651	1991.7	no filter
T2325	1966.7	B	T7693	1991.8	V
T7030	1989.4	B			
T652	1962.8	no filter			
T7056	1989.5	no filter			

TABLE 1. Plate material

Since both dSphs are near the corner of the first epoch Palomar plates, the maximum available overlap field with the later epoch plates was reduced to about 3 (Draco) and 4 (UMi) square degrees compared to the usual 10 square degrees overlap of Tautenburg plates.

3. Plate-to-plate geometry and systematic error removal

In order to avoid magnitude dependent systematic errors we selected only unsaturated images of reference galaxies and stars for further investigation. The saturation of stellar images begins some 3 to 4 magnitudes above the plate limit. In the globular cluster fields with large number densities we excluded the faint objects within one magnitude of the plate limit due to their uncertain classification and much less accurate coordinates. In the cluster fields we selected the reference galaxies only outside a chosen cluster radius.

For the determination of the global plate-to-plate models we used the method of stepwise regression described in Hirte et al. (1990) applied to 3rd order polynomials with only coordinate dependent terms, since the magnitude interval had already been restricted. These calculations were carried out with up to 2000 reference galaxies in the globular cluster fields and about 300 to 600 galaxies in the smaller dSph fields.

The known errors of the APM (cf. Evans 1988) and residual errors of the galaxies' solution dependent on position on the plates were looked for and removed. Corrected

	mean absolute p.m. in x [mas/yr]	mean absolute p.m. in y [mas/yr]
M3 (from 5 plate pairs)		
uncorrected	−2.8±0.4	−1.8±0.6
corrected	−3.1±0.2	−2.3±0.4
M92 (from 5 plate pairs)		
uncorrected	−5.0±0.8	+3.5±2.1
corrected	−4.6±0.6	+2.6±1.6
M92 (from 4 plate pairs)		
uncorrected	−4.5±0.8	+1.4±0.7
corrected	−4.4±0.7	+1.1±0.4

TABLE 2. Mean absolute proper motions of M3 and M92

mean proper motions of the globular clusters were made using the assumption that the systematic errors affected the cluster and the field stars equivalently and that the field star bulk proper motion was independent of plate position. Due to the different plate scales and location of targets, the error functions for the dSphs are more complex. In a first iteration on this problem we have attempted to exclude the systematic errors in the dSph fields by subtracting iteratively the function of the mean galaxies' residuals with x and y dependency.

4. The proper motion of M3 and M92

The main results are listed in Table 2. In the case of M92 we decided not to use the third pair of plates (the second epoch plate T7050 had already been problematic in the global plate matching with high order corrections needed and also showed larger systematic errors than all other pairs of plates). For both clusters the corrected x and y values of the absolute proper motions from all pairs of plates were in a better agreement than the uncorrected ones. The correction improved the solution even if the third M92 pair was included but a residual systematic shift in the y proper motion probably introduced by the high order correction could not be removed.

The results compared with those of Cudworth (1976 and 1979) and with the re-reduction of Cudworth & Hanson (1993) are shown in Figure 1. In the case of M92 we are in good agreement with both Cudworth (1976) using the standard apex and Cudworth & Hanson (1993). Also shown in Figure 1. are the directions of the standard antapex, the old Lick antapex (Klemola and Vasilevskis 1971), the new Lick antapex (Hanson 1987) and the antapex for the Galactic globular cluster system taken from Mihalas and Binney (1981). There is a significant difference between the field and cluster proper motions. As expected there is a general trend of increasing proper motions of the field stars for brighter samples in the M3 field. In the M92 field which is closer to the apex this effect is not present.

More details about the absolute proper motions and the determination of the galactic orbits of M3 and M92 can be found in Scholz, Odenkirchen & Irwin (1993a,b). APM measurements of another 5 globular cluster fields (M53, NGC5466, M5, M13 and M15) including both Palomar and Tautenburg Schmidt plates have already been completed and will be presented in future publications.

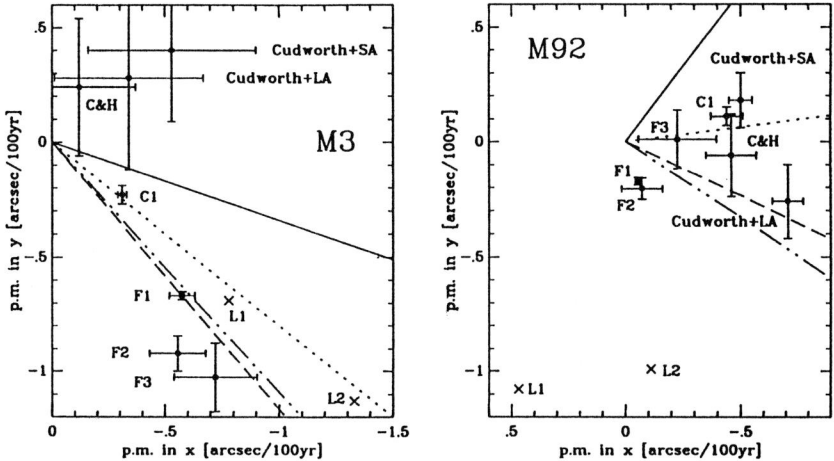

FIGURE 1. Mean absolute p.m. of cluster stars and field stars in comparison to Cudworth (1976, 1979) when he used standard apex (SA) or old Lick apex (LA) and to Cudworth & Hanson (1993) (C&H). Points corresponding to field and cluster stars together with their 1σ error bars are marked. C1 is the corrected cluster motion for 1032 M3 stars and 416 M92 stars with $17 < B < 20$; F1 is the mean p.m. of 2990 and 6466 field stars with corresponding magnitudes; F2 is the mean p.m. of 1121 and 3823 field stars with $14 < B < 17$; F3 is the mean p.m. of 220 and 580 field stars with $B < 14$; respectively in the M3 and M92 fields. The points labelled L1 and L2 are Lick proper motion measurements for about 50 faint and bright stars respectively taken from Klemola and Vasilevskis (1971). The lines represent: solid - standard antapex; dashed - solar antapex with respect to globular clusters; dot-dash - old Lick solar antapex; dotted - new Lick solar antapex.

5. Preliminary results for the dSph

In a first pass at estimating the absolute proper motion of the dSphs we used different plate-to-plate solutions (of similar colour plates) and averaged the results giving higher weights to plate pairs with larger epoch differences. The use of the galaxy residuals in the error analysis leads to a smaller total value of the mean proper motion of the dSphs, which is of the order of 1 mas/yr. (± 0.4 and ± 0.7 mas/yr, respectively for Draco and UMi).

In future work we will use the relative proper motion data of Stetson (1980) and Cudworth, Olszewski & Schommer (1986) to better define the stellar proper motions and then transform these relative proper motions to absolute ones by subtracting the mean "proper motion" of the galaxies. The use of the stellar data seems also to yield better results in the systematic error removal. In addition we intend to use the plate overlap method of Stock (1981) for the determination of the positions of all available reference galaxies on the different plates giving then a higher weight to deeper plates in the reduction of the stellar proper motions from the obtained positions.

Acknowledgements

We would like to thank the Palomar and Tautenburg Observatories for supplying the plates. The APM is a national astronomy facility financed by the Science and Engineering Research Council. The WIP-astrometry project at the Universität Potsdam is being supported by the Koordinierungs- und Aufbau-Initiative für die Forschung in den neuen Bundesländern (KAI e.V.).

REFERENCES

Brosche, P., Geffert, M. & Ninkovic, S. (1983). Lessons from the globular cluster NGC 5466, *Publ. Astron. Inst. Czech. Acad. Sci.*, **56**, 145–155.

Brosche, P., Geffert, M., Klemola, A.R. & Ninkovic, S. (1985). One more space motion of a globular cluster: NGC 4147, *Astron.J.*, **90**, 2033–2038.

Brosche, P, Tucholke, H.-J., Klemola, A.R., Ninkovic, S., Geffert, M. & Doerenkamp, P. (1991). Space motions of globular clusters NGC 362 and NGC 6218 (M12), *Astron. J.*, **102**, 2022–2027.

Cudworth, K.M. (1976). Membership and internal motions in the globular cluster M92, *Astron.J.*, **81**, 975–982.Ω

Cudworth, K.M. (1979). Proper motions, membership, and internal motions in the globular cluster M3, *Astron.J.*, **84**, 1312–1318.

Cudworth, K.M., Hanson, R.B. (1993). Space velocities of 14 globular clusters, *Astron. J.*, **105**, 168–172.

Cudworth, K.M., Olszewski & E.W., Schommer, R.A. (1986). Proper motions and bright-star photometry in the Ursa Minor dwarf galaxy, *Astron. J.*, **92**, 766–776.

Evans, D.W. (1988). Galactic structure and kinematics, Cambridge University D. Phil Thesis.

Hanson, R.B. (1987). Lick Northern proper motion program. II. Solar motion and galactic rotation, *Astron.J.*, **94**, 409–415.

Hirte, S., Dick, W.R., Schilbach, E. & Scholz, R.-D. (1990). Application of stepwise regression in photographic astrometry, In *Errors, bias and uncertainties in astronomy*, ed. C. Jaschek & F. Murtagh, pp. 343–346. Cambridge Univ. Press.

Kibblewhite, E.J., Irwin, M.J., Bridgeland, M.T. & Bunclark, P.S. (1982). A preliminary investigation of proper motions of faint stars in the Hazard 8hr region, *Occ. Rep. R. Obs. Edinburgh*, **10**, 79–89.

Kibblewhite, E., Bridgeland, M., Bunclark, P., Cawson, M. & Irwin, M. (1984). Analysis of images with the APM system at Cambridge, In *Astronomy with Schmidt-Type Telescopes*, ed. M. Capaccioli, pp. 89–97, Dordrecht, Reidel Publishing Company.

Klemola, A.R. & Vasilevskis, S. (1971). A study of solar motion and galactic rotation, *Publ. Lick Obs.*, **XXII**, part III.

Mihalas, D. & Binney, J. (1981). *Galactic Astronomy*, Freeman, San Francisco.

Scholz, R.-D., Odenkirchen, M. & Irwin, M.J. (1993a). Absolute proper motion and galactic orbit of M3, *Mon. Not. R. astron. Soc.*, in press.

Scholz, R.-D., Odenkirchen, M. & Irwin, M.J. (1993b). Absolute proper motion and galactic orbit of M92, *Mon. Not. R. astron. Soc.*, in press.

Stetson, P.B. (1980). The dwarf spheroidal galaxy in Draco. III. Proper motion membership probabilities, *Astron. J.*, **85**, 387–397.

Stock, J. (1981). Block adjustment in photographic astrometry, *Rev. Mex. Astron. Astrofis.*, **6**, 115–118.

Tucholke, H.-J. (1992a). Astrometry of globular clusters I. Relative proper motions for stars in 47 Tuc, *Astron. & Astrophys. Suppl. Ser.*, **93**, 293–309.

Tucholke, H.-J. (1992b). Astrometry of globular clusters II. Relative proper motions and the space motion of NGC 362, *Astron. & Astrophys. Suppl. Ser*, **93**, 311–329.

Webbink, R.F (1988). Kinematics of the galactic globular cluster system, *The Harlow Shapley Symposium on Globular Cluster Systems in Galaxies*, ed. J.E. Grindley & A.G. Davis Philip, Proc. IAU Symp. **126**, pp 49–59.

The central Galactic star cluster and the 'mini-spiral' morphology

By A. M. FRIDMAN[1], O. V. KHORUZUII[1], V. V. LYAKHOVICH[1] AND L. OZERNOY[2,3]

[1] Institute of Astronomy, Russian Academy of Sciences, Moscow, Russia

[2] Inst. for Computational Sciences and Informatics, George Mason University, Fairfax, Virginia 22030, U.S.A.

[3] USRA/Laboratory for Astronomy and Solar Physics, NASA, Goddard Space Flight Center, Greenbelt, Maryland 20782, U.S.A.

We would like to call attention to the important role of optical astrometry as a means of determining the anisotropy of open and globular clusters and to check whether they are on the boundary of their stability. Marginal anisotropy helps to understand the nature of their origin and to construct a model of the interaction of the central star cluster with the gaseous disk at the center of the Milky Way.

---

1. Why must a star cluster be localized on the boundary of the stable region of phase space?

To answer this question let us remember some examples of stellar systems much larger than star clusters. As was shown by Fridman et al. (1990), the stellar disks of galaxies including the Milky Way are near the boundary of gravitational instability. It turns out (Fridman et al. 1990) that gaseous galactic disks possess similar properties: they are located in the stable region near the boundary of gravitational and hydrodynamical instabilities. So we come to the formulation of a universal law concerning the location of dynamical systems in the stable region of parameter space near the boundary of instability. This law is the consequence of the well-known Le Chatelier principle: any physical process acts to cancel its cause. As a result of the development of the radial orbit instability (Polychenko & Shukhman 1972), the transverse velocity dispersion grows until the system arrives at the boundary of instability. The stability region of a spherical stellar system is (Fridman & Polyachenko 1984),

$$0 \leq \xi(r) \leq 2, \qquad (1.1)$$

where

$$\xi(r) \equiv 2T_r(r)/T_\perp(r) \qquad (1.2)$$

and

$$T_{r,\perp}(r) = \int_0^r d\vec{r} \int v_{r,\perp}^2 f_0 d\vec{r}. \qquad (1.3)$$

We assume that open and globular clusters, as dynamical systems, must fit Le Chatelier's principle. This means that, depending on its origin (radial dissipative or collisionless collapse), a star cluster has its value of the parameter $\xi(r)$ inside the interval (1) and close to 0 or 2, respectively.

2. Why is it important to know the value of anisotropy of the central star cluster?

In the literature for many years hypotheses concerning the existence of a black hole in the center of the Galactic central star cluster have been discussed. An estimation of the upper limit of a black hole mass is important to understand the nature of black holes, and for scenarios of the formation of the center of the Milky Way. The equilibrium equation of the central star cluster in the simplest case has the form (Fridman & Polyachenko 1984)

$$\frac{d}{dr}(\rho c_r^2) = -\rho \frac{GM(r)}{r^2} - \rho \left(\frac{2c_r^2 - c_\perp^2}{r} \right), \quad (2.4)$$

where

$$c_{r,\perp}^2 = \frac{\int v_{r,\perp}^2 f_0 d\vec{v}}{\int f_0 d\vec{v}}. \quad (2.5)$$

The presence of a velocity anisotropy, $2c_r^2 > c_\perp^2$, corresponds observationally to the addition of an appropriate additional mass. That is, evidence for velocity anisotropy decreases the allowed upper mass limit of any central black hole.

The total mass of the central star cluster and black hole can be found by means of the rotation curve of the innermost gaseous disk. To measure directly the rotation curve profile is not so easy. But according to our theory of spiral structure generation in gaseous galactic disks, the shape of arms depends on the rotation curve profile. Hence the shape of the observed 'mini-spiral' in the innermost gaseous disk creates the possibility of estimating the total central mass (star cluster + black hole). Taking into account the anisotropy of the star cluster helps further to estimate an upper mass limit to the mass of any black hole in the center of the Milky Way.

3. Estimation of an upper limit to the mass of any black hole in the center of the Milky Way.

We can estimate the mass of the central region of the star cluster of our Galaxy within $r \leq 0.2$ pc using the morphology of the 'mini-spiral' structure, and the hydrodynamical concept of spiral structure generation. The latter was proposed in 1972 (see references in Fridman 1979) and so far is successfully elaborated. The problem was investigated analytically (Morozov, Fainshtein & Fridman 1976; Morozov 1976, 1979; Fridman 1989, 1990) as well as by means of laboratory experiments (Fridman et al. 1985) and numerical simulations (Baev, Makov & Fridman 1986). Kinks and jumps are typical peculiarities of the curves of the angular rotation velocity of gaseous disks of many spiral galaxies (Afanasiev et al. 1988). In addition to these peculiarities the gaseous disk of our Galaxy has also a big jump in surface density 0.35 kpc from the center (Sumin, Fridman & Haud 1991). In recent works we have explored the simultaneous effect of jumps in three gaseous parameters – rotation velocity, surface density $\Sigma(r)$, and sound speed – and show that the observed peculiarities of $\Omega(r)$ in our Galaxy can cause the generation of its main spiral arms (Fridman 1989; Fridman et al. 1993a). In addition we propose a new mechanism of spiral structure generation, caused by either a kink in the curve $\Omega(r)$, that corresponds to a jump of its derivative $d\Omega/dr$, or a jump of the function $\Sigma(r)$ (or both) (Fridman et al. 1993b). We derive elsewhere an important implication: *If, at some radius R_0, there is either a kink in rotational velocity or a jump in density (or both) then (2.4) has a solution in the form of a wave localized in the vicinity of R_0.*

Assuming that the observed 'mini-spiral' in Sgr A West is a density wave generated by hydrodynamical instability due to a jump in density and/or a kink in rotational velocity of gas within the range $0.1 \lesssim R_0 \lesssim 0.3$ pc , we are able to draw several conclusions:

(i) The generated waves can have a variety of different shapes, such as (a) extended spiral waves or (b) local bar-like structures.

(ii) An $m = 1$ mode on its own as the shape of an extended spiral wave can be excited only if a point mass dominates within the central region $r \lesssim 0.5$ pc. Generation of a one-arm spiral having the characteristics proposed by Lacy et al. (1991) does not seem possible for any input parameters of our models.

(iii) Spirals with $m = 2, 3$ can be generated both in the case when a point mass dominates and in the case when the rotation curve is close to flat.

(iv) The observed structure might be a superposition of various modes, namely modes with $m = 2$ and 3 which are extended two- and three-arm spiral density waves; and that with $m = 1$ which is a localized bar-like structure off the center of rotation.

(v) The most substantial constraint on the shape of the gravitational potential at the Galactic center follows from the pitch of the observed mini-spiral. If the observed pattern is either a two-arm or three-arm spiral, its comparatively large pitch suggests that there is no strong central point mass concentration. An estimate of $|dr/d\varphi| \gtrsim 0.1$ pc/rad within $r \sim 1$ pc results in the enclosed total mass within $r \lesssim 0.2$ pc not exceeding $(1.7-4.7) \times 10^5$ M$_\odot$, which leaves for a central point mass a value not exceeding $\sim 10^5$ M$_\odot$. This value is consistent with that derived by other approaches considering a wide variety of physical processes (Ozernoy 1992, 1993).

4. Conclusions

(i) If we assume that the central star cluster was formed in the course of a radial collapse, then depending on whether this collapse is dissipative or collisionless, the cluster may have one of two alternative critical values of velocity anisotropy.

(ii) If one were to assume that the mini-spiral is generated by hydrodynamical instability, then we predict the existence of a peculiarity in the gaseous disk parameters near 0.2 pc from the center. This will be apparent as a jump in surface density or/and a kink in rotation velocity.

(iii) Comparison of the total mass within $r \leq 0.2$ pc of the Galactic center derived from the morphology of the mini-spiral with the distributed mass derived from dynamics allows one to find an upper mass limit for any central black hole.

(iv) Optical astrometry can give important data for dynamical investigations of open and globular clusters, especially in developing stable dynamical models and in understanding the nature of their origin.

REFERENCES

Afanasiev, V.L., Burenkov, A.I., Zasov, A.V. & Silchenko, O.K. 1988, Astrofisika, 28, 343, and 29, 155 (in Russian)

Baev, B.P., Makov, Yu.N., Fridman, A.M. 1986, Pis'ma v Astron. Zh., 13, 964 (in Russian)

Fridman, A.M. 1979, Sov. Phys. Usp. 21, 536

Fridman, A.M. & Polyachenko, V.L. 1984, Physics of Gravitating Systems, Springer Verlag

Fridman, A.M., Morozov, A.G., Nezlin, M.V., Snezhkin, E.N. 1985, Phys. Lett., 109, 228

Fridman, A.M. 1989, Doklady Akad. Nauk, 301, 200 (in Russian)

Fridman, A.M. 1990, Sov. Phys. JETP, 71(4), 627

Fridman, A.M., Khoruzhii, O.V., Lyakhovich, V.V. & Sumin, A&A 1993a (in press)

Fridman, A.M., Khoruzhii, O.V., Lyakhovich, V.V. & Sumin, A.A. 1993b, in Proceedings of the Second International Conference of Fluid Mechanics, 7-10 July 1993, Beijing, China

Lacy, J.H., Achtermann, J.M., & Serabyn, E. 1991, ApJ, 380, L71

Morozov, A.G., Fainshtein, V.G., Fridman, A.M. 1976, Doklady Akad.Nauk, 231, 588 (in Russian)

Morozov, A.G. 1977, Pis'ma v Astron. Zh., 3, 195 (in Russian)

Morozov, A.G. 1979, Astron. Zh., 56, 498 (in Russian)

Ozernoy, L.M. 1992, Testing the AGN Paradigm, Eds. S.S. Holt et al., AIP Conf. Proc., 254, pp. 40 and 44

Ozernoy, L.M. 1993, Proc. 16th TEXAS/3rd PASCO Symp. on Relativistic Astrophysics, Eds. C. Akerlof & M. Srednicki. Ann. New York Acad Sci. (in press)

Sumin, A.A.,Fridman, A.M. & Haud, U.A. 1991, Pis'ma v Astron. Zhurnal, 17, 779 (in Russian).

On the motion of the Magellanic Clouds

By P. KROUPA, S. RÖSER AND U. BASTIAN

Astronomisches Rechen-Institut, Mönchhofstrasse 12-14, D-69120 Heidelberg, Germany

We have measured the proper motion of the Large and Small Magellanic Clouds using Magellanic Cloud stars in the PPM Catalogue. Systematic uncertainties are discussed. Bound and unbound orbits of the Magellanic Clouds around the Galaxy are consistent with our result. A distinction between tidal models and ram-pressure models of the Magellanic Stream cannot be made on the basis of our result. Future astrometry will have to allow measurement of the proper motion of the LMC with an uncertainty no larger than about one tenth mas/yr in order to significantly distinguish between models.

1. Introduction

The two Magellanic Clouds are separated by about 20° on the sky and share a common hydrogen envelope. The associated Magellanic Stream spans a near-great polar circle on the sky nearly crossing the South Galactic Pole and appears to consist primarily of hydrogen gas. The motion of this entire system takes place in the gravitational potential of the Galaxy and so probes the mass of the Galaxy within a distance range of a few tens to a hundred kpc. The Large Magellanic Cloud (LMC) lies at a distance of 52 kpc and recedes with a heliocentric radial velocity of +274 km/s. The corresponding numbers for the Small Magellanic Cloud (SMC) are 63 kpc for the distance and +49 km/s for the heliocentric radial velocity. By contrast, the tip of the Magellanic Stream which is situated about 90° away from the LMC has a large heliocentric velocity of infall of about 380 km/s.

Models of the Magellanic System range from numerical simulations of the LMC-SMC binary or a larger Magellanic Cloud in the gravitational potential of the Galaxy with and without a heavy halo (*e.g.* Murai & Fujimoto 1980, Lin & Lynden-Bell 1982, Gardiner, Sawa & Fujimoto 1993) to a simulation in which a larger Magellanic Cloud partakes in the early Hubble flow, interacts with the Andromeda Nebula and is accelerated towards the Milky Way (Shuter 1992). In these models the Magellanic Stream results from tidal interaction of the LMC-SMC binary, the larger Magellanic Cloud with the Galaxy or with the Andromeda Nebula, respectively. Each scenario is consistent with the observed radial velocities along the Stream. Lin & Lynden-Bell (1982) explicitly predict for the proper motion of the LMC $\mu = 2$ mas/yr for a heavy Galactic halo and $\mu = 1.5$ mas/yr for no halo. The predicted motion is due East (position angle $\theta = 90°$). Gardiner, Sawa & Fujimoto (1993) predict $\mu = 1.8$ mas/yr with $\theta = 80°$. The corresponding galactocentric transverse velocities are 373 km/s (Lin & Lynden-Bell 1982) and 287 km/s (Gardiner, Sawa & Fujimoto 1993). Shuter (1992) requires a smaller Galactic mass and predicts a galactocentric transverse velocity of 355 km/s; the Magellanic Clouds not being bound to the Galaxy. A completely different scenario is discussed by Liu (1992) who argues that the Magellanic Stream may be material from the Magellanic Clouds that was compressed in the wake of the two Clouds owing to the drag exerted by the halo medium of the Galaxy. He uses a galactocentric transverse velocity of 310 km/s.

The aim of this paper is to investigate whether the Magellanic Clouds have proper motions significantly different from the values expected from current theory. We keep in mind that in order to differentiate between models we need a proper motion with an uncertainty not larger than about one tenth mas/yr. A more detailed account of the work presented here can be found in Kroupa, Röser & Bastian (1993).

2. The data and their systematic errors

The southern portion of the PPM Catalogue has been completed recently (Bastian et al. 1992). It is based on observational catalogues spread over almost one century. The proper motions in the FK5 sytem are based on the IAU (Lieske et al. 1977) value for the constant of precession. There are indications (Williams et al. 1991) that the IAU constant of precession should be corrected by -2.7 mas/yr. This yields a correction to the proper motions of only $+0.1$ mas/yr in right ascension and -0.1 mas/yr in declination for the LMC. For the SMC the corrections are -0.5 mas/yr and -1.0 mas/yr, respectively.

The change of the constant of precession affects the rotation of the FK5 system as a whole. Systematic errors of the FK5 proper motion system at the position of the Magellanic Clouds are a second cause of uncertainty. As a third cause of uncertainty we have to consider how well the PPM Catalogue approximates the FK5 system. We note that mainly OB-stars of the LMC are in the PPM Catalogue. The proper motions of these may suffer from systematic errors owing to the different spectral response of the astrographic systems at short wavelengths (i.e. "colour equations" in the jargon of photographic astrometry).

It is impossible now to quantify any possible systematic difference of the PPM Catalogue from an inertial reference frame. However, once the systematic uncertainties for the PPM are known it will be straightforward to transfer the proper motions measured here to an inertial frame. Our purpose here is to investigate if the Magellanic Clouds show motions that differ significantly from the values predicted by current theory. For this purpose it suffices to assume that the PPM represents an inertial reference frame.

The statistical uncertainty of an individual proper motion in the PPM Catalogue is approximately 3 mas/yr. We need about ten common motion stars to reduce the uncertainty to about 1 mas/yr which, at a distance of 50 kpc, corresponds approximately to the circular velocity of the isothermal halo if it extends that far.

Sanduleak (1968, 1969, 1970) provides a deep objective-prism survey of the LMC and SMC and lists definite or probable members. We identify the PPM Catalogue stars in Sanduleak's list by matching coordinates, coincidence of HD numbers and/or spectral types and magnitudes, and obtain a list of 35 LMC stars and 8 SMC stars (Kroupa, Röser & Bastian 1993).

3. Results

The mean of all proper motion values of the list of LMC member stars is:

$$15\cos\delta\ \mu_\alpha = +1.3 \pm 0.6\,\text{mas/yr}; \qquad \mu_\delta = +1.1 \pm 0.7\,\text{mas/yr}.$$

The uncertainties are the standard errors of the mean values. This corresponds to a heliocentric transverse velocity of 420 ± 220 km/s and a galactocentric transverse velocity of 335 ± 220 km/s adopting 220 km/s as the circular velocity of the sun and a distance of 8.5 kpc from the Galactic centre.

For the SMC we obtain:

$$15\cos\delta\ \mu_\alpha = +1.7 \pm 1.1\,\text{mas/yr}; \qquad \mu_\delta = -0.8 \pm 1.3\,\text{mas/yr}.$$

3.1. Other studies

A preliminary result of an independent measurement of the proper motion of the LMC has been reported (Tucholke & Hiesgen 1991). They find $15\cos\delta\,\mu_\alpha = +0.91 \pm 2.34$ mas/yr and $\mu_\delta = -0.23 \pm 2.77$ mas/yr by measuring absolute proper motions relative to background galaxies. They measure photographic plates with an epoch difference of 15 years but do not explicitly define their sample of LMC members. The preliminary report by Jones, Klemola & Lin (1989) states that they have measured a proper motion of the LMC that is "less than half of that predicted" by Lin & Lynden-Bell (1982).

Indirect methods based on the difference in position angle between the kinematical and photometric line of nodes allow a determination of the transverse velocity which leads to a gradient in radial velocity across the LMC. However, Meatheringham et al. (1988) and Prevot, Rousseau & Martin (1989) find a heliocentric transverse velocity of 275 ± 65 km/s and a galactocentric transverse velocity of 150 km/s, respectively.

All four measurements thus suggest a proper motion for the LMC of ≈ 1 mas/yr due East.

3.2. Interpretation of our result

The modulus of the proper motion of the LMC is 1.7 ± 0.9 mas/yr pointing within 40° of the opposite direction of the Magellanic Stream. The proper motion is consistent with the value predicted by Murai & Fujimoto (1980), Lin & Lynden-Bell (1982), Shuter (1992), Liu (1992) and Gardiner, Sawa & Fujimoto (1993). We emphasize that in order to distinguish between models, the proper motion of the LMC (and preferably also of the SMC) has to be known to about one tenth mas/yr. This corresponds to 25 km/s and is about one half to one third of the maximum rotation velocity of the LMC (Meatheringham et al. 1988, Prevot, Rousseau & Martin 1989), complicating the task further.

REFERENCES

Bastian, U., Röser, S., Yagudin, L., Nesterov, V. V., Polochentsev, D. D., Potter, Kh. I., Wielen, R., Yatskiv, Ya. S. 1993, *PPM South Star Catalogue*, Astronomisches Rechen-Institut, Heidelberg, Germany (magnetic tape version)

Gardiner, L. T., Sawa, T. & Fujimoto, M. 1993, preprint

Jones, B. F., Klemola, A. R., Lin, D. N. C., 1989, BAAS, 21, 1107

Kroupa, P., Röser, S. & Bastian, U. 1993, MNRAS, in press

Lieske, J. H., Lederle, T., Fricke, W., Morando, B. 1977, A&A, 58, 1

Lin, D. N. C., Lynden-Bell, D. 1982, MNRAS, 198, 707

Liu, Y. 1992, A&A, 257, 505

Meatheringham, S. J., Dopita, M. A., Ford, H. C., Webster, B. L. 1988, ApJ, 327, 651

Murai, T., Fujimoto, M. 1980, PASJ, 32, 581

Prevot, L., Rousseau, J., Martin, N. 1989, A&A, 225, 303 Sanduleak, N. 1968, AJ, 73, 246

Sanduleak, N. 1969, AJ, 74, 877

Sanduleak, N., 1970. *Contr. Cerro Tololo Interamerican Obs.*, 89

Shuter, W. L. H. 1992, AJ, 386, 101

Tucholke, H.-J., Hiesgen, M., 1991. In: *The Magellanic Clouds*, IAU Symp. No. 148, eds: Haynes, R., Milne, D., (Kluwer Academic Publishers), p.491

Williams, J. G., Dickey, J. O., Newhall, X X, Standish, E. M., 1991. The Orientation of the Dynamical Reference Frame, *Reference Systems*, IAU Coll. No. 127, [eds: Hughes, J. A., Smith, C. A. & Kaplan, G. H.,] p.146.

ROEMER satellite project: the first high-accuracy survey of faint stars

By E. HØG[1] AND L. LINDEGREN[2]

[1] Copenhagen University Observatory, Ostervoldgade 3, DK-1350 Copenhagen K, Denmark

[2] Lund Observatory, Box 43, S-22100, Lund, Sweden

The proposed ROEMER mission aims at accurate astrometric and photometric measurements of 100 million stars down to 18th magnitude. Based on the well-proven concepts of the ESA Hipparcos mission, but employing a vastly superior CCD detection system, it would give the positions, parallaxes and annual proper motions to 0.1–1.5 milli-arcsec, depending on magnitude, and multi-colour photometry to 0.001–0.02 mag. The programme would include a complete survey of the 50 million stars brighter than $V = 15.5$. Some basic considerations for the design of the ROEMER payload are given.

1. Introduction

The highly successful Hipparcos mission has demonstrated the astrometric and photometric capabilities of a small scanning satellite using a very modest ($\simeq 30$ cm) optical telescope. But Hipparcos was designed in the early eighties, before the maturing of solid-state detectors for use in space. The realization that modern CCDs could be applied to such a project, with an enormous gain in performance, led one of us to propose a concept for a mission called ROEMER (Høg 1993). A more elaborated version of this proposal was recently submitted to the European Space Agency (Lindegren et al. 1993).

The proposed ROEMER mission would be the first high-accuracy astrometric and photometric survey of faint stars, measuring all the 50 million stars brighter than $V = 15.5$ and an additional 50 million stars down to 18th magnitude. A 2.5-year mission would yield the positions, absolute trigonometric parallaxes and annual proper motions with an accuracy in the range of 0.1 milli-arcsec ($V \leq 11$) to 1.5 milli-arcsec ($V = 17$), and magnitudes in five wavelength bands to an accuracy of 0.001 to 0.02 mag.

The scientific objectives include a thorough improvement of the current knowledge of stellar structure and evolution, stellar dynamics, star clusters and associations, unseen companions, the galactic and cosmological distance scales, and solar-system dynamics. The mission would establish an inertial optical reference frame several times better than currently achieved with radio observations and directly tied to the extragalactic frame. With an average density of 70 stars in a $10' \times 10'$ field, this astrometric reference frame will be directly accessible to large ground-based telescopes. Since the same stars will obtain accurate multi-colour photometry, they will also form a dense net of photometric standards.

2. Summary of the proposed mission

Basic mission concepts are derived from the Hipparcos mission (Perryman and Hassan 1989): like its predecessor, ROEMER is conceived as a free-flying, three-axis controlled satellite in geostationary orbit, with a modest optical telescope continuously sweeping across the sky in a pre-defined pattern. Two widely separated viewing directions provide the large-angle measurement capability necessary to build a globally rigid astrometric system.

ROEMER is, however, much more than a mere replica of Hipparcos. The use of several CCDs directly in the focal plane, with their higher quantum efficiency, smaller transmission losses and capability to observe thousands of stars strictly simultaneously, provides a measuring system which is at least 10^5 times more efficient than Hipparcos in terms of photon utilization. This translates into 20 times smaller errors, five magnitudes fainter limit, and a thousand times more programme stars—noting that Hipparcos is already achieving orders-of-magnitude improvement over previous knowledge. The inclusion of colour filters for several wavelength bands, from near-UV to near-IR, adds tremendously to the astrophysical value of the mission.

In order to fit into the cost envelope of an ESA Medium Size Mission, the instrument must be of about the same size as that of the Hipparcos payload. For cost reasons as well as light efficiency, the optical and mechanical design is simplified as much as possible. Much higher demands can on the other hand be tolerated in terms of on-board data storage, computing, and telemetry. Main mission characteristics are summarised in Table 1

3. Considerations for the payload design

The Hipparcos mission has shown that a beam combiner mirror can provide the very stable angular reference needed for global astrometry. The attitude control system used by Hipparcos, based on cold gas thrusters, is also suited to give the required long periods of smooth free rotation of the satellite, provided the torques from solar radiation pressure, gyros and structural non-rigidity are decreased. Finally, Hipparcos has hinted at the enormous photometric potential of a scanning instrument, provided it is equipped with proper colour filters to define the wavelength bands — something which could not be fully exploited in that project, mainly because of the low sensitivity of its main detection system.

For the ROEMER mission there are strong reasons to incorporate a multi-colour photometric capability for all its observed stars. Firstly, because such observations are scientifically extremely valuable as a complement to the astrometric measurements; secondly, because a colour index must be obtained for all the stars in order to determine and correct for an expected astrometric effect of chromaticity in the optical system; and thirdly in order to increase the dynamic range of astrometry (while filters are not used for the astrometry of the faint stars, they push the bright limit by at least two magnitudes before saturation of the detector). The wavelength bands should be carefully selected to optimize the astrophysical information content of the measurements for different types of objects. In this context, the possible use of polarization filters should also be considered.

The detection system is based on CCDs in the focal plane, capable of measuring thousands of stars simultaneously with high quantum efficiency. The charge image of each CCD is electronically shifted at the same rate as the optical star images move across the detector. The accumulated charges of selected pixels are then read out as they reach the edge of the chip. This replaces the Hipparcos measurement of one star at a time through a modulating grid by a low-efficiency ($\sim 8\%$) image dissector tube. Special-purpose CCDs with rectangular pixels can be manufactured today as required to match the precision requirement and the resolving power of the telescope, which differ in the two dimensions (Figure 1). High positional stability is needed even though cooling to $-70°$ C necessarily introduces large temperature differences between the CCDs and their connection to the telescope. A dimensional stability inside a CCD of better than one

Expected results

number of stars	100 million
limiting magnitude	$V = 18$
completeness limit	$V = 15.5$ (50 million stars)
accuracy of positions and parallaxes .	0.1 mas ($V \leq 11$)
	to 1.3 mas ($V \simeq 17$)
accuracy of proper motions	0.1 mas/yr ($V \leq 11$)
	to 1.5 mas/yr ($V \simeq 17$)
systematic errors	< 0.1 mas
photometric accuracy	0.006 mag ($V \leq 11$)
	to 0.016 mag ($V \simeq 17$)
photometric colour bands	(e.g.) U, B, V, R, I
	and polarization filters ($P+, P-$)
	plus a wide band (W)

Payload

optical system	all-reflective Wright-Väisälä
telescope aperture diameter	336 mm
telescope focal length	4700 mm
angle between viewing directions	$\sim 54.5°$
detection system	20 CCDs cooled to $-70°$ C
number of pixels	$4096 \times 1024 + 256 \times 64$
pixel size	$4 \times 16~\mu m^2 + 4 \times 256~\mu m^2$
wavelength band	300–950 nm

System

mission duration	2.5 yr (consumables: 5 yr)
orbit	geostationary
power source	fixed solar panels (7 m^2)
attitude control	cold gas thrusters
spin rate	170 arcsec/s (127 min period)
angle from spin axis to the Sun	55°
precession of spin axis about the Sun	73 day period
nominal motion of spin axis on sky ..	4.1° per day
scientific telemetry rate	50 kbytes/s
external dimensions of satellite	Ø3.0 m×3.5 m (cylindrical)
launch mass	about 1100 kg

TABLE 1. Main characteristics of the ROEMER mission.

part in a million when first-order bulk motion and magnification changes are removed, has been reported by Buffington *et al.* (1990) for periods of 10 hours. Such stability makes a calibration of CCD irregularities feasible as required for the present purpose.

The optical telescope has two half-circular openings with a diameter of 34 cm. Matching the diffractional resolution to the smallest available CCD pixel width (about 4 μm)

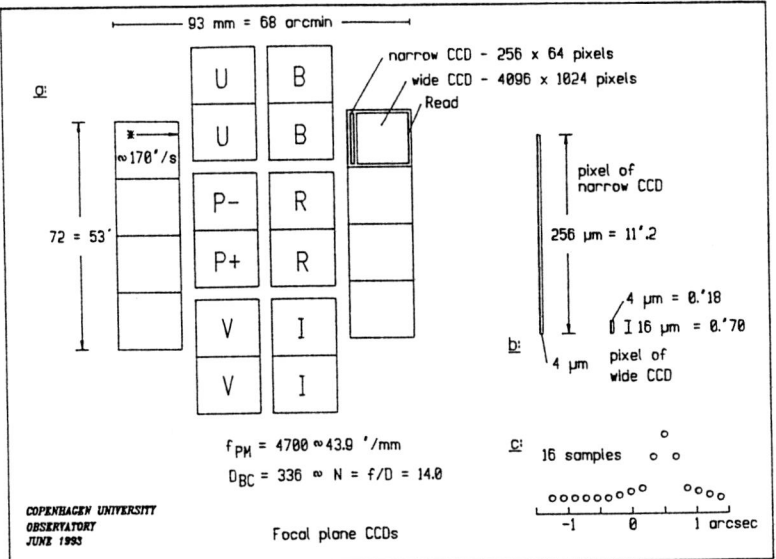

FIGURE 1. Focal plane arrangement of ROEMER. **a:** The stars drift across 20 chips, each containing a narrow and a wide CCD. Reading of the number of accumulated electrons (counts) takes place in a special register at the right edge of each CCD. Eight of the chips, at left and right, measure in a wide spectral band W and twelve in the photometric bands $UBVRI$ and, perhaps, in polarized light ($P-$, $P+$). **b:** The rectangular pixels. **c:** A star record of 16 samples.

implies a focal length of nearly 5 m. This can be accommodated in a 3 m diameter cylindrical spacecraft by means of two folding mirrors (Figure 2). A Wright–Väisälä optical system provides diffraction-limited images in a flat field of one degree diameter. A very smooth rotation with a period of about 2 hours is assured by a rigid spacecraft structure with minimum perturbations from solar radiation pressure and gyroscopes—the latter not being used at all during science operations. The attitude control system is based on cold-gas thrusters to be fired at intervals of approximately 30 min. A rotationally symmetric, compact shape of the satellite will best satisfy these requirements (Figure 3).

The proposed attitude control for ROEMER uses observations of reference stars for all normal attitude determination and acquisition without use of gyroscopes. If gyros are nevertheless required for special manoeuvres or emergency situations, they should normally be turned off or kept at a low rate. To ensure sufficient mechanical stiffness, the solar panels should be mounted on the satellite surface instead of hinges. The apogee boost motor should be jettisoned after use, lest jitter arise from the loose solid residue particles inside the chamber.

4. Predicted performance

The following estimate of the astrometric and photometric mean errors for the ROEMER mission is based on photon-statistical errors, since it is expected that the other significant errors in Hipparcos from 'non-rigidity' (Makarov et al. 1993) and from parasitic stars can be neglected in ROEMER. The measurement of a programme star could be disturbed if another 'parasitic' star is situated within about 6×4 pixels = 3.0 arcsec^2

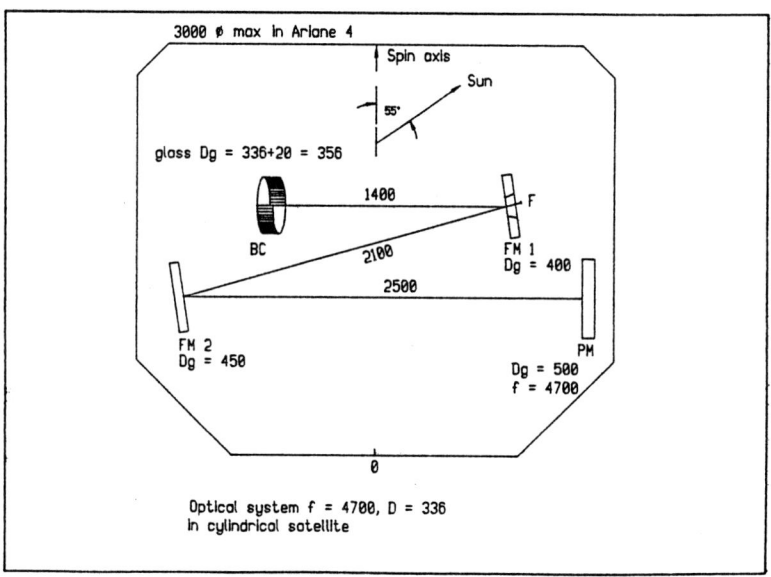

FIGURE 2. Optical system with focal length 4700 mm and aperture diameter 336 mm. The beam combiner mirror BC has a basic angle of (say) 54.5°. A telescope with two flat folding mirrors, FM1 and FM2, and the primary mirror PM, can be placed in a cylindrical spacecraft of 3000 mm diameter.

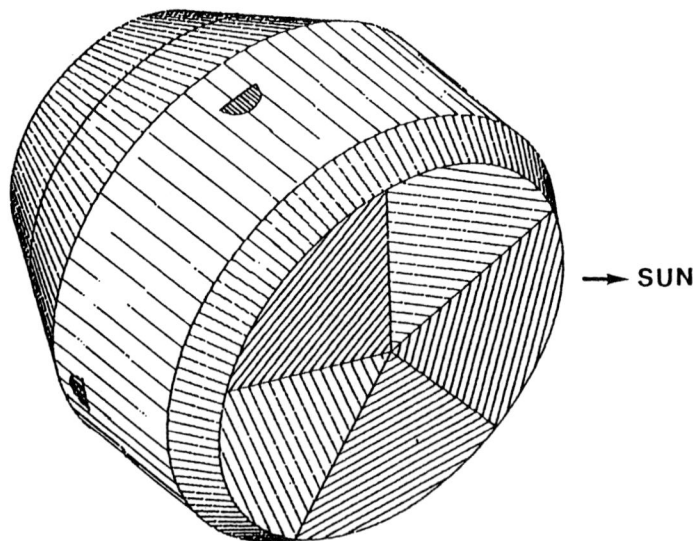

FIGURE 3. A view of the ROEMER satellite. The cylindrical/conical envelope and internal mass distribution are designed to minimize perturbations due to solar radiation pressure. The solar panels cover the flat top end of the satellite.

	Astrometry		Photometry				
V	parallax	p. m.	U	B	V	R	I
mag	mas	mas yr^{-1}	mag	mag	mag	mag	mag
11	0.08	0.09	0.003	0.001	0.001	0.001	0.001
12	0.12	0.14	0.005	0.002	0.002	0.001	0.001
13	0.19	0.22	0.008	0.003	0.002	0.002	0.002
14	0.30	0.35	0.013	0.004	0.004	0.004	0.004
15	0.48	0.56	0.022	0.007	0.006	0.006	0.006
16	0.77	0.91	0.037	0.011	0.010	0.009	0.010
17	1.29	1.50	0.069	0.018	0.016	0.015	0.016
18	2.25	2.63	—	0.032	0.028	0.026	0.028

TABLE 2. Predicted mean errors in astrometry and photometry for a 2.5 year ROEMER mission.

centered on the programme star (cf. top Figure 3c). The probability of having a star brighter than $V = 21$ within this area in either field-of-view is 0.024 for the mean sky and 0.10 if both fields are in the galactic plane. Therefore, parasitic stars will give negligible disturbance in astrometry and photometry, compared with photon noise.

Predicted mean errors for a 2.5 year ROEMER mission are given in Table 2. The table refers to a star of spectral type G0 and is based on realistic assumptions concerning optical transmittance and quantum efficiency, background and read-out noise, as well as image smearing resulting from the pixel width, discrete sampling and scan velocity errors; in addition a margin factor of 1.2 has been applied to all photon noise errors, both for astrometry and photometry. (The similar table by Høg (1993) gave much smaller mean errors mainly because a five-year mission was assumed for a satellite with two telescopes, while here only 2.5 years and one telescope with the same collecting aperture are assumed.)

The mean errors in this Table are nevertheless too optimistic for the bright stars. An asymptotic mean error must be added, especially to the astrometric errors. This is very difficult to estimate, but based on the experiences with Hipparcos data, asymptotic errors of 0.1 mas for the parallaxes, positions and annual proper motions would seem conservative. It should be noted that an extension of the mission length from 2.5 to five years, which is technically feasible, would reduce the mean errors in position, parallax and photometry by a factor 0.7, and the proper motions by a factor 0.35.

Acknowledgements

This work was supported by the Danish Space Board and the Swedish Space Board.

REFERENCES

Buffington, A., Hudson, H.S. & Booth, C.H., 1990, *Publications of the Astronomical Society of the Pacific*, **102**, 688

Høg, E., 1993, Astrometry and photometry of 400 million stars brighter than 18 mag, in: I.I. Mueller and B. Kołaczek (eds.) *Developments in Astrometry and Their Impact on Astrophysics and Geodynamics*, IAU Symp. No. 156

Lindegren, L. (ed.), Bastian, U., Gilmore, G., Halbwachs, J.L., Høg, E., Knude, J., Kovalevsky, J., Labeyrie, A., van Leeuwen, F., Pel, W., Schrijver, H., Stabell, R. & Thejll, P., 1993, ROEMER, Proposal for the Third Medium Size ESA Mission (Lund)

Makarov, V.V., Høg, E. & Lindegren, L., 1993, Accuracy of star abscissae in the ROEMER project, Submitted to *Experimental Astronomy*

Perryman, M.A.C. & Hassan, H., 1989, *The Hipparcos mission, Vol. I, The Hipparcos Satellite*, ESA Publications Division, ESA SP-1111.

Meridian circle observations of the planets

By E. M. STANDISH

Jet Propulsion Laboratory 301-150, Pasadena, CA 91109, U.S.A.

Throughout the nineteenth and much of the twentieth centuries, planetary ephemerides were based entirely upon meridian circle observations from a number of different observatories. For much of that time, these observations represented possibly the most accurate measurements of anything in the history of mankind. Even at present, meridian circle observations form an important part of the data fit by the JPL ephemerides; especially the photoelectric observations which have given a further improvement in accuracy over the past decade. For the present study, however, some earlier meridian circle observations have been analyzed, with the intention of possibly incorporating their information also into the data set. These, however, show varying degrees of usefulness because of some surprising reasons. Observations by Lalande in 1795 are of special interest for ephemerides, because they contain two (unintentional) observations of the planet Neptune. Observations by Robertson in the early 1800's are also of interest for ephemerides, because they include measurements of Uranus; however, there are some suspicious features of Robertson's early measurements which render their validity questionable.

1. Introduction

This paper presents some investigations and experiences with a few sets of meridian circle observations taken over the past two centuries. Some of the findings are curious from a historical aspect; others have a more practical bearing upon present-day ephemerides, which initially was the main purpose for the investigations.

The observations of Michel Lalande are of special interest, for he measured the position of Neptune fifty-one years before the "discovery" of that planet in 1846. Importantly, these two positions, on May 8 and May 10, 1795, have seemed not to be consistent with modern-day ephemerides, even though they are quite consistent with each other. Other than these two observations and the sightings of Neptune in 1612 and 1613 by Galileo himself (Drake and Kowal, 1979), all of the pre-discovery observations of Neptune are within a year or two of its discovery; therefore, these other observations do not add much dynamical information about Neptune's orbit. Thus, Lalande's observations assume added importance. They are discussed in §2.

Pages from the 1811 and 1812 observing notebook of Abram Robertson were copied and sent by CA Murray (1991) with permission from the Royal Astronomical Society's Manuscripts Division. In a letter, Murray stated, "It was the late David Thackeray who drew my attention to the fact that the Radcliffe Observations from the death of Hornsby (1810) to the appointment of Manuel Johnson (1839) had never been reduced. Abram Robertson succeeded Hornsby and Stephen Peter Rigaud followed Robertson in 1827..." Evidently, there are tens of thousands of these unreduced observations. Observations of Uranus and/or Neptune could be of interest if contained in these pages. However, Murray noted that Neptune would not appear; its magnitude is too faint, and, furthermore, Robertson was observing only known stars and planets, listing the names of each object in the margin. He measured only a few stars per night, covering a large range in declination, interspersed with planets, including "Georgian's Center"; i.e., Uranus. This was in contrast to Lalande, who selected a very narrow declination zone and then tried to measure every star as it transited. Robertson's observations are discussed in §3.

A large amount of transit observations from many sources has been collected over the past couple of decades, both at MIT and at the US Naval Observatory. In the early 1970's, JPL received from MIT, a series of magnetic tapes containing their collection of planetary optical observations. The observations had been taken from many sources and had been put onto the tapes in exactly the form in which they had been published. In the 1980's the author also obtained a set of Uranus observations and residuals from members of the US Naval Observatory. These observations, also taken at a number of different observatories, had been transformed, by the USNO, first onto the FK4 reference frame by the means of catalogue corrections, and then onto the FK5 reference frame by the application of the IAU-adopted equinox correction (Fricke, 1982). It is this set, in particular, that has been widely distributed to astronomers and others who have eventually used it to predict a tenth planet in the solar system. There are some differences between the tapes; these are discussed in §4.

Section 5 discusses the transit data presently being used at JPL for the adjustment of the planetary ephemerides. The series of 6-inch and 9-inch Washington transit observations were the only sources used at JPL until the Royal Greenwich Observatory (Herstmonceux) observations were obtained in order to fill the gap in the early 1970's when the USNO 6-inch meridian circle was being refurbished. During the past decade, the newer photoelectric observations from the Carlsberg Automatic Meridian Circle (La Palma) and Bordeaux Observatory Meridian Circle have provided a further increase in accuracy. Also, the first meridian circle observations of Pluto have been obtained with the CAMC, starting in 1989. These are discussed by Morrison et al. (1992), and the improvement which they provide to the ephemeris of Pluto is shown by Standish (1993b).

2. The Observations of Lalande, 1795.

Can the inconsistency between Lalande's observations and the other observations of Neptune be real, since Neptune's longitude is very uncertain back in 1795 (see e.g., Standish, 1993a)? Are the large residuals really indicators of "incompleteness in the present model of the outer solar system" (see, e.g., Seidelmann and Harrington, 1988)? The issue was pressed by Marsden (1992):

" ... I would like to see ... a set of osculating elements for Neptune that does in fact accommodate the 1795 observations. We owe it to Lalande to fit them ..."

Rawlins (1970) provided the reduction of Lalande's observations which has been used in recent times. He, however, used a very limited set of reference stars: only stars within 40′(minutes) of Neptune, only timings from the middle wire, and only stars between magnitudes 6 and 8.5. As a result, there were only six reference stars, three of which were in common over the two nights.

Could the problem with the Neptune residuals merely be the result of an inadequate reduction by Rawlins? It was decided to use all of the measurements taken by Lalande over the two nights, copies of Lalande's hand-written observing records being kindly provided by Rawlins (1992). The fitting of the observations included seven unknowns for each of the two days: five unknowns for the timing (right ascension) observations: a clock offset for each of the three wires, a correction to the clock rate and a dependence upon the star's declination, and two unknowns for the altitude (declination) observations: a constant offset and a dependence upon the star's declination. Most of the observations looked reasonable, but only after a fair number of them had been corrected for obvious mis-recordings, mistakes and illegible or ambiguous-looking characters. Two solutions, one including all residuals less than 10 seconds of time and the other including all residuals less than 100 seconds of time, gave similar results: with respect to DE200, the right ascension residuals were nearly $-8''$ in right ascension and effectively 0 in declination,

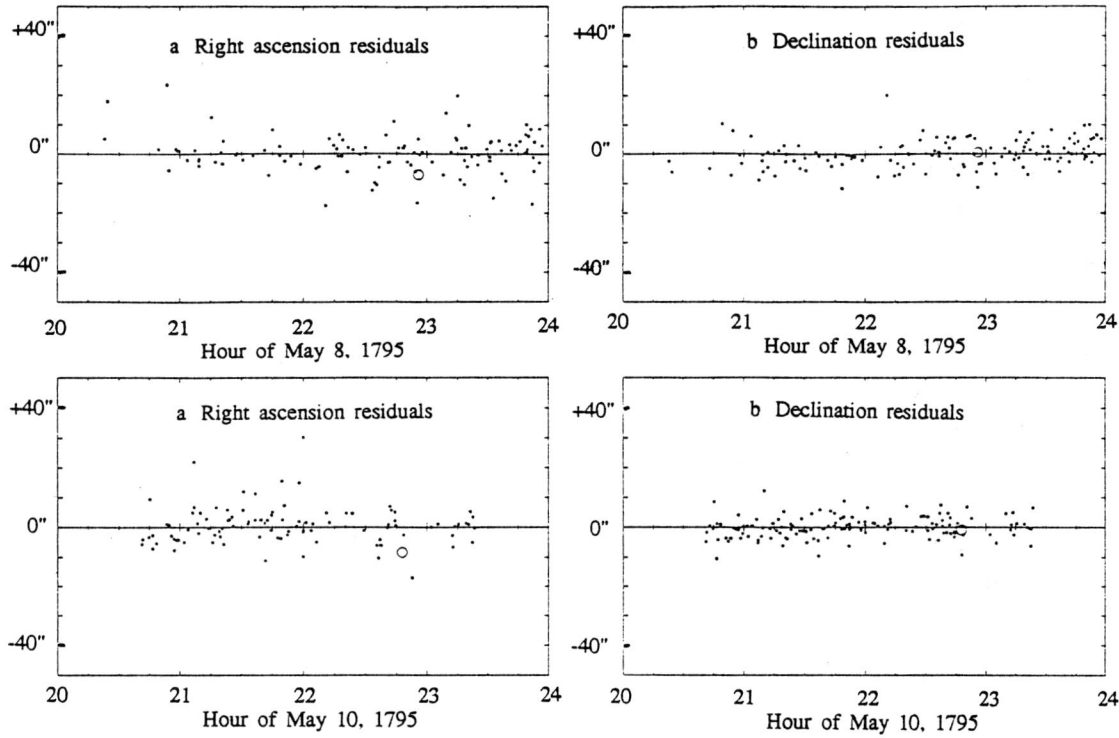

FIGURE 1. The May 8 and May 10 residuals of Lalande, including those of Neptune, shown as a circle: (a) right ascension (central wire only), (b) declination.

with uncertainties of about 4″ in both components. Figure 1 shows the post-fit residuals from only the central wire readings on the two nights; Neptune is represented by a circle, the stars by dots.

These results are very similar to those of Rawlins.

If the observations and the reductions of them seem certain, could they possibly be accommodated by an ephemeris as suggested by Marsden? Three different ephemeris least-squares adjustments were performed, fitting to all of the Neptune transit observations covering 1846-1992 as well as to the Lalande observations. The purpose was to see what was required in order to force the ephemerides to fit Lalande's observations. Consequently, different *a priori* sigmas were assigned to the Lalande observations for each solution: for the first solution, Lalande's observations were excluded completely; for the second, the sigmas were 1″; and for the third, 0.″08. Figure 2 shows all of the post-fit residuals of Neptune, in right ascension and declination. One can see that all of the observations are well-fit in the first case, except for the Lalande points which were −9.″2 and −8.″4 (the Lalande residuals are a bit worse on a present-day ephemeris than they are w.r.t. DE200). As shown by the second case, when the Lalande points were included in the solution with weights of 1″, all of the residuals changed, but only slightly, even those of Lalande. However, only when the *a priori* sigmas for the Lalande points were reduced to the unrealistically small value of 0.″08, did the post-fit residuals (−3.″6 and −4.″4) approach the true 4″ uncertainty of the observations themselves. Moreover,

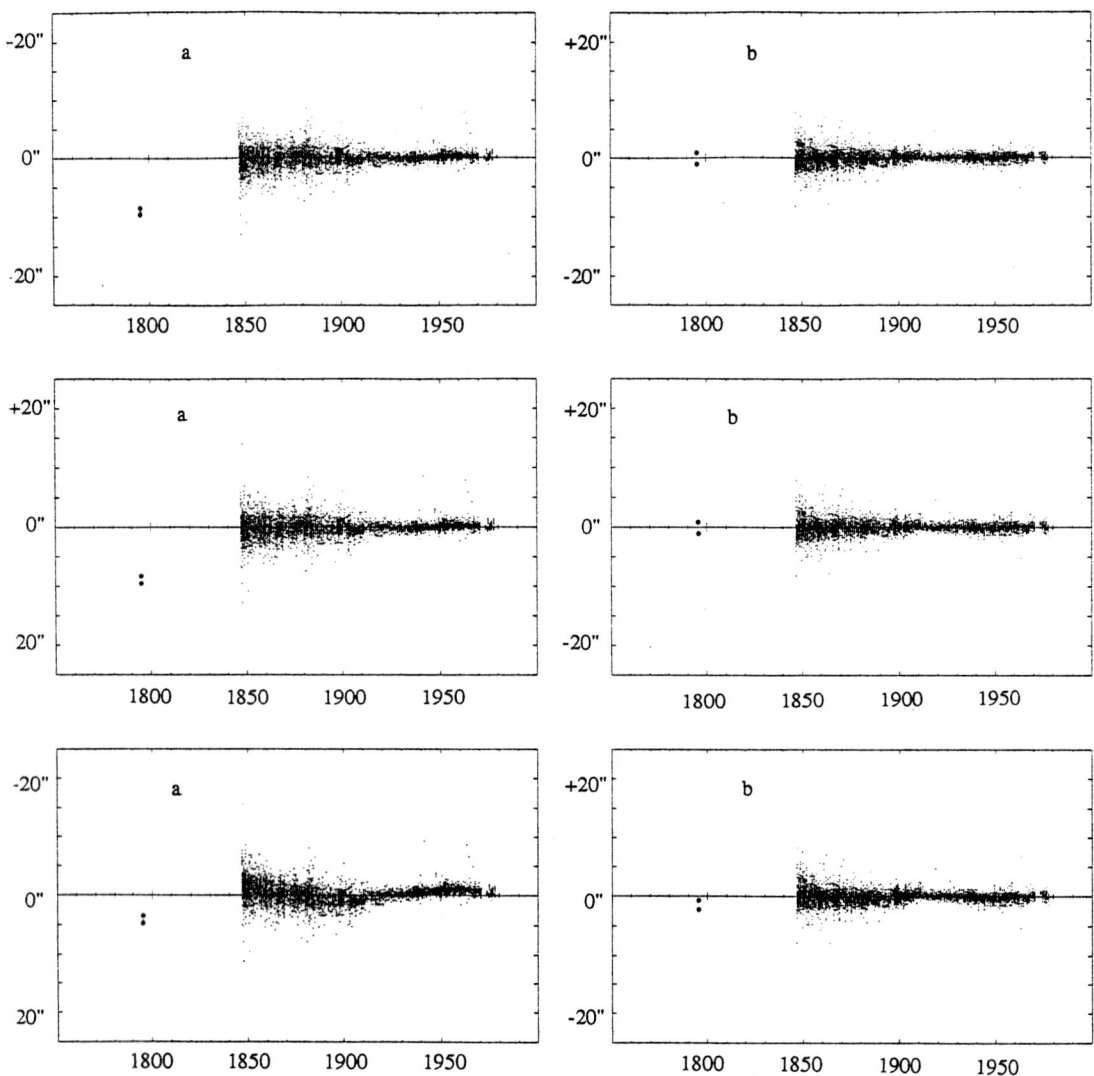

FIGURE 2. The full set of Neptune residuals in (a) right ascension and (b) declination computed against three ephemerides that were fitted to all the observations apart from Lalande's which were treated as follows: *top*, Lalande's excluded; *middle*, Lalande's weighted with $\sigma=1''$; *bottom*, Lalande's weighted with $\sigma=0''.08$.

the figure(bottom) shows the obvious signature forced into the other observations by the distortion of the ephemeris in order to even partially fit Lalande's observations.

Thus, at face value, Lalande's observations are not consistent with the other observations of Neptune.

Could Lalande's treatment of Neptune have been different from his treatment of the stars? Certainly, not intentionally, for he thought Neptune was an ordinary star when he measured it. However, it has been suggested (Franz, 1993) that Neptune indeed would have appeared different in Lalande's telescope - a bluish-green, non-scintillating image - possibly just enough ghost-like to cause Lalande to hesitate for a fraction of a second - enough to make the difference, for nearly all of Lalande's observations are recorded to the nearest half-second of time ("xx" or "xx.5"). A shift of one recording unit would change the residual by 7″.5. Whatever, this discrepancy has no certain explanation.

As a sidelight, it is not widely known that someone noticed something strange about the Neptune observations. J. J. Lalande, the uncle of Michel Lalande, later assembled his nephew's observations into the published *Histoire Celeste*. At some time, by someone, there was written in the margin of the notebook next to the May 8th measurement of the star preceeding Neptune, "see the 10th of May... there is a transposition of altitude and error on the passage of the other ... of the following star". And in the margin next to the May 10th measurement of Neptune, there is "see the 8th of May... there is a transposition of altitude with the star which is at 59° 54′ 40″. Whoever it was, he noticed something unusual (the order of transit between Neptune and the mentioned star had indeed changed over the two nights), but he evidently assumed it to be a mere recording mix-up. If only... Of course, had this anomaly been pursued and had Neptune been consequently discovered, we would have been deprived of one of astronomy's great triumphs - Leverrier, Adams and the Berlin Observatory.

3. The Observations of Robertson, 1811 and 1812

The 1811 observations of Robertson show timings at five wires; they are sometimes listed to an integer number of seconds of time; sometimes they are listed to a half-second of time ("xx.5"). There are also many observations which are recorded to a tenth of a second of time. Using just the central wire (since the others are virtually dependent), one may plot the timing (right ascension) residuals as is done in Figure 3. The reference stars are shown as stars, the sun as a pair of dotted circles for the 1st and 2nd limbs, and Jupiter as a concave square. The severe slopes and the discontinuities in the data are obvious. These occur at the times where, in the notebook, Robertson says something like, "After this observation I put the Clock forward 2′ " or "After this observation I screwed up the Bob of the Pendulum 3 divisions of the Nut". Robertson was evidently making major clock adjustments, and it seems that he eventually became successful toward the middle of March (near day #70), since the slopes and discontinuities are greatly reduced. If one solves for these clock offsets and rates using only the reference stars and then applies the results to all of the objects, the residuals become as they appear in Figure 4. The standard deviation is about 0.65 seconds of time (10 arcseconds). There is a hint that the ephemeris objects are offset a bit from the stars; possibly, this is a magnitude effect or simply the fact that the sun and Jupiter were further south than the majority of the reference stars, though among the reference stars, there seems to be no dependence of the right ascension residuals upon the declinations.

However, Robertson's observations cannot be trusted! Starting on January 7, 1811 and continuing for a full month, the sum of the 1st and 5th wire readings (modulo 60) is always exactly equal to the sum of the 2nd and 4th readings, and these are both always exactly equal to the twice the reading of the 3rd wire. Further, in an extraordinary number of cases, the reading on the 5th wire is merely an integer while the other four are recorded to a tenth of a second of time; but still the means are exactly equal. Occasionally, the second wire is recorded to one-quarter/three-quarters of a second, and in all such cases, the fourth wire is recorded to three-quarters/one-quarter of a second, so that their sum is an integer. A few lines from the notebook are reproduced in Table 1, with only the seconds parts listed, showing the precision written

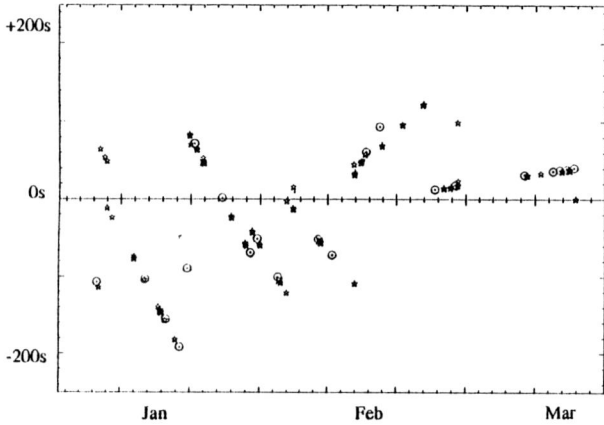

FIGURE 3. The 1811 observations of Robertson before correcting for clock offsets and rates. The solar residuals are indicated by a dotted circle; those of Jupiter by a concave square.

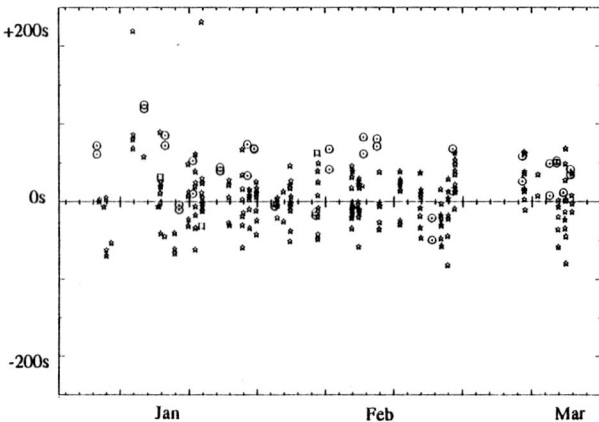

FIGURE 4. The 1811 observations of Robertson after correcting for clock offsets and rates.

by Robertson. Can it possibly be that Robertson's observations are accurate below the level of 0.1 seconds? The answer is no, for he observed many of the same stars night after night, and the *rms* about the mean for the residuals of any given star is reasonable: well over half a second of time.

After February 9, virtually all of Robertson's observations are listed to only an integer number of seconds. For the observations in 1812, analysis again shows evidence that the individual wire measurements are not independent. A few of these planetary residuals are listed in Table 2. The correlation between wires observed on the same day is extremely high, but the correlation from day to day is non-existent. The second observation is an extreme example; it is an observation of Uranus. All four observed wires show a residual of minus five seconds! This is not coincidence; it is not observational error, or a misprint,

	wire#1	wire#2	wire#3	wire#4	wire#5
	4	26.75	49.5	12.25	35
	58.6	20.7	42.8	4.9	27
	3.8	30.1	56.4	22.7	49
	32.4	56.8	21.2	45.6	10
	4.2	26.4	48.6	10.8	33
	23	56	29	2	35
	33.4	56.3	19.2	42.1	5
	52.3	14.6	36.9	59.2	21

TABLE 1. Some of the 1811 listing of Robertson. Only the seconds parts of the observations are listed, but those are given to the precision as recorded by Robertson. The sum of the 1st and 5th wires (modulo 60) is exactly equal to the sum of the 2nd and 4th wires, and these are exactly equal to twice the 3rd wire. Note, also, the preponderance of integer values in the 5th column. One must conclude that all five readings are not independent.

Planet	Date	wire#1	wire#2	wire#3	wire#4	wire#5
Ven	Apr 20	1.4	1.6		1.3	1.5
Ura	Apr 24	−5.2	−5.4		−5.4	−5.5
Ven	Apr 26	1.1	0.9		1.1	0.8
Ura	Apr 27	2.3	2.7		2.2	2.5
Ura	May 04	1.0	0.9	1.0	1.1	0.8
Jup	May 04		0.3		0.5	0.6
Ura	May 05	1.0	0.9		0.9	0.5
Ura	May 07	0.1	0.5		0.5	0.2
Jup	May 07	−0.1	−0.1		−0.2	−0.4
Ura	May 08	0.7	0.3		0.3	0.2
Ven	May 08	−1.7	−2.1		−1.8	−1.9
Ven	May 11	−1.6	−1.4		−1.2	
Ven	May 12	−0.7	−0.8	−0.9	−0.7	−0.7
Ura	May 12	1.7	1.6		1.6	1.5

TABLE 2. Planetary residuals of Robertson covering three weeks in 1812. The high correlations between the wires during a single transit suggests that these observations were not observed independently.

or a transcription error. Also, it is not an ephemeris error, for in subsequent days, the residuals of Uranus are reasonable and seem to scatter around a near-zero mean.

4. The MIT and USNO observation tapes of Uranus

The tape from MIT contained the observations as published. The USNO tape, on the other hand, contained not only the observations, transformed first to the FK4 and then to the FK5 reference system, but it also contained residuals. Evidently, these were computed with respect to JPL's planetary ephemerides, DE200, since similar computations by the author give nearly exact agreement with the USNO residuals. It is basically this set of observations and residuals that has been widely distributed by the USNO. However,

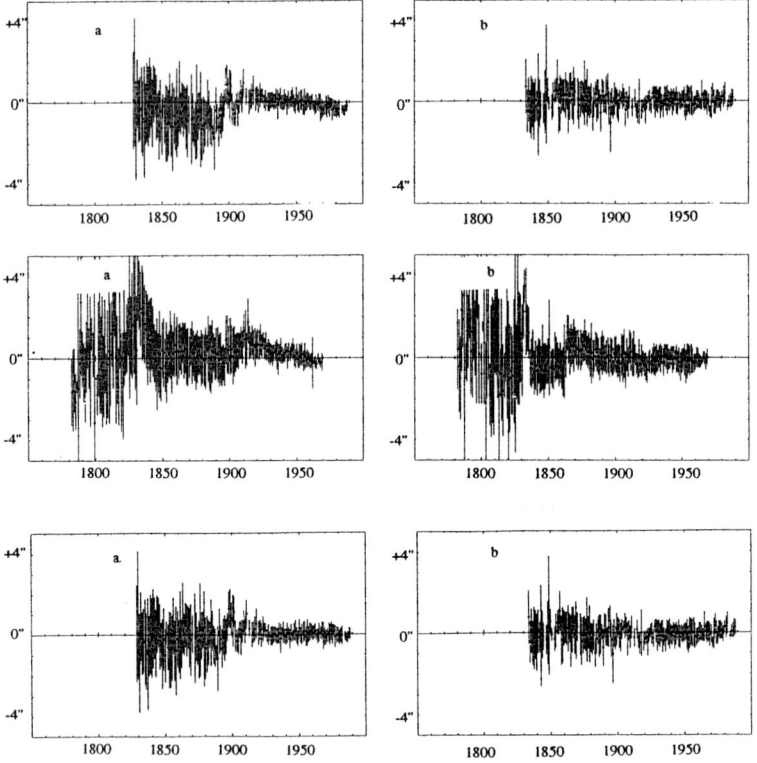

FIGURE 5. The normal points (means and standard deviations of the residuals in (a) right ascension and (b) declination for each opposition taken from the following tapes of Uranus observations: *top*, USNO; *middle*, MIT; *bottom*, USNO after accounting for the correct mass of Neptune and then re-adjusting the orbit of Uranus.

the set is far from complete. It contains no Greenwich data and only a scattering of Paris data, among others. Possibly, catalogue corrections for these were not available at the time of the tape's creation and, as such, it was deemed preferable to exclude them. Certainly, such a choice would be wise, for it would be incorrect to do analyses upon the uncorrected data.

Normal points of Uranus for each collection are plotted in Figure 5 (a normal point is the mean and standard deviation of all residuals within a given opposition). The differences in the two sets represent, not "unexplained signatures in the motion of Uranus", but rather some of the observational errors known to exist. Still remaining in the USNO residuals, Figure 5(top), are signatures due to the incorrect mass of Neptune which was used in DE200 and due to the need to re-adjust the orbit of Uranus. Corrections for these have been shown by Standish (1993a) to eliminate nearly all of the remaining signature, as shown in Figure 5(bottom). Does some signature remain? Certainly. Consider the observations from the Radcliffe Observatory in the late 1700's shown in the MIT plot only (Figure 5(middle)). These are reported to be highly inaccurate (Seidelmann and Harrington, 1988); they are probably uncorrectable; their use could do more harm than good. Note also the discontinuities in the USNO residuals (top) around 1900; Uranus cannot jump like that; those are observational errors.

5. Twentieth century transit observations

Until the last decade, the only optical observations used for orbit adjustment at JPL (Standish, 1990) were the post-1911 transit observations from the 6-inch and the 9-inch meridian circles of the USNO. Earlier observations were suspected of being severely biased (see, e.g., Fricke 1973, 1975; Duncombe and van Flandern, 1976); the later observations seem to be greatly improved in quality due to the introduction onto the telescope of the impersonal micrometer. In addition, it was possible to transform the USNO observations consistently onto a single fundamental catalogue, the FK4. Nevertheless, even these observations are not free from systematic errors. The literature abounds with different known effects in these data - day/night corrections, the theory of refraction, phase effects, the polar caps of Mars, solar heating of the telescope, rings of Saturn, etc. (Seidelmann et al., 1985; Seidelmann, 1986; Standish et al., 1976; etc.), and there is still suspicion that the modern optical reference system may have errors in the proper motion system as large as $1''$/century (Stumpff and Lieske, 1984; Standish and Williams, 1990). Nevertheless, over the centuries, the optical transit observations have represented some of the most accurate scientific measurements in existence until the advent of electronics. The scatter of the residuals for the daytime objects (sun, Mercury and Venus) is about $0''.8 - 0''.9$; that for the other planets is about $0''.4 - 0''.5$.

Within the last decade, photoelectric meridian circles on La Palma (CAMC) and at Bordeaux have been observing stars, planets, satellites and some selected minor planets with a substantial increase in accuracy over the preceeding visual observations. The scatter for the planetary positions now approaches the $0''.2$ range for both instruments, and there are strong indications that systematic errors are well below $0''.05$ (see, Morrison et al., 1991). The overall improved quality of these observations has already been of great benefit: they provided last-minute positions of Uranus prior to the Voyager encounter in 1986 - determination of an ephemeris which was a bit questionable at that time (Standish, 1985). The Neptune ephemeris for 1989 received a similar benefit.

Since 1989, the CAMC has been observing Pluto (Morrison et el., 1992). Here the quality gain is even more pronounced, since former observations of Pluto were strictly photographic, taken with narrow field cameras and reduced with secondary stellar catalogues. The great improvement in the Pluto ephemerides made available by the CAMC data is discussed by Standish (1993b). However, such a small portion of Pluto's orbit has been observed that extrapolation into the future still deteriorates extremely rapidly. Thus, for the ephemerides, continuing these observations is a necessity.

6. Conclusions

Meridian circle observations continue to improve, even in the present day. The oldest observations have certainly been useful throughout the past two centuries, but their accuracy is limited. Use of these for ephemeris improvement must be done with caution. The observations of the present century show a substantial improvement over their predecessors, due mainly to the development and use of the impersonal micrometer. During the past decade, the CAMC and Bordeaux photoelectric meridian circles have provided yet a further advance in accuracy and have already been of benefit to spacecraft navigation.

7. Acknowledgements

The research in this paper was carried out by the Jet Propulsion Laboratory, California Institute of Technology, under a contract with the National Aeronautics and Space Administration.

REFERENCES

Carlsberg Meridian Catalogue, La Palma, #4, Copenhagen University Observatory, Royal Greenwich Observatory, Real Instituto y Observatorio de Marina.

Drake,S. and Kowal,C.T. (1980). Scientific American, 243, 74–81.

Duncombe,R.L. and Van Flandern,T.C. (1976). Secular Variation of the Obliquity of the Ecliptic, Astron. J., 81, #4, 281–284.

Franz,O. (1993), private communication.

Fricke,W. (1973). Pro and Contra Changes in the Conventional Values of Precession, in Highlights in Astronomy, ed. G.Contopoulos, (Sydney), 211–219.

Fricke,W. (1975). Effects of the Introduction of a new Fundamental Reference Coordinate System on Time–Scales in Time Determination, Dissemination and Synchronization, in Proc. Second Cagliari International Meeting, eds. H.Enslin and E.Proverbio, (Cagliari), 235-240.

Fricke,W. (1982). Determination of the Equinox and Equator of the FK5, Astron. Astrophys., 107, L13–L16.

Marsden,B.G. (1992), private communication.

Morrison,L.V., Buontempo,M.E., Fabricius,C., and Helmer,L. (1992). First meridian circle observations of Pluto. Astron Astrophys, 262, 347–349.

Morrison,L.V., Gibbs,P., Helmer,L., Fabricius,C., Einicke,O., Requieme,Y., and Rapaport,M. (1991). Evidence of Systematic Errors in FK5, Astrophys & Space Sci, 177, 31-34.

Rawlins,D. (1970). The Great Unexplained Residual in the Orbit of Neptune, Astron J, 75, #7, 856- 857.

Rawlins,D. (1992), private communication.

Seidelmann,P.K. (1986). Unsolved Problems of Celestial Mechanics, Cel Mech J, 39, 141-146.

Seidelmann,P.K. and Harrington,R.S. (1988). Planet X - The Current Status, Cel Mech, 43, 55-68.

Seidelmann,P.K., Santoro,E.J. and Pulkkinen,K.F. (1985). Systematic Differences between Planetary Observations and Ephemerides, Dynamical Astronomy, eds. V.G.Szebehely and B.Balazs, (U.Texas Press, Austin) 55-78.

Standish,E.M., Keesey,M.S.W., and Newhall,XX (1976). JPL Development Ephemeris Number 96, JPL Technical Report, 32-1603.

Standish,E.M. (1985). Planetary and Lunar Ephemerides, DE125/LE125, JPL Inter Office Memorandum, 314, 6-591.

Standish,E.M. (1990a). The observational basis for JPL's DE200, the planetary ephemerides of the Astronomical Almanac, Astron Astrophys, 233, #1, 252-271.

Standish,E.M. (1990b). An approximation to the outer planet ephemeris errors in JPL's DE200. Astron Astrophys, 233, #1, 272-274.

Standish,E.M. and Williams,J.G. (1990). Dynamical Reference Frame in the Planetary and Earth-Moon Systems, in Inertial Coordinate System on the Sky, eds. J.H.Lieske and V.K.Abalakin, (Kluwer Academic Publishers, Dordrecht), 173-181.

Standish,E.M. (1993a). Planet X: No Dynamical Evidence in the Optical Observations. Astron J, 105, #5, 2000-2006.

Standish,E.M. (1993b). Improved Ephemerides of Pluto. Submitted to Icarus.

Stumpff,P. and Lieske,J.H. (1984). The Motion of the Earth-Moon System in Modern Tabular Ephemerides, Astron Astrophys, 130, 211-226.

Astrometric observations of minor planets and comets: present and future needs

By B. G. MARSDEN

Harvard-Smithsonian Center for Astrophysics, Smithsonian Astrophysical Observatory, 60 Garden St, Cambridge, MA 02138, U.S.A.

The numerous provisional designations that have been given to minor planets refer to perhaps no more than 30 000 individual objects, of which 5610 had received permanent numbers by June 1993. There are more than 2000 known or suspected apparitions of comets, referring to 857 individual objects. The total number of observations in the files is currently increasing by about 10 percent per year, and the better photographic positions from the early 1980s are comparable in accuracy to the best micrometric positions of a century earlier. Re-reduction of the micrometric data can nowadays be accomplished relatively easily, but photographic re-reductions are more of a problem. The introduction of the CCD and its steady replacement of photography is resulting in a significant increase in accuracy, despite the inadequacies of the *Guide Star Catalogue*. The increase in the number of objects for which meridian observations are now being made is also noteworthy. Use of Bayesian probability for orbit determination, as in the recent Muinonen-Bowell procedure, has merit for minor planets observed at several oppositions, but care needs to be exercised in accepting its predictive capability for comets affected by nongravitational forces and minor planets with poor observational coverage.

1. Quantity

As of the beginning of June 1993, the IAU Minor Planet Center and its predecessors have given 70 258 provisional minor-planet designations, up from 69 977 one month earlier. To these one must add the 1000 or so minor planets that were given their permanent numbers before the modern designation system came into use in 1925, as well as the 5000 and more objects discovered during the collaborative examination by Leiden astronomers of plates taken with the 1.2-m Palomar Schmidt between 1960 and 1977. Some kind of positional information is needed for all these entries. With the help of such information, one can then subtract almost 18 000 from the list to account for multiple designations involving the 5610 minor planets that now have permanent numbers, the occasional comet or outer satellite of Jupiter. More than 1000 designations also need to be subtracted because multiple designations are known to have been given to the same object during a particular opposition, and there are perhaps 8000 more multiple designations involving identifications of the same object at different oppositions. One cannot say how many unknown identifications will then remain in the data, but it is not impossible that there will be as many as 20 000 more. Removing also a few obvious mistakes and cases where the observer has withdrawn the data, we can see that the total number of individual minor planets for which observations have been recorded may be no more than about 30 000.

In addition to the numbered minor planets, there are 3595 objects with reasonably satisfactory orbits linking the observations at more than one opposition and for which few additional multiple designations are likely to be found (Marsden and Williams 1993a). Inclusion of almost all of the Palomar-Leiden survey objects and an undetermined number, perhaps between 5000 and 6000, of other objects with general orbits computed from observations at a single opposition yields a total number of discrete objects with orbit determinations approaching 20 000. The difference between this number and the minimum estimate of the number of individual minor planets is more than compensated by the 15 000 available "Väisälä" orbits (i.e., orbits computed from two observations on the assumption that the objects were at perihelion or aphelion, a circle being therefore a special case), but these clearly have numerous duplications, and in some cases the Väisälä orbits are from observations on a single night. There are also a large number of cases where a provisional designation was given to a single observation.

Likewise, there are upwards of 2000 apparitions of comets recorded in the literature. In 1397 cases there exist orbital elements representing the observed positions during those apparitions (Marsden and Williams 1993b). As with the provisional designations for minor planets, these cases include multiple designations of the same periodic comet, and the number of individual comets for which orbits have been computed is 857—of which the 86 recorded before the end of the seventeenth century can scarcely be considered of astrometric quality.

In terms of the number of separate observations, it should be noted that when Paul Herget was directing the Minor Planet Center he aimed at completing the machine-readable files of observations (precessed, if necessary, to 1950.0 using the expressions given by Newcomb) back to the beginning of the year 1939, although many earlier observations were included, particularly those from the original photographic observing program at Heidelberg (Reinmuth 1953, 1960), which dates back to the end of 1891. After the Minor Planet Center moved to Cambridge (Mass.) in 1978 it was decided that there should be at least some observational record for each provisional designation, which therefore augmented the collection back to 1925. More recently, pre-1939 observations of lower-numbered minor planets have been added. In the case of comets, the plan has been that the computer file should be complete back to the beginning of 1964, but again, there are some earlier data. Following the recommendation of IAU Commission 20, all the positions were converted to J2000.0 (Yeomans 1991), and when a full version of the file was last made available for public consumption, in July 1992, it contained 655 673 observations of minor planets and 40 183 of comets. The annual rate of increase is on the order of 10 percent for both minor planets and comets. There is also a file of some 16 000 older cometary observations, many of them from the late-nineteenth century, that do not contain references or magnitudes and have not been properly checked.

2. Quality

It can be said that the bulk of the better astrometric observations of comets and minor planets being made and published in the early 1980s were comparable in accuracy to the best observations published a century earlier. This rather provocative statement requires some explanation. The observations in the 1880s were mainly micrometric, with a smattering of meridian data. The filar micrometer had been perfected, and the series of Astronomische Gesellschaft catalogues (AGK_1) was available as a source for reference stars. As a general rule, only a single reference star was used, and the principal astronomical journals published the observations in a convenient manner, including some useful redundancy. The object's apparent position was supplied, together with the offset from the reference star, for which a mean position was given (referred to the equinox of the beginning of the year), together with the reduction necessary to obtain the corresponding apparent position. The time of observation was generally in local mean time, and it was usual also to give the correction—or, rather, its logarithm—necessary to reduce the position of the comet or minor planet to the geocenter, although this correction needed first to be divided by the object's distance, that being computed from an available orbit.

Each position was produced from several comparisons, the number usually being stated, and allowance for the object's motion had to be made in deriving what was published; this would occasionally result in the publication of the right ascension and the declination for different times.

As micrometry gave way to photography, generally around the 1920s and 1930s, the redundancy started to vanish. Perhaps the catalogue used for the positions of the reference stars (the AGK_2, Yale zones, later the AGK_3 and the very popular $SAOC$ or Smithsonian Astrophysical Observatory catalogue) would be mentioned, and sometimes even the particular stars used were indicated. In a few cases the reference stars were listed together with their "dependences", following a recommendation by Comrie (1925). These dependences, which essentially allowed the object's right ascension and declination to be obtained as a weighted mean of the individual right ascensions and declinations of, say, three to six comparison stars (Schlesinger 1911), have nowadays largely been forgotten in favor of the more flexible plate-constants computation, but they were very convenient to use in the days of mechanical calculating machines. For the small-field plates obtained with long-focus reflectors one usually had recourse to the AC (*Astrographic Catalogue*) for the reference stars. It was customary to use the AC plate constants to obtain the right ascensions and declinations of these stars from the AC xy measures. This process was decidedly non-uniform from one part of the sky to another, although for some zones (e.g., Heckmann et al. 1954) redeterminations of the plate constants have been obtained in a more standard manner. Another way of achieving uniformity has been to make individual redeterminations from the AC data as needed, using $SAOC$ positions for the reference stars involved in producing the AC. This was standard practice in the photographic observing program carried out with the 1.5-m reflector at the Oak Ridge Observatory in Massachusetts between 1973 and 1989.

As star catalogues, e.g., the PPM (*Positions and Proper Motions*), $ACRS$ (*Astrographic Catalogue Reference Stars*) and GSC (Space Telescope Institute *Guide Star Catalogue*), began to become available in the FK_5/J2000.0 system, IAU Commission 20 recommended that positions of minor planets and comets should be reduced directly in this system starting 1992 Jan. 1. By this time CCD astrometry was rapidly superseding photographic astrometry, except for global reductions of fields several degrees across. The extreme linearity of the CCD, and the ease and speed with which CCD exposures could be obtained and relative positions of celestial objects established to a fraction of a pixel, brought the potential for the first significant improvement in the accuracy of the bulk of the astrometric data on minor planets and comets in more than a century. The distribution of the GSC, with its average of some 400 stars per square degree, on compact disks in May 1989 was a tremendous boon to the CCD observers, and the use of the inconvenient AC (e.g., in the Oak Ridge program, which converted from photography to CCD a few months later) ceased almost immediately. Cometary positions have suffered in the past from the difficulty of the manual centering of the wires of a micrometer on the visual image or the crosshairs of a measuring engine on the over-exposed photographic image of a diffuse object. The algorithms for automatic CCD centroiding seem able to cope even with this problem, and repeated tests on bright cometary images (and the same reference stars) over intervals as long as 24 hours frequently show relative consistency in setting on the best-condensed region of a comet to better than $0\rlap{.}''1$.

Meridian observations of minor planets also have the potential for high accuracy, but they have historically been restricted to a handful of the brightest objects. In recent years there has been much more extensive meridian activity, using the automatic meridian circles at La Palma (Carlsberg Automatic Meridian Circle) and Bordeaux (Helmer and Morrison 1985, Rapaport et al. 1989). Tens of thousands of meridian observations have been obtained of more than 100 of the brighter minor planets. Among these minor planets are the 48 selected for observations by HIPPARCOS (Bec-Borsenberger 1990). It has been traditional to publish meridian observations as apparent positions referred to the geocenter. Although this practice can be criticized (when one considers that the vast majority of the available data are mean, topocentric positions), reduction to the geocenter will clearly be helpful in attending to observations made from the orbiting HIPPARCOS.

3. Re-reduction

Use of the redundancy included in the publication of the old observations went far beyond the correction of typographic or simple arithmetic errors, although that was certainly a significant consideration. The principal value of the redundancy was that it allowed the user to make a re-reduction of the object's position if an improved position could later be obtained for the reference star, or if it could be demonstrated that the reference star had been misidentified. Partly because of the HIPPARCOS project, there has recently been increased interest in making very precise determinations of the orbits of low-numbered minor planets. Although most of these determinations have simply made use of the observations in the Minor Planet Center's files, those by Edwin Goffin, in particular, have benefited by the investigator's extraction and re-reduction of old micrometric data. A computer program written by Gareth Williams at the Minor Planet Center automatically provides efficient re-reductions in terms of the FK_5/J2000.0 positions in the *PPM*. As time permits, the data pertaining to old micrometric observations of comets and minor planets in the *Astronomische Nachrichten*, *Astronomical Journal*, *Monthly Notices*, *Bulletin Astronomique* and *Journal des Observateurs* are being systematically prepared for and subjected to this computer re-reduction at the Minor Planet Center. It should also be noted that the input data are also being saved for potential *further* re-reduction using future catalogues of greater accuracy and star density than the *PPM*.

An example of the re-reduction is shown in the top part of Table 1. This is a measurement (Abetti 1917) of the minor planet (41) Daphne with the 0.28-m refractor at the Arcetri Observatory in Florence at 1916 Apr. $26^d 10^h 16^m 33^s$ Arcetri Mean Time (= 1916 Apr. 26.89690 UT). There is obviously a discordance of exactly 1' among the declination data, and this is automatically noted. The 1916.0 position specified for the comparison star is precessed to 2000.0, and comparison with the on-line *PPM* yields an obvious match within a few seconds of arc. The *PPM* FK_5/J2000.0 position and proper motion of the star are then converted into an apparent position for the true equinox at the time of observation of the minor planet, the offset is applied, and the resulting position of the minor planet is reconverted to the FK_5/J2000.0 system. This position compares very well with the Arcetri topocentric position derived by integrating the orbit of (41) Daphne (available on-line for a nearby epoch) to the time of observation (a computation that in the interests of efficiency is done separately at a later stage, and for this purpose it is generally sufficient to ignore the perturbations over an interval of ±100 days), showing that the discordance was in the object's published apparent declination, which should have been 1' further to the south. One could also have confirmed that the 1916.0 position given for the stated reference star, Leipzig II 6201 = BD +8°2632 (and this is indeed PPM 159085), is precisely consistent with the AGK_1 1875.0 position and first- and second-order precession terms, but this is a time-consuming exercise that would have to be done manually—and it is, fortunately, completely unnecessary.

The lower part of Table 1 shows what would happen if the 1' error were instead in

	α	δ	
(41) Daphne	12ʰ 38ᵐ 54ˢ.31	+7° 45′ 06″.8	apparent
Offset from reference star	+ 1 19.04	− 5 26.6	apparent
Reference star	12 37 32.32	+7 49 51.5	1916.0
Reduction to apparent place	+ 2.95	− 18.1	
Discordance		+ 1 00.0	
Transformed reference star	12 41 47.976	+7 22 12.85	2000.0
PPM 159085	12 41 47.725	+7 22 13.55	2000.0
Annual proper motion	− 0.0015	+ 0.002	
Re-reduced (41) Daphne	12 43 06.84	+7 16 48.5	2000.0
Computed (41) Daphne	12 43 06.81	+7 16 49.3	2000.0
(41) Daphne	12 38 54.31	+7 44 06.8	apparent
Offset from reference star	+ 1 19.04	− 5 26.6	apparent
Reference star	12 37 32.32	+7 48 51.5	1916.0
Reduction to apparent place	+ 2.95	− 18.1	
Discordance		+ 1 00.0	
Transformed reference star	12 41 47.976	+7 21 12.85	2000.0
No star; transformed (41)	12 43 06.94	+7 16 47.5	2000.0
Search over wider area			
PPM 159082	12 41 36.772	+7 18 56.37	2000.0
Annual proper motion	− 0.0022	+ 0.037	
Re-reduced (41) Daphne	12 42 55.94	+7 13 28.4	2000.0
PPM 159085	12 41 47.725	+7 22 13.55	2000.0
Annual proper motion	− 0.0015	+ 0.002	
Re-reduced (41) Daphne	12 43 06.84	+7 16 48.5	2000.0
PPM 159088	12 41 50.133	+7 23 48.70	2000.0
Annual proper motion	− 0.0051	− 0.064	
Re-reduced (41) Daphne	12 43 09.54	+7 18 29.2	2000.0

TABLE 1. Re-reduction of the Arcetri observation of (41) Daphne on 1916 Apr. 26.

the 1916.0 position of the reference star. In this case there would be no match, and the apparent position of (41) Daphne (which is now free from the 1′ error) would be transformed to FK_5/J2000.0. The search for the reference star would then be made over a larger area, three candidates being revealed within several minutes of arc of the indicated position. The offset is used to derive a position for the minor planet that corresponds to each of the stars, and comparison with the transformed position (or, if necessary, the position derived later from the orbit) quickly reveals the one that is correct.

It is not necessary to utilize all the redundancy. The program works if only an apparent position (which could even be derived from an ephemeris) and an offset are available. Although errors in the offset are more difficult to uncover, the offset can be derived as long as the apparent position of the reference star is either provided directly or can be obtained from the mean position and reduction to apparent place—or if same-equinox mean positions are supplied both for the object and the reference star.

In deciding to publish dependences, the early photographic observers felt that subsequent users would be able to improve the results by correcting the positions of the

	α		δ	
Published	$15^h 44^m 00^s.26$		$-13°22'46''.8$	
Relative xy measures				
	-33865		-07786	
	-24021		-21183	
	-18431		$+22396$	
	-01639		-23282	
	$+33695$		$+23097$	
	$+38589$		-04549	
Derived reference-star positions				
	15 40 08.19		-13 35 45.2	
	15 41 15.65		-13 58 05.0	
	15 41 53.96		-12 45 27.4	
	15 43 49.03		-14 01 34.9	
	15 47 51.16		-12 43 17.4	
	15 48 24.90		-13 30 21.3	
PPM 230557	15 40 08.324	$-0^s.0008$	-13 35 55.05	$0''.000$
PPM 230594	15 41 15.975	$+0.0001$	-13 58 13.14	-0.003
PPM 230610	15 41 53.570	-0.0001	-12 45 35.66	$+0.009$
PPM 230657	15 43 49.800	-0.0008	-14 01 36.78	-0.018
PPM 230763	15 47 49.771	-0.0003	-12 43 55.77	-0.016
PPM 230775	15 48 25.003	0.0000	-13 29 57.42	$+0.003$
Improved	15 44 00.30		-13 22 47.9	
	15 44 00.30		-13 22 48.4	
	15 44 00.30		-13 22 47.4	
"Schmidt"	15 44 00.31		-13 22 48.3	

TABLE 2. Re-reduction of Palomar observation of 1976 GD$_2$.

reference stars in much the same way as for the micrometric data. In retrospect, it might have been more desirable to give the xy measurements instead, and if these were suitably averaged, scaled, rotated and specified with respect to the object of interest, it would not in fact have been necessary to identify the stars, so the amount of space utilized would have been the same as was occupied by the dependences.

An example, showing what might have been done in the case of the Palomar observation of 1976 GD$_2$ on 1993 Apr. 23.46493 UT (Helin 1993), is given in Table 2. In performing plate reductions at the Minor Planet Center it is the practice first to adjust the xy measures so that the scale and orientation are precisely compatible with two of the reference stars, and the xy measures of the minor planet or comet are then subtracted from the figures for each star. The published J2000.0 position of 1976 GD$_2$, shown at the top of the table, had been adjusted from an original 1950.0 position in the manner already discussed, and the relative xy values for the six reference stars succinctly listed beneath it were obtained using the 1950.0 *SAOC* positions (corrected for proper motion) of the first and last stars. These numbers, given here simply as integers, therefore roughly correspond to the offsets of the reference stars from the minor planet in right ascension and declination in units of 0.1 second of arc. Approximate J2000.0 values of the positions of the reference stars can thus be derived from the published values for the minor planet (the right ascension differences being divided by the cosine of the declination). As in the micrometric case, inspection of the on-line *PPM* then allows the stars to be identified, and their *PPM* positions and proper motions can be used with the xy figures to perform a full plate-constants solution and obtain FK_5/J2000.0 coordinates for the minor planet. The position labeled "improved" is this linear solution. The following lines show the effect of eliminating the fifth and the third stars, respectively, the elimination of the fifth giving the most plausible residuals.

In retrospect, it is rather unfortunate that xy data of this type have not been routinely supplied with each photographic observation, although to be useful the position of the plate center (assumed to be at the centroid of the star configuration in the 1976 GD_2 example) should also be supplied. (Measurements could in fact more usefully be published with respect to a point close to this center, so that the *same* values for the stars would be used when several minor planets are measured on the same plate.) Not only would the process of conversion from a generally unspecified 1950.0 system to FK_5/J2000.0 have thereby been rendered more rigorous, but there would also have been the opportunity of replacing the large number of averaged three-star-dependence solutions with the use of least-squares plate constants. If measures were available for enough reference stars, further improvement might be achieved by solving for the plate constants of the second order and the third-order terms important for a Schmidt observation. The last line in Table 2 is this Schmidt solution for the observation of 1976 GD_2, although there is no check on it because it is exactly defined by the six stars (and the position of the plate center is critical).

The implicit assumption has been, of course, that photographic plates will always be available for remeasurement at some later date. The wisdom of this has to some extent been borne out as the processes of measurement and reduction have steadily become more automated, and some astronomers like to envisage the day when one can work through and fully reduce the images on a pile of plates as quickly as it takes to remove the plates from their envelopes. In practice, remeasurement (and, for many plates taken more than a quarter of a century ago, the *first* measurement of positions for which only rough approximations have been published) is currently being carried out on a more-or-less routine basis only at the Lowell Observatory, which has in its possession the plates obtained at Indiana University during 1949-1966, as well as its own even more extensive collection. But it is quite clear that photographic observations tend to become less useful as they age. There are obviously many exceptions to this, notably cases where deliberate searches for critical past observations of specific objects are being made in the Schmidt sky surveys (Palomar, La Silla and Siding Spring), where the measurements are actually made from the widely available copies of the original plates. Otherwise, plates do break, get misfiled or generally deteriorate with age, and even in the best-run observatories the logbooks in which the times of the beginning and end of each exposure were recorded have been known to disappear. While the loss of notebooks can also render impossible the complete reproducibility of micrometric data, errors in the published observation times (and dates) are—for reasons that are not entirely clear—much less likely to occur for micrometric than for photographic data.

If there are difficulties in archiving photographic plates, CCD observations suffer from the problem that the magnetic tapes on which they are recorded have to be reused, sometimes after as short a time as a month. The times of exposure can be more permanently recorded (and generally more reliably than in the old photographic logbooks), and some observers have thought to store separately and more concisely the CCD imagery of the objects of interest and the reference stars—but not of the blank spaces between them. In any case, it seems foolhardy not to retain a computer record of the xy measurements obtained from the CCD images, and perhaps such records should be collected and made accessible under some central control.

4. Accuracy

For the photographic observations of minor planets and comets from the past century or so, the general level of accuracy (involving both external and internal errors) is $\sim 2''$-$3''$. Many observations, particularly those reduced by hand long ago, are significantly worse than this. Some are better, but they are usually not numerous enough to make much difference, even if one can isolate which ones they are. For the bulk of the orbital computations that are made, the level stated above has been considered quite adequate. It is adequate to permit the ephemerides of some 98 percent of the numbered minor planets to be predicted to within $\sim 5''$ for a decade or more into the future, and this is an extraordinary improvement over the situation little more than two decades ago (Marsden 1971). It allows some analysis of the effects of nongravitational forces on short-periodic comets, as well as the determination of the "original" orbits of long-period comets and their membership in the Oort Cloud (Marsden and Williams 1993b), provided that the observations also cover a sufficiently long time span. It has even allowed a good determination of the mass of Ceres (Goffin 1991, Williams 1992).

As already noted, however, we are starting to witness a significant improvement in the general accuracy of the observations. The principal agent of this revolution is the CCD, although some of the more sophisticated procedures used to perform global solutions for Schmidt plates, for example, have played a role. The introduction of the new, uniform, inertial, all-sky catalogues for the reference stars is also helping, particularly for the widespread re-reduction of micrometric data for which old-epoch AGK_1 star positions (and no proper motions) have hitherto been used. As an indication of this improvement, the Minor Planet Center is nowadays generally using $\pm 2''\!.0$ for the maximum residuals accepted in an orbit solution. In the past the cutoff was more like $\pm 2''\!.5$—and frequently higher; furthermore, it was not considered a sin to ignore the elliptic-aberration terms that were present in mean star positions before the introduction of J2000.0.

Of course, there are some purposes for which an accuracy of $\pm 2''\!.0$ is still inadequate. The most obvious of these is perhaps the prediction of occultations by minor planets, where last-minute adjustments to the predicted track have to be made with the help of differential measures of the minor planet and the star to be occulted. Accurate predictions are also required for objects that are to be the targets of space missions. In such cases an individualized reference-star catalogue might be used, and there is sometimes also the possibility of incorporating the very precise fix that a radar detection can offer. In terms of optical astrometry the most consistent positions currently being obtained are those with the automatic meridian circles; Leslie Morrison reports that since 1988 the Carlsberg Automatic Meridian Circle has been giving positions of minor planets brighter than 14th magnitude with external errors that are typically $\pm 0''\!.2$.

The *internal* consistency of many of the CCD data currently being obtained (for much fainter objects) is actually comparable to this—and often better. The problem here is the inadequacy of the *GSC*. Like the *SAOC* before it, the *GSC* was prepared—and very conveniently so—for a specific purpose that is somewhat removed from the highest astrometric principles, and it is well known that the positions of the stars near the edges of the Schmidt plates on which it is based were improperly reduced. Figure 1, provided by James Scotti, shows this inadequacy quite dramatically. The abscissa corresponds to a typical right-ascension CCD scan with the University of Arizona's 0.9-m Spacewatch telescope. The ordinates show the residuals of the *GSC* positions in right ascension and declination. The field is some 10° wide, and the discontinuities corresponding to the edges of the 6° Schmidt plates are very obvious.

A proper re-reduction of the *GSC* must therefore be considered as the prime need for the future of astrometry of minor planets and comets—if there is to be further improvement in the external accuracy of the bulk of the observations made in this CCD era. It is gratifying to know that such a re-reduction is essentially now complete (Bucciarelli et al. 1993), with, further, the *PPM*, rather than the old *SAOC*, used for the reference stars. The absence of proper motions can be expected to become a problem by the end of the century, and it is hoped that, in due course, this point can be addressed, presumably by making use of the first-epoch Palomar Sky Survey and the *AC*. Star catalogues that are even more dense than the *GSC* would, of course, reduce below the present 5-8 arcmin the field size where it becomes necessary to make reductions using field transfers.

Yeomans and Chodas (1993) have made the point that orbits of minor planets and comets established with the help of the positions of the perturbing planets given by the J2000.0 standard DE200 should incorporate the effects of relativity. Since the DE200 planetary positions were themselves fitted to the observations with the help of a relativistic model, the point has a certain validity. But the nut-and-sledgehammer analogy does rather come to mind when one considers that any relativistic effect must be completely masked by the uncertainty of the Newtonian orbital solution for more than 99.98 percent of the known members of the solar system—even when the relativistic computation is carried out correctly for the remainder (cf. Yeomans 1992). Furthermore, although the perturbative effects of some of the larger minor planets were taken into account in its production, DE200 does not in fact include coordinates for these bodies. Taken as a whole, the orbit computations for the numbered minor planets are far more likely to be deficient because the effect of Ceres was neglected—or even because erroneous values are used for the masses of Uranus and Neptune (Standish 1993). Simon Newcomb was well aware of the need for consistency in orbital computations, and he produced a superb system of planetary constants and Newtonian ephemerides (relativistic corrections subsequently being supplied for use, if desired, in the case of Mercury) that served the whole astronomical community tremendously well for almost a century. The principal motivation for the DE series of planetary ephemerides is to assist space operations at the Jet Propulsion Laboratory—a very necessary and worthy endeavor. But the needs of the astronomical community as a whole are somewhat different, and the production of an long-lasting update to Newcomb that meets those needs in the framework of more modern accuracy standards would seem highly desirable.

5. Coverage

The accuracy of the ephemeris of a member of the solar system depends, not only on the innate accuracy of the astrometric observations on which it is based, but also on the distribution of the available observations and the time span covered by them. For a minor planet that is reasonably well observed and does not make particularly close approaches to the earth, a useful rule of thumb is that a properly constructed ephemeris is extrapolable with accuracy approaching that of the observations for a comparable interval of time into the future (or past). In achieving such ephemeris accuracy, one must be aware of the possibility of contamination by observations that do not meet the supposed accuracy standards, either because they do not refer to the same minor planet, or because they are affected by gross errors.

Minor planets are particularly prone to the problem of misidentification. The simplest way to avoid misidentification is by progressively making observations in pairs (Marsden

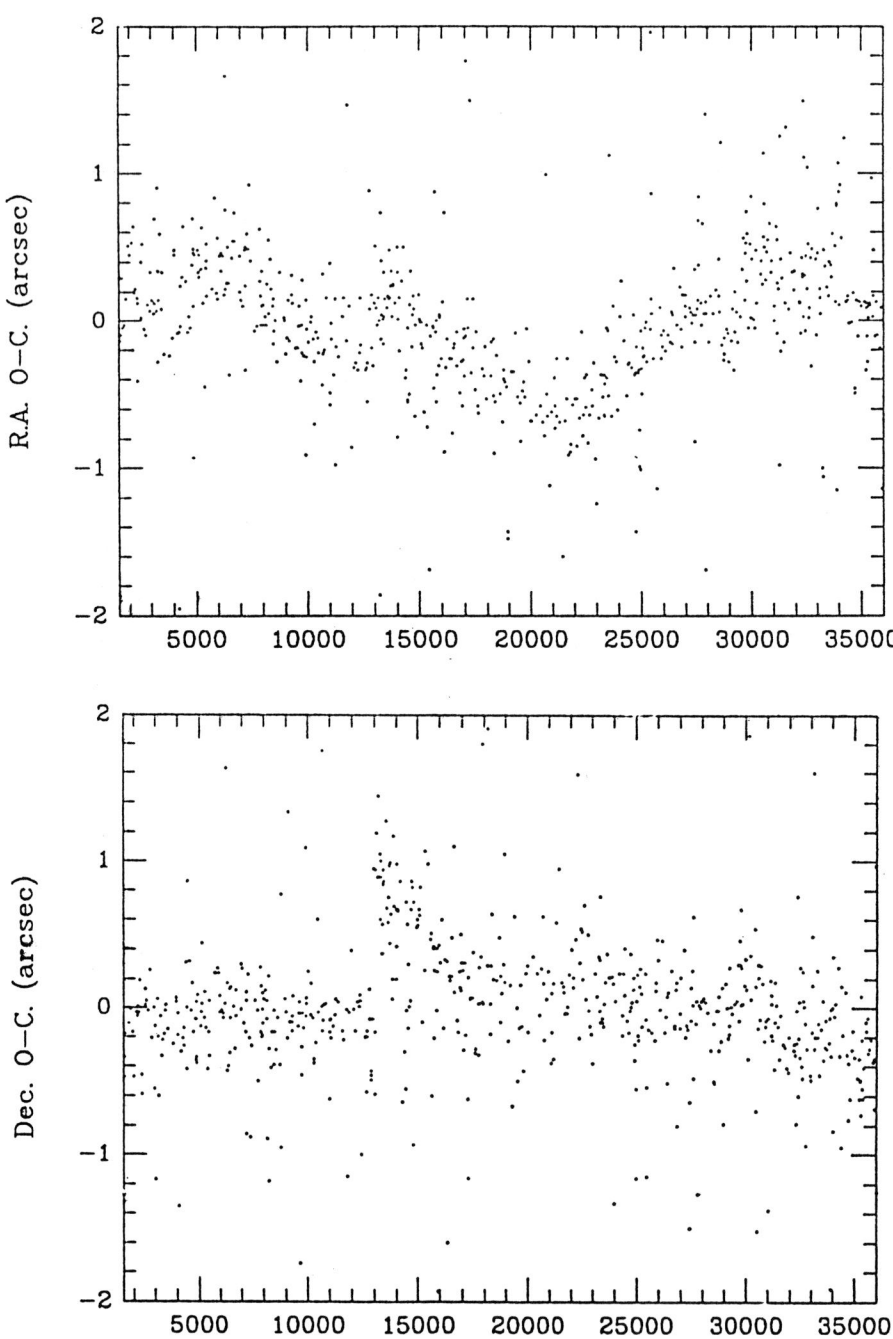

FIGURE 1. *GSC* reduction by James Scotti of a Spacewatch field observed on 1993 May 24. The 36 000 pixels along the abscissa correspond to $\sim 10°$. The upper plot shows the residuals in right ascension, the lower those in declination.

1986). Two observations made a short while apart on the discovery night should be accompanied by another pair a few nights later. There is little point in observing a new near-opposition discovery of a main-belt minor planet at the same time on several nights over an interval of much less than a week, for the small departure of the motion from a great circle renders a general orbit solution indeterminate. However, since it is the amount by which this departure exceeds the accuracy of the observations that is important, it will perhaps become possible to make useful orbit determinations from somewhat shorter time spans than in the past. Indeed, one can seriously consider the possibility of determining initial two-night general orbits using observations separated by several hours on the first or second night (Marsden 1992), where even for a minor planet in the main belt the diurnal parallax usefully increases the departure from a great circle.

The Minor Planet Center generally refrains from giving recognition to a discovery until accurate measurements are received for two nights. Bowell et al. (1989) have discussed in detail an appropriate regimen for month-to-month and year-to-year follow-up, and observational coverage that adheres to this regimen should, while maximizing efficiency, allow misidentifications to be precluded, observations greatly in error to be tagged, and the discovery eventually to receive a permanent minor-planet number.

Muinonen and Bowell (1993) have recently extended the normal "deterministic" concept of orbit improvement to utilize covariance matrices and Bayesian probabilistic analysis. As a result, they have developed a more precise formulation that enables one to establish when follow-up observations of a particular object would be particularly useful, when the orbit is well-enough known that the object should qualify for numbering, and when further astrometry would be an unnecessary waste of effort. There is much merit to this formulation, although the Minor Planet Center prefers to use an *a priori* criterion, rather than this necessarily *a posteriori* process, to examine when an object might be considered for numbering. Nevertheless, the Center does nowadays generally publish new orbit improvements for multiple-opposition orbits whenever new observations become available.

The Minor Planet Center's criterion for consideration for numbering uses a scoring system that counts the total time span covered by the observations, the number of oppositions and the coverage at those oppositions (single-night observations at the first or last oppositions being ignored), as well as the distribution of the observations during different months of the year. Of course, an object that scores well could still be rejected for numbering if the actual orbit computation shows deficiencies in the observations. In practice, this scoring system gives results that are quite comparable to those of the Muinonen-Bowell method. One exception is that the Minor Planet Center insists that observations should have been obtained—and with reasonable coverage—at a recent opposition. The other is that Trojans and other objects of long-period are numbered earlier than Muinonen-Bowell would stipulate; this could easily be fixed by changing the scoring so that the time-span is measured in revolution periods as well as in years, but it was decided to balance the accuracy against the patience of observers who may already have provided excellent coverage at three or four oppositions. Accuracy is not really compromised by this, in the sense that there is no possibility that an object will be lost, and the Minor Planet Center continues to observe Herget's boast that no minor planet numbered after he took over the reins (at number 1565) in 1947 has subsequently been lost; furthermore, of those lost previously, all but (719) Albert—which has an unusual, eccentric orbit that can render the object bright only infrequently—have now been found.

If the recommended observing regimen has been compromised, however, gross errors and misidentifications can go unnoticed, and caution should also be exercised in accepting the conclusions of Bayesian probabilities for comets whose motions are affected by nongravitational forces. The problem can be particularly troublesome for minor planets observed at only one opposition, as in a study by Muinonen and Bowell (1992) of earth-crossing objects. Although the method's authors are careful to dismiss observations that are made on a single night at any opposition, it is not so easy to deal objectively with isolated observations of a single-opposition object. Potential observers can therefore be misled into attempting a later recovery when the uncertainty is greatly underestimated. A case in point is 1989 ML, where the positional uncertainty in November 1992 was computed to be $\sim 1°$. As discussed in more detail by Marsden (1993), the object was then found by accident some $28°$ from the prediction. Advocates of the Bayesian approach would like to see covariance matrices published as a matter of course. For single-opposition orbits covariance matrices are obviously much less useful than knowledge of the distribution of the observations and how well they are represented; individual residuals are more basic than covariance matrices anyway, even for multiple-opposition objects.

REFERENCES

Abetti, A. 1917 Journal des Observateurs, **2**, 20.

Bec-Borsenberger, A. 1990 A&AS, **86**, 299.

Bowell, E., Chernykh, N. S. & Marsden, B. G. 1989 In Asteroids II, ed. R. P. Binzel, T. Gehrels and M. S. Matthews, 21. Tucson: University of Arizona Press.

Bucciarelli, B., Doggett, J. B., Sturch, C. R., Lasker, B. M., McLean, B. J., Lattanzi, M. G. & Taff, L. G. 1993 GSC 1.2: A new astrometric reduction of the HST Guide Star Catalog. This conference.

Comrie, L. J. 1925 Popular Astronomy, **33**, 382.

Goffin, E. 1991 AA, **249**, 563.

Heckmann, O., Dieckvoss, W. & Kox, H. 1954 AJ, **59**, 143.

Helin, E. 1993 Minor Planet Circular No. 22149.

Helmer, L. & Morrison, L. V. 1985 Vistas in Astronomy, **28**, 505

Marsden, B. G. 1971 In Physical Studies of Minor Planets, ed. T. Gehrels, 639. Washington: National Aeronautics and Space Administration.

Marsden, B. G. 1986 In Asteroids, Comets, Meteors II, ed. C.-I. Lagerkvist, B. A. Lindblad, H. Lundstedt and H. Rickman, 1. Uppsala: Uppsala University Observatory.

Marsden, B. G. 1992 In Observations and Physical Properties of Small Solar System Bodies, ed. A. Brahic, J.-C. Gérard and J. Surdej, 251. Liège: Institut d'Astrophysique.

Marsden, B. G. 1993 IAU Symposium No. 161, in preparation.

Marsden, B. G. & Williams, G. V. 1993a Catalogue of High-Precision Orbits of Unnumbered Minor Planets, Cambridge, Mass.: Smithsonian Astrophysical Observatory.

Marsden, B. G. & Williams, G. V. 1993b Catalogue of Cometary Orbits, 8th edition. Cambridge, Mass.: Smithsonian Astrophysical Observatory.

Muinonen, K. and Bowell, E. 1992 Bulletin of the American Astronomical Society, **24**, 642.

Muinonen, K. and Bowell, E. 1993 Icarus, in press.

Rapaport, M., Requième & Mazurier, J. 1989 Minor Planet Circular, Nos. 14443–14459.

Reinmuth, K. 1953 Katalog von 6500 genauen photographischen Positionen Kleiner Planeten. Veröffentlichungen der Landessternwarte Heidelberg-Königstuhl, **16**.

Reinmuth, K. 1960 Katalog von 6000 genauen photographischen Positionen Kleiner Planeten. Veröffentlichungen der Landessternwarte Heidelberg-Königstuhl, **17**.

Schlesinger, F., 1911 ApJ, **33**, 161.

Standish, E. M. 1993 AJ, **105**, 2000.

Williams, G.V. 1992 In Asteroids, Comets, Meteors. 1991, ed. A. W. Harris and E. Bowell, 641. Houston: Lunar and Planetary Institute.

Yeomans, D. K. 1991 Report of the Commission 20 ad hoc System Transition Committee. Pasadena: Jet Propulsion Laboratory.

Yeomans, D. K. 1992 AJ, **104**, 1266.

Yeomans, D. K. & Chodas, P. W. 1993 In Hazards Due to Comets and Asteroids, ed. M. S. Matthews, in press. Tucson: University of Arizona Press.

Astrometry and space missions to asteroids and comets

By DONALD K. YEOMANS

Jet Propulsion Laboratory, 4800 Oak Grove Drive, Pasadena, CA 91109, U.S.A.

To a significant degree, the success of spacecraft missions to comets and asteroids depends upon the accuracy of the target body ephemerides. In turn, accurate ephemerides depend upon the amount, time span and accuracy of the astrometric dataset used to generate the ephemerides. Special reference star catalogs and astrometric quality CCD detectors, when used by experienced observers, can provide observations that are comparable with the accuracy of the reference stars. The observing campaigns are described for the comets and asteroids that have been mission targets and lessons are drawn from these experiences for future target body ephemeris development efforts.

1. Introduction

To date, there have been three comets and two main-belt asteroids visited by spacecraft. Space missions to comets and asteroids began when the International Cometary Explorer (ICE) spacecraft flew approximately 7800 km tail-ward of comet Giacobini-Zinner on September 11, 1985. A few months later in March 1986, six separate spacecraft flew sunward of comet Halley and one of these craft (Giotto) flew as close as 600 km from the comet and was later re-targeted for a 200 km tail-side flyby of comet Grigg-Skjellerup on July 10, 1992. Although its primary science objectives will take place during its rendezvous with Jupiter in late 1995, the Galileo spacecraft flew within 1620 km of the main-belt asteroid 951 Gaspra on October 29, 1991 and within 2400 km of the main-belt asteroid 243 Ida on August 28, 1993. Each of these missions required special efforts to ensure that the orbit and ephemeris of the target body was sufficiently accurate to satisfy the requirements of the various on-board scientific instruments. In general, those missions requiring imaging sequences had the most stringent demands upon the accuracy of the target body ephemerides (comet Halley, and asteroids Gaspra and Ida). Section 2 will discuss some problems that are peculiar to cometary orbit determination and §3 will outline the special efforts undertaken to provide accurate astrometric observations for the ephemeris development of comets and asteroids that are mission targets. Section 4 will preview the Clementine mission that is scheduled to fly within 100 km of the near-Earth asteroid 1620 Geographos in late August 1994 and §5 will draw together some of the most important conclusions resulting from these efforts to improve the ephemeris accuracies of comets and asteroids that are mission targets.

2. Cometary orbit determination

Even after all the gravitational perturbations of neighboring planets are taken into account, the observations of many active comets cannot be well represented without the introduction of additional so called nongravitational effects into the dynamic model. These nongravitational effects are brought about by cometary activity when momentum is transferred to the nucleus by the sublimating mass ices. Whipple (1950, 1951) noted that for an active icy cometary nucleus, a thermal lag between cometary noon and the time of maximum outgassing would introduce transverse nongravitational accelerations in a comet's motion. A rocket-like thrust acting to increase the comet's orbital energy (increase its period) would result from a nucleus in direct rotation while an outgassing nucleus in retrograde rotation would suffer a decrease in its orbital energy. In an attempt to model the effects of these nongravitational forces upon the motions of active comets, Marsden et al. (1973) introduced what has become the standard, or symmetric, nongravitational acceleration model for cometary motions; a rapidly rotating cometary nucleus is assumed to undergo vaporization from water snow that acts symmetrically with respect to perihelion. That is, at the same heliocentric distance before and after perihelion, the cometary nucleus experiences the same nongravitational acceleration. This model was used to successfully represent the nongravitational accelerations affecting the motions of comets Giacobini-Zinner and comet Halley. For the very active comet Halley, it was necessary to model the offset between the observed photometric center of the comet's image and the comet's unseen center-of-mass. This offset was assumed to vary along the comet-sun line with an inverse square dependence on the heliocentric distance. A recent estimate for this offset at one AU from the sun was determined to be 848 km sunward (Yeomans 1993). This result is in accordance with Medvedev's (1993) theoretical estimate of about 1000 km for Halley's post-perihelion offset in the photocenter. Comet Halley produces far more gas and dust than most short-periodic comets so this modelled offset between the comet's observed center-of-light and its center-of-mass may only be appropriate for comet Halley.

3. Astrometric observations for comet and asteroid mission targets

3.1. *The Encounters with Comets Giacobini-Zinner, Halley and Grigg-Skjellerup*

Although not originally intended as a cometary spacecraft, the International Sun Earth Explorer 3 (ISEE-3) spacecraft was re-targeted with a series of Earth and lunar swingbys to fly through the tail of comet Giacobini-Zinner (Von Rosenvinge et al 1986). Before encountering the comet on September 11, 1985 this spacecraft was renamed the International Cometary Explorer (ICE). A group of experienced astrometric observers who were already cooperating within the Astrometry Network of the International Halley Watch provided astrometric data in support of the encounter and while the ICE spacecraft was not equipped with a camera, the magnetometer clearly showed a magnetic field signature indicating that the spacecraft passed directly through the tail of comet Giacobini-Zinner. The ICE spacecraft subsequently made a distant flyby of comet Halley on March 25, 1986. Through the efforts of the Astrometry Network of the International Halley Watch, there were a number of special efforts undertaken to refine the ephemeris accuracy of comet Halley prior its March 1986 encounter with six separate spacecraft (see Figure 1). For much of the older 1909-1911 data, Röser (1987) re-reduced the existing astrometric positions using modern reference star catalogs and for the 1986 return, Burton Jones, Arnold Klemola of the Lick Observatory and William Owen, Jr. of the Jet Propulsion Laboratory generated a special reference star catalog based upon the AGK3R that was distributed to a large group of international observers. Each observer was then asked to reduce their astrometric measurements of comet Halley with respect to this special catalog.

Since the Soviet VEGA 1 and VEGA 2 spacecraft flew past comet Halley a few days prior to ESA's Giotto spacecraft, a knowledge of the inertial position of the Soviet spacecraft from ground-based radio tracking and the VEGA camera angles necessary to image the comet were used to improve the comet's ephemeris just prior to the March 14, 1986 flyby of Giotto. Together with the accurate ephemeris provided by the ground-based as-

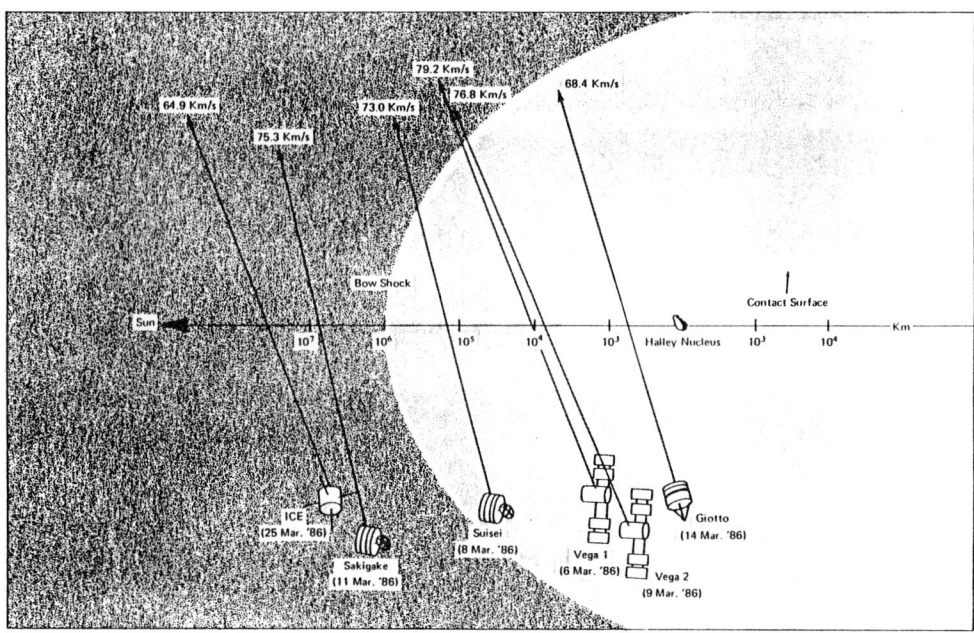

FIGURE 1. A schematic diagram showing the flyby circumstances of each of the six spacecraft that encounter comet Halley in March 1986.

trometry, this pathfinder effort (Munch et al. 1986) allowed the Giotto spacecraft to fly on the sunward side of comet Halley to within 620 km. Using an Earth close approach in July 1990, the Giotto spacecraft was re-targeted for a July 10, 1992 tail side flyby of comet Grigg-Skjellerup. Although Giotto's Multicolor camera was disabled by a particle hit just prior to the Halley close approach in March 1986, many of the particles and fields instruments were operational during the 1992 Grigg-Skjellerup encounter.

3.2. *The Galileo spacecraft encounters with main-belt asteroids 951 Gaspra and 243 Ida.*
Launched from the Kennedy Space Center aboard the Shuttle Atlantis on October 18, 1989, the Galileo spacecraft will arrive at Jupiter in December 1995. En route to Jupiter, the spacecraft passed by Venus once (Feb. 10, 1990) and the Earth twice (Dec. 8, 1990 and Dec. 8, 1992) to pick up orbital energy. Enroute to Jupiter, the Galileo spacecraft was scheduled to fly past the two main-belt asteroids, 951 Gaspra and 243 Ida, on October 29, 1991, and on August 28, 1993 (see Figure 2). The number of images required to ensure capture of a small asteroid during an encounter depends upon the accuracy of both the asteroid's ephemeris and that of the spacecraft. A spacecraft's position in space can be determined from radiometric tracking to within about 20 km, while asteroid position uncertainties are often more than an order of magnitude larger. Pre-launch analyses of the Gaspra encounter (Murrow et al. 1989) showed that the number and quality of the spacecraft images would depend primarily upon the accuracy of Gaspra's ground-based ephemeris. The subsequent failure of the spacecraft's high gain antenna to open properly limited the spacecraft's optical navigation capability, which further increased the importance of an accurate ephemeris. Under the direction of Arnold Klemola (Lick

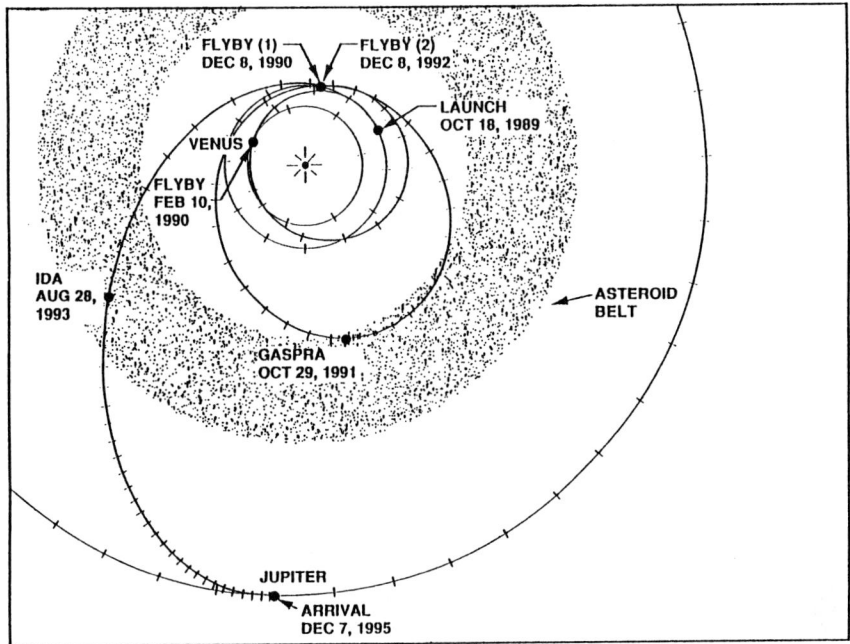

FIGURE 2. This schematic diagram shows the interplanetary Venus - Earth - Earth Gravity Assist (VEEGA) trajectory of the Galileo spacecraft. The time ticks represent 30 day intervals on the Galileo trajectory and the orbits of Venus and Earth; they represent 100 day intervals on the orbit of Jupiter.

Observatory) and William Owen, Jr. (JPL) special reference star catalogs were prepared for two oppositions prior to the Gaspra encounter and for three oppositions prior to the Ida encounter. A small group of experienced observers agreed to reduce their astrometric data with respect to these reference star catalogs. Significantly, the same reference star catalog was used to support the final spacecraft terminal navigation; in the few weeks prior to encounter, the cameras on-board the spacecraft were used to image the asteroids on the star background and these optical navigation images were employed to refine the asteroid ephemerides just prior to the encounters. In general the selection of reference stars for the special Lick catalogs was guided by two criteria. Within one degree of Gaspra's celestial track, about 25 stars per square degree were selected while within 2 arc minutes of this track, all available stars were selected. The second group of stars was intended to provide the necessary density of stars for reducing data taken with CCD detectors - systems that normally have narrow fields of view. The magnitudes of the selected stars were generally in the range from 10 to 17. Before being distributed to the various observers, the star positions were brought into the coordinate system defined by the JPL planetary ephemeris (DE-125) that was standard for the Galileo flight project. Table 1 presents the characteristics of the special Gaspra and Ida star catalogs. The details of the Gaspra astrometric observation campaign are presented in Yeomans et al. (1993) along with a list of the experienced observers who provided the critical data for Gaspra's orbital improvement.

The root mean square orbital residual (observed minus computed) for those Gaspra

	Gaspra				
Observ. period	No. of plates	Epoch	Ref. cat.	No. of stars	Mag. range
10/1989 - 4/1990	5	1990.09	PPM	4970	9 - 16
2/1991 - 9/1991	3	1990.46	P70	5628	9 - 16
	Ida				
5/1990 - 1/1991	3	1990.56	P70	3439	9 - 16
11/1991 - 3/1992	3	1991.93	ACRS	1781	9 - 16
3/1993 - 8/1993	4	1992.34	ACRS	3692	9 - 16

TABLE 1. Special Reference Star Catalogs for asteroids 951 Gaspra and 243 Ida. The time interval for which each catalog was employed is followed by the number of yellow plates taken with the Lick Observatory's 50.8 cm dual astrograph, the epoch of the catalog, the reference star catalog used in the reductions, the number of stars and their magnitude range

and Ida observations made with respect to the special star catalogs was less than $0''\!.3$. CCD observations generally had residuals less than $0''\!.2$ and a limited set of Ida CCD astrometric data that was taken by the personnel at the Naval Observatory's Flagstaff Station achieved a RMS residual of $0''\!.06$ - an unprecedented value for asteroid astrometry. These latter data were reduced by W.M. Owen using the special Ida star catalog, two HIPPARCOS star positions and a image overlap technique. An analysis of on-board optical images from the Galileo spacecraft has shown that the final ground-based ephemeris accuracy for asteroid 951 Gaspra was about 80 km. At the time of this writing, only a single optical navigation image of Ida has been received and processed but a preliminary analysis suggests that the ground-based ephemeris accuracy for ida will be about 80 km as well. Figure 3 presents the ephemeris uncertainties of Gaspra in Gaspra's orbit plane for three cases. The largest error ellipse represents the expected uncertainty prior to encounter based upon an expected, but conservative, observation set. The post-encounter error ellipse (1σ) is based upon the observational data actually received and the smallest ellipse represents the final position uncertainties of Gaspra after all the ground-based astrometric data and the spacecraft optical navigation images were processed.

4. The upcoming encounter with 1620 Geographos – an Earth-approaching asteroid

The Deep Space Program Science Experiment (DSPSE) is a joint project of the U.S. Department of Defense and NASA to first orbit the moon with a small spacecraft and then de-orbit and fly closely past asteroid 1620 Geographos in late August 1994. The DSPSE spacecraft (also known as Clementine) will launch in January 1994, orbit the moon for three months and then use lunar and Earth swingbys to re-shape the spacecraft's orbit and fly sunward of Geographos about August 31, 1994. The nominal flight plan calls for a very close flyby to within 100 km of the asteroid so that an accurate asteroid ephemeris is essential for the success of this portion of the mission. The accuracy of an asteroid's orbit is primarily dependent upon the accuracy and size of the dataset. As a rule of thumb, optical astrometric observations for asteroids are accurate to the $1''$ level because the reference star positions from commonly used star catalogs are not much more accurate than this level. However, when special efforts are made to construct

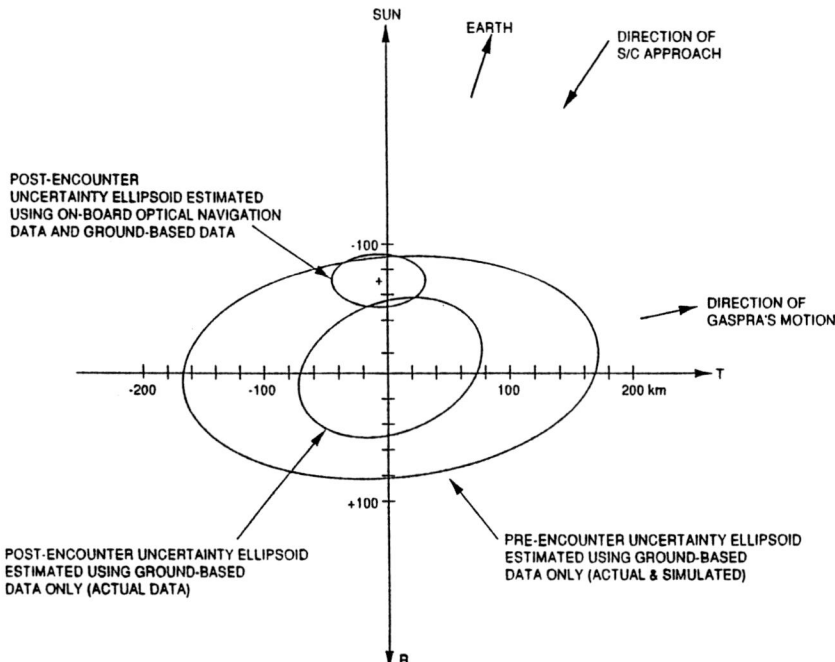

FIGURE 3. The Gaspra position error ellipsoids at the Galileo spacecraft close approach point are viewed projected onto Gaspra's orbital plane. The largest ellipse denotes Gaspra's 1σ position uncertainties resulting from the pre-encounter, ground-based ephemeris development effort. The smaller ellipse denotes the corresponding Gaspra position uncertainties resulting when additional ground-based observations became available. The smallest ellipse represents the final, post-encounter position uncertainties using both on-board and ground-based data.

a more accurate reference catalog of stars along the asteroid's path and these catalogs are used by experienced observers to reduce their CCD observations, the astrometric accuracy of optical measurements can be reduced to the $0.''3$ level or less. In fact, this level of accuracy was achieved for the astrometric observations of asteroid 951 Gaspra and 243 Ida taken during the opposition intervals prior to the respective flyby of the Galileo spacecraft on October 29, 1991 and August 28, 1993. For the 1994 astrometric observations of asteroid 1620 Geographos, current plans call for experienced observers to use reference stars from the Hubble Space Telescope Guide Star Catalog (version 1.2) together with additional stars from the PPM southern reference star catalog. As a rule, the along-track component of an asteroid's position error is the largest because the object's mean motion, or orbital semi-major axis is often poorly known. Orbits based upon a uniform set of optical astrometric data over a long time interval normally result in an orbital semi-major axis that is well determined. Asteroid Geographos has a lengthy optical observation interval extending back to August 1951.

In terms of improving the asteroid's ephemeris accuracy, its proximity to the Earth just prior to the DSPSE spacecraft flyby on August 31, 1994 is particularly fortuitous (see Figure 4). Six days prior to the spacecraft encounter, the asteroid reaches a minimum Earth - asteroid distance of 0.033 AU - a circumstance that should allow the asteroid's ephemeris uncertainties to be driven down to very small levels. As previously noted, a typical accuracy for optical asteroid astrometric observations is $1''$, an angle that subtends

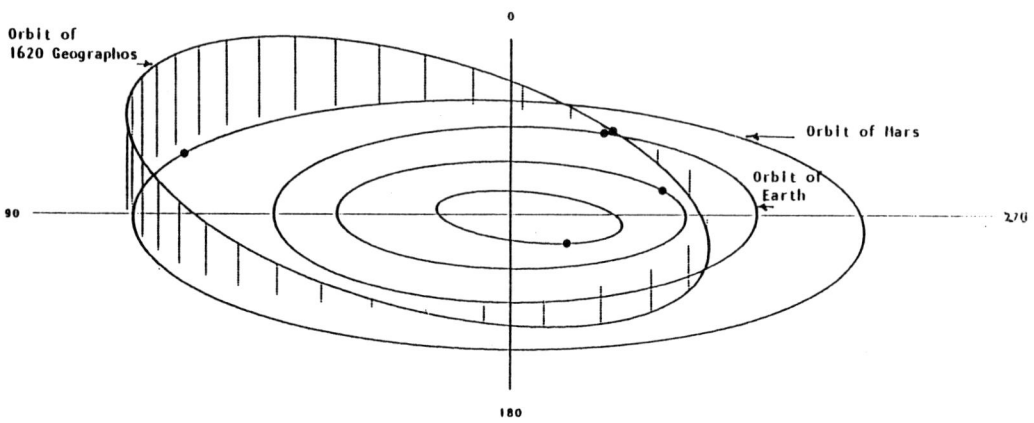

FIGURE 4. The orbit of asteroid 1620 Geographos is displayed together with those of the planets Mercury through Mars. The view is 20 degrees from the ecliptic plane and 180 degrees from the vernal equinox. The positions of the planets and the asteroid are given for the time of the DSPSE encounter on August 31, 1994.

some 725 km at one AU from the Earth. Thus, for optical astrometric observations made at a distance of one AU, individual astrometric observations define the asteroid's position to a only a few hundred kilometers. However, during the Geographos-Earth close approach, a single astrometric observation, accurate to the 1" level, can define the asteroid's position in the plane-of-sky to about 25 km. The close Earth approach is responsible for the asteroid's peculiar motion on the celestial sphere (see Figure 5) whereby the asteroid nearly passes over the south celestial pole on August 23 and moves rapidly northward at several degrees a day. As a result of the asteroid's motion on the sky, the most powerful optical astrometric data will have to be taken in late August by experienced observers in the southern hemisphere.

The asteroid-Earth close approach in August 1994 will also allow radar astrometric observations to be taken at the Goldstone 70-meter radar telescope in California's Mojave desert. When used in combination with optical plane-of-sky observations, radar time delay (range) and Doppler shift (range-rate) measurements are extremely powerful for reducing ephemeris errors because these radial Earth-asteroid observations nicely complement the optical plane-of-sky measurements. In addition, the radar measurements have a fractional precision two orders of magnitude better than optical angle measurements. For example, recent X-band radar observations of asteroid 4179 Toutatis during its close Earth approach in December 1992 provided radar data resolution of $\frac{1}{8}$th microsecond in time delay and less than one Hertz in Doppler shift. Thus, these measurements had a range resolution of a few tens of meters and a radial velocity resolution less than 2 cm/s.

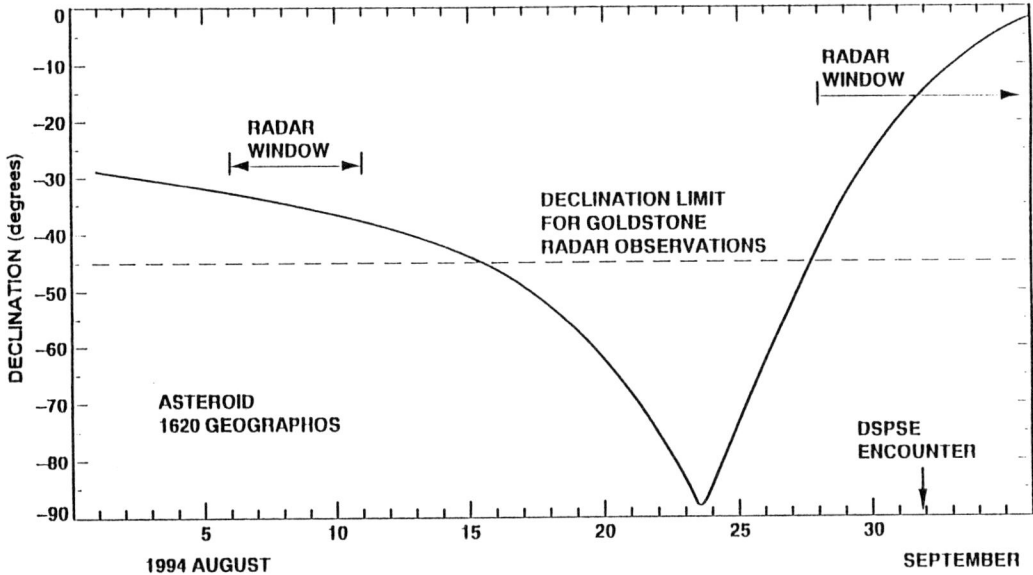

FIGURE 5. The declination of asteroid Geographos is plotted as a function of date in August - September 1994. Radar observations are possible from the Goldstone facility in California when the asteroid's declination is above –45 degrees and the asteroid is less than 0.15 AU from the Earth. In the eight days prior to the DSPSE encounter, the asteroid's declination changes rapidly from –88 degrees to –15 degrees.

Because the radar data were used in near real-time to update the orbit and ephemeris of 4179 Toutatis, the ephemeris error during much of the observation interval was less than the 4-km object itself (Ostro 1993). The Goldstone radar facility is located at 35 degrees north latitude and the antenna is not normally allowed to operate below a 20 degree elevation limit. To extend the time interval over which radar data can be taken, special permission has been requested to operate the antenna with a 10 degree horizon limit. Thus if the asteroid's declination is below 45 degrees south, it will not rise above the observing horizon at Goldstone. In addition, the asteroid's size demands that it be within about 0.15 AU of the Earth before radar observations can be successfully carried out; Geographos satisfies this constraint from the second week of August through the second week in September 1994. With this distance constraint in mind, one notes from Figure 4 that the asteroid's declination allows pre-encounter radar observations of more than a few hours only during the intervals August 6-11 and August 28-31. During the first interval, only Doppler measurements are possible while during the second interval, both Doppler shift measurements at about 1 Hz resolution and sub-microsecond time delay measurements are anticipated.

5. Conclusions

Successful spacecraft encounters have been achieved for three comets and two asteroids. The ephemerides of each of these target bodies was of sufficient accuracy so that all of these missions successfully achieved their scientific objectives. In part, these successes are due to the astrometric observation campaigns and preparations that preceded the arrival of each of the target bodies. Lessons learned from these observation campaigns include the following:

(i) Special reference star catalogs were generated and distributed to a few experienced observers for comet Halley and asteroids Gaspra and Ida. These star catalog efforts paid off handsomely. Observations reduced with respect to these special catalogs had orbit residuals significantly lower than those observations that were not reduced with respect to these special catalogs.

(ii) Experienced observers, who employed the special reference star catalog, generated observations that were near the accuracy level of the star catalog positions themselves. Data reduced with respect to the special reference star catalogs were weighted at least twice the level of the remaining data.

(iii) The final pre-encounter orbits and ephemerides for target bodies should be based upon lengthy data intervals to reduce the uncertainty in the object's mean motion. For example, the final orbit of Gaspra was based upon 342 observations spanning the lengthy interval from 1913 December 19 through 1991 October 5.

(iv) In terms of reducing the position uncertainty of target bodies at encounter, those data taken nearest the Earth are the most powerful. For main-belt asteroids, these Earth close approaches occur during opposition. For Earth approaching asteroids, radar Doppler and time delay measurements are powerful data types for significantly reducing ephemeris errors down to levels comparable to the target's physical size.

(v) The critical role played by dedicated and experienced astrometric observers cannot be overstated. The participation of these observers, who often reduce and transmit their data within hours of actually making the observations, is one of the primary reasons for the string of successful space missions to comets and asteroids.

Acknowledgements

The research described in this paper was carried out by the Jet Propulsion Laboratory, California Institute of Technology, under contract with the National Aeronautics and Space Administration.

REFERENCES

Marsden, B.G., Sekanina, Z., and Yeomans, D.K. 1973. Comets and Nongravitational Forces. V. Astron. J. 78, 211 - 225.

Medvedev, Yu.D. 1993. On the Dust Cloud in a Comet's Head and its Role in Forming the Photocentre Shift Phenomena. Asteroids, Comets, Meteors 1993 poster paper, Belgirate, Italy, June, 1993.

Munch, R.E., Sagdeev, R.Z., and Jordan, J.F. 1986. Pathfinder: Accuracy Improvement of Comet Halley Trajectory for Giotto navigation. Nature, 321, 318 - 320.

Murrow, D.W., Chodas, P.W., and Kallemeyn, P.H. 1989. Galileo Asteroid Encounter Navigation. AIAA paper AAS 89-406.

Ostro, S.J. 1993. Personal communication.

Röser, S. 1987. Catalogue of Astrometric Observations of Comet P/Halley at its Apparition 1909-1911. Astronomy and Astrophysics Supplement Series, 71, 363 - 395.

Von Rosenvinge, T.T., Brandt, J.C., and Farquhar, R.W. 1986. The International Cometary Explorer Mission to Comet Giacobini-Zinner. Science, 232, 353 - 355.

Whipple, F.L. 1950. A Comet Model. I. The Acceleration of comet Encke. Astron. J. 111, 375 - 394.

Whipple, F.L. 1951. A Comet Model II. Physical Relations for Comets and Meteors. Astrophys. J. 113, 464 - 474.

Yeomans, D.K. 1993. A Review of Comets and Nongravitational Forces. In Asteroids, Comets, Meteors, 1993 (A. Milani, M. Di Martino, and A. Cellino, editors), Kluwer publishers, in press.

Yeomans, D.K., Chodas, P.W., Keesey, M.S., Owen, W.M., and Wimberly, R.N. 1993. Targeting an Asteroid: The Galileo Spacecraft's Encounter with 951 Gaspra. Astronomical Journal, 105, 1547 - 1552.

Occultation astrometry: predictions and post-event results

By C. B. OLKIN[1,2] AND J. L. ELLIOT[1,2,3]

[1]Department of Earth, Atmospheric, and Planetary Sciences, Massachusetts Institute of Technology, Cambridge, MA 02139, U.S.A.

[2]Lowell Observatory, Flagstaff, Arizona 86001, U.S.A.

[3]Dept. of Physics, Massachusetts Institute of Technology, Cambridge, MA 02139, U.S.A.

Stellar occultations potentially provide the highest precision data for relating solar system ephemerides to the stellar reference frame. For example, occultations by the Uranian rings can define the position of the occulted star relative to the rings to better than 0.02 mas (equivalent to a few hundred meters at the distance of Uranus). Occultations by atmospheres can be less precise than occultations by symmetrical solid bodies, like rings and large asteroids, with a precision on the order of 1.0 mas. Planetary astronomers have published the results of about 80 occultation observations, including events involving all known planets (with the exception of Mercury) and more than 40 asteroids. In order to improve the relation between solar system ephemerides and stellar reference frames, an astrometric program is needed to accurately determine the positions and proper motions of previously occulted stars. Since only the brightest stars are currently selected for occultation observations – in order to probe the physical properties of planetary systems – many observations that involve fainter stars are not observed. However, these events would be useful for improving planetary astrometry in the stellar reference frame. Also, improvement of the ephemerides relative to FK5 would benefit the occultation prediction process, since most of the effort in the prediction refinement involves modeling differences between the observed planet and its ephemeris.

1. Introduction

In order to adequately predict an occultation for deployment of ground based observers and the Kuiper Airborne Observatory, one needs accurate relative astrometry of the star and occulting body. The accuracies required are a fraction of the angular body radius – to ensure the observer is within the occultation shadow. Small outer solar system bodies like Pluto and Triton (with radii in the range $0\rlap{.}''06 - 0\rlap{.}''07$) require accuracies of $0\rlap{.}''02$. The prediction process has been refined over the years to identify fainter candidates and to achieve these levels of accuracy.

Not only is astrometry important to the prediction of occultations, the resulting occultation timings provide the relative position of the star and occulting body. These data are very accurate (on the order of milli- to micro-arcseconds), especially if the star is occulted by a sharp ring feature or gasless limb.

2. Candidate searches and predictions

Early occultation predictions were performed by comparing star catalogs against planetary ephemerides (Taylor 1963; Taylor 1970; Taylor 1974). This task has been accelerated with the use of modern computing facilities. Currently occultations have been predicted for every planet and nearly 200 asteroids using catalogs like the Astrographic Catalog Reference Stars (ACRS), Hubble Space Telescope Guide Star Catalog (GSC) and Carlsberg Meridian Catalogues (CAMC) (Wasserman et al. 1990; Bosh & McDonald 1992).

This technique of catalog searches is useful for bright planets, because a star of comparable magnitude to the occulting body is needed to provide an adequate signal-to-noise ratio to satisfy most scientific objectives (Elliot 1979).

With the advent of CCD detectors, occultations of fainter bodies–such as Pluto and Triton–could be observed, so that stars fainter than catalog limits could be considered. This was the incentive for observational searches of occultation candidates. The earliest observational searches were performed by A. Klemola & B. Marsden, who found numerous occultations by the Uranian rings using the Double Astrograph at Lick Observatory (Klemola & Marsden 1977). Klemola and his colleagues have expanded this effort to include Neptune and Pluto (Klemola et al. 1981a; Klemola et al. 1981b; Klemola & Mink 1991; Mink & Klemola 1985; Mink et al. 1991). Searches for occultations by asteroids have been carried out at the Lowell Observatory Astrograph (Wasserman et al. 1981; Wasserman et al. 1987; Wasserman et al. 1990). Recently CCD strip scans have been used for these searches (Dunham et al. 1991; McDonald & Elliot 1992) for three reasons: (i) they can record fainter stars than photographic plates for a telescope of the same size, (ii) the raw data are ready for reduction (eliminating the scanning process needed for plates) and (iii) CCDs are now more readily available to observers.

Following the identification of a candidate, further astrometry of the star and occulting body is needed to generate occultation predictions for various sites, if the occulting body has a small angular size. Fortunately, only the relative positions are relevant for predicting the path of the occultation shadow.

As an example of how occultations are currently determined, we shall describe the prediction process used for the 1993 July 10 occultation of Tr60 by Triton. First, a search for Triton occultation candidates was carried out using CCD strips scans from Wallace Astrophysical Observatory (McDonald & Elliot 1992). In that search, the candidate Tr60 was identified along with 129 other candidates through 1994. CAMC positions confirmed the star's location within $0\overset{''}{.}15$ in declination, so plans began in earnest to refine the prediction and to observe the occultation. Strip scans containing both objects began two months prior to the occultation at Lick Observatory and Lowell Observatory. Stare frames from the USNO also began at the same time, however, they did not contain both objects until 4 days before the occultation. Having both objects on the same stare frame or strip scan greatly reduces the errors in their relative positions because both the star and occulting body are reduced with the same set of reference stars. All data were sent to MIT by the fastest means possible, where they were reduced relative to a standard network determined with the Lick Double Astrograph.

The positions of Triton relative to its ephemeris were fit to a model that projected its position relative to Tr60 at the occultation time. This model had three components in both right ascension and declination: (i) a constant offset that accounted for RA and Dec errors of Neptune relative to the star, (ii) a linear time variation that absorbed any rotational or scale error of the reference system, and (iii) a term that accounted for an in-track error in Triton's ephemeris about Neptune. The total number of fitted parameters in this model was five. The result of this procedure was a successful prediction (McDonald et al. 1993), which allowed several teams of ground-based observers and the KAO to be deployed within Triton's shadow (Olkin et al. 1993). However, due to weather and instrument problems, only the KAO observed the occultation (Elliot et al. 1993).

3. Post-event results

Occultation timings could become a valuable tool for linking the FK5 frame and the solar system frame. Over eighty occultations have been observed using modern detectors

Planet or satellite	Number of published occultations	References
Mercury	0	
Venus	1	(deVaucouleurs & Menzel 1960; Taylor 1963)
Mars	1	(Elliot et al. 1977; Hubbard 1979; French & Taylor 1981)
Asteroids	>40	(Millis & Elliot 1979; Millis & Dunham 1989)
Jupiter	3	(Baum & Code 1953; Hubbard & Van Flandern 1972; Veverka et al. 1974)
Saturn	2	(French et al. 1993; Hubbard et al. 1993; Elliot et al. 1993)
Uranus	20	(Elliot & Nicholson 1984; French et al. 1991)
Neptune	8	(French et al. 1985; Hubbard et al. 1987)
Pluto	1	(Elliot et al. 1989; Millis et al. 1993)
Io	1	(Taylor et al. 1971)
Ganymede	1	(Carlson et al. 1973)
Titan	1	(Hubbard et al. 1992)
Triton	1	(Olkin et al. 1993; Elliot et al. 1993)
Charon	1	(Walker 1980; Elliot & Young 1991)

TABLE 1. Occultations observed

and timing methods (see Table 1). Some of these events provide very accurate relative positions between the star and the occulting body. For example, solutions for Uranian ring orbits have given formal errors considerably less than a kilometer in the shadow center (French et al. 1991; Mason 1992). This translates to an error of about 0.02 mas between the star and the center of Uranus. The Saturn ring models are not as precise, but Elliot et al. (1993) find uncertainties in the closest approach between the star and Saturn to be 0.5 mas for the 28 Sgr occultation and 0.6 mas for the GSC6323-01396 occultation. For objects without sharp edges, such as Pluto's atmosphere, accurate astrometry is still possible. The uncertainty in the mid-time of the atmospheric occultation of P8 by Pluto is 0.046 seconds, which translates to 0.04 mas at Pluto (Elliot et al. 1989; Millis et al. 1993).

4. Conclusions

Astrometric networks of faint stars near the outer planets would aid the current methods of occultation prediction. However, even with current techniques, occultations by bodies with an angular size similar to Triton and Pluto can be accurately predicted.

In order to use occultation timings to improve planetary ephemerides, three things need to be accomplished: (i) past and future occultation candidate stars need to be measured accurately in the FK5 system, (ii) the occultation light curves need to be reduced relative to modern ephemerides, and (iii) the results would have to be included in new calculations of planetary ephemerides. Additional astrometric data could be obtained from the many occultations not observed because their signal-to-noise is too low for physical studies, but observations of these events would provide sufficient signal-to-noise for accurate relative astrometry.

Occultation timings could provide improved planetary ephemerides, which would in turn facilitate occultation predictions and improve the tie between solar system ephemerides and the FK5 system.

Acknowledgments

This work was supported in part by NASA Grants NAGW-1494, NAG 2-836 and NAG 2-811.

REFERENCES

Baum, W. A., & A. D. Code, 1953, AJ, 58, 108
Bosh, A. S., & S. W. McDonald, 1992, AJ, 103, 983
Carlson, R. W., J. C. Bhattacharyya, B. A. Smith et al., 1973, Science, 182, 53
deVaucouleurs, G., & D. H. Menzel, 1960, Nat, 188, 28
Dunham, E. W., S. W. McDonald, & J. L. Elliot, 1991, AJ, 102, 1464
Elliot, J. L., 1979. Ann. Rev. Astron. Astrophys., 17, 445
Elliot, J. L., A. S. Bosh, M. L. Cooke et al., 1993, AJ (in press).
Elliot, J. L., E. W. Dunham, A. S. Bosh et al., 1989, Icarus, 77, 148
Elliot, J. L., E. W. Dunham, & C. B. Olkin, 1993, Bull. Amer. Astron. Soc. (in press).
Elliot, J. L., R. G. French, E. Dunham et al., 1977, ApJ, 217, 661
Elliot, J. L., & P. D. Nicholson, 1984, in Planetary Rings (R. Greenburg & A. Brahic, Ed.), p. 25, University of Arizona Press, Tucson.
Elliot, J. L., & L. A. Young, 1991, Icarus, 89, 244
French, R. G., P. A. Melroy, R. L. Baron et al., 1985, AJ, 90, 2624
French, R. G., P. D. Nicholson, M. L. Cooke et al., 1993, Icarus, 103, 163
French, R. G., P. D. Nicholson, C. C. Porco et al., 1991, in Uranus (J. T. Bergstralh, E. D. Miner & M. S. Matthews, Ed.), p.327, Univ. Az. Press, Tucson.
French, R. G., & G. E. Taylor, 1981, Icarus, 45, 577
Hubbard, W. B., 1979, ApJ, 229, 821
Hubbard, W. B., P. D. Nicholson, E. Lellouch et al., 1987, Icarus, 72, 635
Hubbard, W. B., C. C. Porco, D. M. Hunten et al., 1993, Icarus, 103, 215
Hubbard, W. B., B. Sicardy, R. Miles, A. J. Hollis et al., 1992, AA, 269, 541
Hubbard, W. B., & T. C. Van Flandern, 1972, AJ, 77, 65
Klemola, A. R., & B. G. Marsden, 1977, AJ, 82, 849
Klemola, A. R., & D. J. Mink, 1991, AJ, 102, 389
Klemola, A. R., D. J. Mink, & J. L. Elliot, 1981a, AJ, 86, 135
Klemola, A. R., D. J. Mink, & J. L. Elliot, 1981b, AJ, 86, 138
Mason, E. C., 1992, The Rings of Uranus. B.A. Thesis, Wellesley College.
McDonald, S. W., & J. L. Elliot, 1992, AJ, 104, 862
McDonald, S. W., C. B. Olkin, E. W. Dunham et al., 1993, Bull. Amer. Astron. Soc. (in press).

Millis, R. L., & E. W. Dunham, 1989, in Asteroids II (R. P. Binzel, T. Gehrels & M. S. Matthews, Ed.), p.148, The University of Arizona Press, Tucson.

Millis, R. L., & J. L. Elliot, 1979, in Asteroids I (T. Gehrels, Ed.), pp. 98, The University of Arizona Press, Tucson.

Millis, R. L., L. H. Wasserman, O. G. Franz et al., 1993, Icarus (in press).

Mink, D. J., & A. Klemola, 1985, AJ, 90, 1894

Mink, D. J., A. R. Klemola, & M. W. Buie, 1991, AJ, 101, 2255

Olkin, C. B., J. L. Elliot, E. W. Dunham et al., 1993, Bull. Amer. Astron. Soc. (in press).

Taylor, G. E., 1963, Royal Obs. Bull. E355

Taylor, G. E., 1970, MNRAS, 147, 27

Taylor, G. E., 1974, Tech. Note No. 34. Nautical Almanac Office, Royal Greenwich Obs., Hailsham, Sussex, England.

Taylor, G. E., B. O'Leary, T. C. Van Flandern et al., 1971, Nat, 234, 405

Veverka, J., L. H. Wasserman, J. Elliot et al., 1974, AJ, 79, 73

Walker, A. R., 1980, MNRAS, 192, 47p

Wasserman, L. H., E. Bowell, & R. L. Millis, 1981, AJ, 86, 1974

Wasserman, L. H., E. Bowell, & R. L. Millis, 1987, AJ, 94, 1364

Wasserman, L. H., E. Bowell, & R. L. Millis, 1990, AJ, 99, 723.

Methods for development of satellite theories

By P. J. MESSAGE

Dept of Applied Mathematics and Theoretical Physics, The University, PO Box 147, Liverpool L69 3BX, U.K.

Methods are briefly reviewed for the development of theories for the prediction and analysis of natural satellite motions, considering a few examples from the many recent researches in this field, including both analytical developments and numerical integrations. These techniques can be used in mutually supporting ways, each having its own appropriate contributions to make.

1. Introduction

Before considering some of the methods in use for the construction of ephemerides to predict the future positions of planetary satellites, let us first bring to mind the three main types of perturbative influence on satellite motion:

First, the effect of the fact that the primary planet is not exactly spherically symmetrical (the "oblateness" terms). These are dominant in the motion of Jupiter's inner satellite Amalthea, and make significant contributions (though are not dominant) in the cases of all but the outermost satellites of Saturn (in the case of Mimas and Tethys, and Enceladus and Dione, they help to govern the nature of the long-term perturbations), and the Galilean satellites of Jupiter, but they are readily treated by relatively simple analytical methods (though second-order terms are necessary for the secular motion of the node of Amalthea).

Second, the perturbative effect of the Sun. This of course underlies the complexity of the motion of the Moon, and causes the motion of the outer satellites of Jupiter to be so very far from even approximately elliptical. For all but the Moon, and the outermost satellites of Jupiter, Saturn and Neptune, however, the solar terms are not dominant, and succumb to first-order analytical perturbation theory, with a relatively small number of terms.

Third, the mutual perturbations of the satellites. In many cases these take on a dominance as a result of near-commensurabilty of the orbital periods. One may characterise resonance as "shallow", where, while it is close enough for the correspondingly small linear combinations of the mean orbital motions in longitude, appearing as denominators of the appropriate terms in the perturbation series, to cause the appropriate terms in those series to have markedly enhanced amplitudes, (and markedly long periods), the usual methods of calculating the perturbations in powers of the disturbing masses nevertheless still succeed. Let us call "deep" resonance those cases where these methods fail, and the effect of the resonance is to induce a different type of motion, related to periodic solutions of the three-body problem (of Poincaré's first, second or third sorts), or oscillatory motion about them. This of course dominates the motion of the three innermost Galilean satellites of Jupiter, and many pairs amongst the satellites of Saturn. One may further classify deep resonance according to whether the associated small linear combination of mean motions, as a fraction of either of the mean motions, is still a few powers of ten larger than the mass of the perturbing satellite, as a fraction of the mass of the primary planet (let us call this "moderately close" resonance), or whether the first of these two ratios is not appreciably larger than the second (let us call this "very close" resonance). Examples of "moderately close" resonance are, the pairs Mimas and Tethys, and Enceladus and Dione, in Saturn's system, and, amongst the Galilean satellites of Jupiter, the pairs Io and Europa, and Europa and Ganymede. The pair Titan and Hyperion in Saturn's system provides an example of "very close" resonance, in this terminology, and this underlies the particular complexity in the motion of Hyperion, in which the associated periodic solution of the three-body problem is of Poincaré's second (high eccentricity) sort (see Message 1966).

2. Development of Theories

Many varieties of method are currently in use for the construction of theories of satellite motion, ranging from completely analytical perturbation theories, through theories in which particular numerical values of some parameters have been used at some stages in the development, to direct numerical integration of the equations of motion in rectangular co-ordinates. The choice of method used is governed by the nature of the dominant features of the particular satellite motions being studied, or by the purpose of the particular investigation in hand, or by the predilection of the investigator. This review cannot be extensive enough to be in any way exhaustive, but let us attempt to sample from amongst the main varieties. Where, to the precision sought, first-order perturbations arising from a relatively few terms in the appropriate disturbing function are all that are significant, an analytical development of the Lagrange differential equations governing the rates of change of the parameters of the instantaneous Keplerian orbit will usually provide the most effective way to proceed. The analytical expressions obtained for the perturbations as multiple Fourier series in the appropriate linear functions of time may be readily re-evaluated for changed values of the parameters of the system, i.e. masses of satellites, the co-efficients (J_n) in the non-spherical harmonics in the gravitational potential of the primary planet (oblateness parameters), pole vector, or of the parameters of the solution, i.e. mean values of the elements, etc.

It is where the perturbations are larger than can be treated as first-order perturbations of Keplerian motion that a variety of choices of method is usually found, and the increased precision to be expected from future space missions, e.g. Cassini, will make this true for more satellites. In a few cases, development of analytical perturbation theory as far as the second order gives what is needed, but in the frequent cases of deep resonance, if analytical methods are used, special solutions are needed for the long-term part of the motion, taking account of the particular features of the type of resonance in question, as pioneered by H. Struve for the orbital inclination-based resonance of Mimas and Tethys, and the eccentricity-based resonance of Enceladus and Dione, and by Newcomb and, later, Woltjer, for the very close resonance with Titan in the motion of Hyperion. The large mean eccentricity in this latter case leads to the need to use expansions in powers of the difference of the eccentricity from a value near its mean rather than of the eccentricity itself (and it proves also convenient to use expansions in powers of the difference of the major semi-axis from a value near its mean) (see Message, 1989).

Many investigators use the numerical solution of the differential equations of motion, written in terms of rectangular co-ordinates, frequently choosing a central difference integration scheme of the Gauss-Jackson type (as was in fact recommended by Cowell and Crommelin in their pioneering of this approach to predict the 1910 apparition of Halley's comet). This of course has the advantage of giving the positions of the satellites directly, and makes it simple to use a full model of the system, though the results are valid only for the particular set of parameters and initial conditions chosen. From the positions and velocity components given by the numerical integration at each time step, parameters of the instantaneous Kepler orbit may of course be calculated, and analysis of the resulting run of values carried out to give multiple Fourier expressions for the perturbations of the orbital elements of the kind resulting from an analytical perturbation theory (e.g. Taylor 1992).

A further approach is to carry out numerical solution of the Lagrange equations for the Keplerian orbital parameters, or their equivalent, and once again to follow this by analysis of the run of values to give multiple Fourier expressions for the perturbations of the orbital elements (e.g. Vienne and Duriez, 1991).

3. Comparison with Observations

No theory can be better as a predictor than the observational data set on which it is founded allows, through the particular observationally-determined parameters of the theory. In the case of numerical solution of the equations for the rectangular co-ordinates, these will be the initial values of the co-ordinates and velocity components, and, in the case of analytical perturbation theories, often the mean values of the elements, though often, especially where near resonance is important, other parameters of the theory. The observations made during the 19th. century were mostly filar micrometer measures, either of a satellite relative to the primary planet, or relative to another satellite. In the case of the Galilean satellites of Jupiter, timings of mutual eclipses and occultations of the satellites were made, and also the series of heliometer measures of Gill at the Cape. From about the turn of the century, observations began to be made photographically, relative positions being taken by use of a measuring machine. More recently, there has been increasing use of charged coupled device images at many observatories. For a survey of observational techniques and discussion of difficulties, see Pascu (1977). For description of comprehensive prepared catalogues of collected observations of satellites, see Strugnell and Taylor (1990), and Morley (1989).

It is important to choose the parameters to fit the observations in the closest way, to make full use of the information contained in the observational data set. Consider first the numerical solution of the equations of motion for the rectangular co-ordinates, and write these schematically:

$$\frac{d^2 \mathbf{x}}{dt^2} = \mathbf{F}(\mathbf{x}, \mathbf{c}, t) \tag{3.1}$$

where $\mathbf{x} = (x_1, x_2, \ldots, x_n)$, and $\mathbf{c} = (c_1, c_2, \ldots, c_m)$, the x_i being the co-ordinates of all the satellites under study, and the c_i the parameters of the system (masses and oblateness parameters, etc).

Let the solution be

$$\mathbf{x} = \mathbf{X}(\mathbf{a}, \mathbf{u}, \mathbf{c}, t) \tag{3.2}$$

with

$$\mathbf{a} = \mathbf{X}(\mathbf{a}, \mathbf{u}, \mathbf{c}, t_0), \tag{3.3}$$

and

$$\mathbf{u} = \frac{\partial \mathbf{X}}{\partial t}(\mathbf{a}, \mathbf{u}, \mathbf{c}, t_0) \tag{3.4}$$

i.e. the components of $\mathbf{a}$ and $\mathbf{u}$ are the values taken at the initial time t_0 of the positions and velocity components x_i and dx_i/dt, respectively. Then the set of provisional estimates, $(\mathbf{a}_0, \mathbf{u}_0, \mathbf{c}_0)$, lead to the *"calculated"* solution,

$$\mathbf{x}_C(t) = \mathbf{X}(\mathbf{a}_0, \mathbf{u}_0, \mathbf{c}_0, t) \tag{3.5}$$

and we seek improved estimates $(\mathbf{a}_0 + \delta\mathbf{a}, \mathbf{u}_0 + \delta\mathbf{u}, \mathbf{c}_0 + \delta\mathbf{c})$, which correspond to the solution

$$\mathbf{x}_C(t) + \xi(t) = \mathbf{X}(\mathbf{a}_0 + \delta\mathbf{a}, \mathbf{u}_0 + \delta\mathbf{u}, \mathbf{c}_0 + \delta\mathbf{c}, t),$$

so that

$$\xi(t) \approx \frac{\partial \mathbf{X}}{\partial \mathbf{a}} . \delta\mathbf{a} + \frac{\partial \mathbf{X}}{\partial \mathbf{u}} . \delta\mathbf{u} + \frac{\partial \mathbf{X}}{\partial \mathbf{c}} . \delta\mathbf{c} \tag{3.6}$$

We will need to have solved,

$$\frac{d^2}{dt^2}\left(\frac{\partial \mathbf{X}}{\partial \mathbf{a}}\right) = \frac{\partial \mathbf{F}}{\partial \mathbf{X}} . \frac{\partial \mathbf{X}}{\partial \mathbf{a}}, \tag{3.7}$$

$$\frac{d^2}{dt^2}\left(\frac{\partial \mathbf{X}}{\partial \mathbf{u}}\right) = \frac{\partial \mathbf{F}}{\partial \mathbf{X}} . \frac{\partial \mathbf{X}}{\partial \mathbf{u}} \tag{3.8}$$

and

$$\frac{d^2}{dt^2}\left(\frac{\partial \mathbf{X}}{\partial \mathbf{c}}\right) = \frac{\partial \mathbf{F}}{\partial \mathbf{X}} . \frac{\partial \mathbf{X}}{\partial \mathbf{c}} + \frac{\partial \mathbf{F}}{\partial \mathbf{c}} \tag{3.9}$$

simultaneously with the equations of motion, taking, as initial values at $t = t_0$

$$\frac{\partial \mathbf{X}}{\partial \mathbf{a}} = \mathbf{I}_n, \tag{3.10}$$

$$\frac{d}{dt}\left(\frac{\partial \mathbf{X}}{\partial \mathbf{u}}\right) = \mathbf{I}_n, \tag{3.11}$$

where $\mathbf{I}_n$ is the $n \times n$ identity matrix, and

$$\frac{d}{dt}\left(\frac{\partial \mathbf{X}}{\partial \mathbf{a}}\right) = \mathbf{0}_n, \quad \frac{\partial \mathbf{X}}{\partial \mathbf{u}} = \mathbf{0}_n, \quad \frac{\partial \mathbf{X}}{\partial \mathbf{c}} = \mathbf{0}_n, \quad \frac{d}{dt}\left(\frac{\partial \mathbf{X}}{\partial \mathbf{c}}\right) = \mathbf{0}_n, \tag{3.12}$$

where $\mathbf{0}_n$ is the $n \times n$ zero matrix.

Ideally the solution $\mathbf{x}_c + \xi$ will pass through all the observed positions $\mathbf{x}_{obs}(t)$, giving us an equation of condition,

$$\mathbf{x}_{obs}(t) - \mathbf{x}_c(t) = \frac{\partial \mathbf{X}}{\partial \mathbf{a}} . \delta\mathbf{a} + \frac{\partial \mathbf{X}}{\partial \mathbf{u}} . \delta\mathbf{u} + \frac{\partial \mathbf{X}}{\partial \mathbf{c}} . \delta\mathbf{c} \tag{3.13}$$

for each value of the time t for which an observed position is available. Solution by least squares then provides estimates of $\delta\mathbf{a}$, $\delta\mathbf{u}$ and $\delta\mathbf{c}$ as required. Iteration of this process, using the improved estimates of the parameters $\mathbf{a}$, $\mathbf{u}$ and $\mathbf{c}$ from one least squares solution as the provisional estimates for the following one, is now feasible with the availability of modern computers. Convergence of this iteration is found in practise to provide a very keen test of the adequacy of the model and procedures being used.

Suppose now that Lagrange's planetary equations or their equivalent are being used:

$$\frac{d\mathbf{x}}{dt} = \mathbf{F}(\mathbf{x}, \mathbf{c}, t) \tag{3.14}$$

where $\mathbf{x} = (x_1, x_2, \ldots, x_n)$, $\mathbf{c} = (c_1, c_2, \ldots, c_m)$, being here the elements of the instantaneous Keplerian elliptic orbits of all the satellites under study, and the c_i the parameters of the system as before. Let the solution be,

$$\mathbf{x} = \mathbf{X}(\mathbf{a}, \mathbf{c}, t) \tag{3.15}$$

where $\mathbf{a} = (a_1, a_2, \ldots, a_n)$ the a_i being the parameters of the solution (e.g., mean values of the elements, or their initial values, or other parameters, according to the nature of the theory being used, e.g. special parameters of a near-commensurability theory). Here the set of provisional estimates, $(\mathbf{a}_0, \mathbf{c}_0)$ lead to the *"calculated"* solution,

$$\mathbf{x}_C(t) = \mathbf{X}(\mathbf{a}_0, \mathbf{c}_0, t) \tag{3.16}$$

and we seek improved estimates $(\mathbf{a}_o + \delta\mathbf{a}, \mathbf{c}_o + \delta\mathbf{c})$ which correspond to the solution $\mathbf{x}_c(t) + \xi(t) = \mathbf{X}(\mathbf{a}_0 + \delta\mathbf{a}, \mathbf{c}_0 + \delta\mathbf{c}, t)$ so that

$$\xi(t) \approx \frac{\partial \mathbf{X}}{\partial \mathbf{a}}.\delta\mathbf{a} + \frac{\partial \mathbf{X}}{\partial \mathbf{c}}.\delta\mathbf{c} \tag{3.17}$$

and we have an equation of condition,

$$\mathbf{x}_{obs}(t) - \mathbf{x}_c = \frac{\partial \mathbf{X}}{\partial \mathbf{a}}.\delta\mathbf{a} + \frac{\partial \mathbf{X}}{\partial \mathbf{c}}.\delta\mathbf{c} \tag{3.18}$$

for each value of the time t for which a set of orbital elements derived from observations is available. This has traditionally been at each time of opposition of the primary planet with the Earth, giving a mean value of each element derived from the observations during that opposition appearance of the planet. The availability of modern computing power makes it more often possible to carry out complete reductions in which each individual observation is treated separately in the comparison with theory (Harper and Taylor, 1993 and Dourneau, 1993). If a completely analytical theory is being used, then the partial derivatives $\frac{\partial \mathbf{X}}{\partial \mathbf{a}}$ and $\frac{\partial \mathbf{X}}{\partial \mathbf{c}}$ will be available from it to be used in the equations of condition. Not all the relevant parameters c_i may be solved for, since some may be better determined from another source, in which case the corresponding term is omitted from each of the equations of condition.

4. Criteria governing interval of validity

The success of a theory in predicting future positions will, as has been said above, be limited by the accuracy of the estimates of the parameters of the solution (initial values in the case of a numerical integration) from the observational data. These estimates will inevitably only have a finite accuracy, and, in the case of a numerical integration, the unavoidable errors in, especially, the values of the mean motions and other frequencies implicit in the initial conditions will lead to a "run-off", or increasing discrepancy between the positions given by the numerical integration and the actual motion, which will limit the usefulness of the predictions to a time interval of the same magnitude as the interval over which the observational data has been fit. In the case of an analytical theory, errors in the estimates of the frequencies will also lead eventually to a drift-off of the accuracy of the predictions it gives (noticeable for example in the case of Saturn's satellite Phoebe), though improvement in the estimates of these parameters will enable the same theory to continue to be useful. Clearly, the success of an analytical theory in prediction, and in indicating the nature of the longer-term evolution of the system, will depend crucially on how successfully it has modelled the important resonances which underly the motion over time intervals of the order of magnitude of the interval over which the predictions are sought.

Analytical theories and numerical integrations are not only exclusive alternative methods; they can be used together, not only for mutual testing, but also by using one in the process of developing the other. For example, in use of a numerical integration which has been fit to the observational data, from which, by Fourier analysis of values of orbital elements derived from it, are provided values of co-efficients of appropriate periodic terms and other parameters, to provide key data in the construction of an analytical theory. In this way, the numerical integration serves as a dynamically consistent carrier of the information from the observations to the analytical theory (see Sinclair and Taylor, 1985, and also Message, 1993, using results from Taylor, 1992, in the case of Hyperion).

REFERENCES

Dourneau, G., 1993, AA, 267, 292
Harper, D. , & Taylor, D.B. , 1993 , AA, 268, 326
Message, P.J., 1966, Proceedings of I.A.U. Symposium No. 25, 197
Message, P.J., 1989, Cel. Mech., 45, 45
Message, P.J., 1993, Cel. Mech., 56, 277
Morley, T.A., 1989, AA, 77, 209
Taylor, D.B., 1992, AA, 265, 825
Pascu, D., 1977, in Planetary Satellites, (ed. J. Burns, Univ. of Arizona Press), 63
Sinclair, A.T., & Taylor, D.B., 1985, AA, 147, 241
Strugnell, P.R., & Taylor, D.B., 1990, AAS, 48, 289
Vienne, A, and Duriez, L., 1991, AA, 257, 331.

CCD observations at the Bureau des Longitudes: analysis of the positions of satellites

By J.-E. ARLOT, F. COLAS, W. THUILLOT AND D. T. VU

Bureau des Longitudes, URA 707 du CNRS, 77 Avenue Denfert-Rochereau, Paris, F-75014, France

CCD observations were started in 1988 at the Bureau des Longitudes. Most of the planetary satellites have been observed and we now know how well CCDs perform from the point of view of astrometry. As with photographic observations, we chose long focus telescopes for better measurement accuracy. The main drawback of the CCDs is their size: therefore we have developed special techniques for astrometric reduction. Moreover, the use of an anti-blooming CCD allows us to observe bodies very close to their primary.

We obtained extensive observations of the Martian satellites with a similar accuracy to the best photographic plates. We also observed Thebe (J XIV) with our CCD camera, and we now have more than 1000 positions referred to Amalthea (J V). Mimas is known to be a difficult body to observe, nevertheless we are carrying out systematic measurements. Uranian satellites are also routinely observed with the objective of fitting a new dynamical theory. Lastly, the satellite of Neptune, Proteus (N VIII), discovered on Voyager's images, was observed for the first time from a ground-based observatory with our CCD camera on the ESO 2.2 m telescope.

1. Observing satellites with CCD cameras

Starting in 1988, Bureau des Longitudes began to make a series of astrometric observations using CCD cameras. Whilst this technique has many advantages, some problems did occurr which had to be solved before obtaining reliable results. The most important problem arises from the small size of the CCD combined with the long focal length of the telescopes that we use: a focal length of 10 to 20 m is necessary to get accurate positions, but the field decreases as the focal length increases (Table 1). Therefore, often no calibration stars are available in the field. In order to calibrate the field, we use the centre of a globular cluster which provides us with the constants of the CCD. (*cf.* Figure 1). These constants are calculated for each night, allowing us to use a telescope not dedicated to astrometry. We note that catalogued globular clusters are not equally distributed over the sky and we require many more such clusters to be catalogued for astrometric use.

The large differerence in magnitude between close-in satellites and their primary poses another problem. In order to solve this problem and to obtain measurable images, we could use either a focal coronograph, a neutral density filter, or an anti-blooming target. We adopt the last solution because of its ease of use and efficiency, allowing us to observe objects closer to the bright objects than with other techniques. The main advantage of the CCD detector is its ability to provide directly numerical images which can be analyzed and corrected using image-processing software.

Figure 2 shows the raw and processed images before and after numerically removing the background, thereby producing measurable images of faint objects. Another attribute of the CCD detector is its ability to perform photometric measurements, and recent results show the application of this technique to the observation of the Galilean satellites (Descamps et al. 1992). The ability to get photometric measurements at the same time

Focal length:	15 m	Pixel size		Field	No. of stars (GSC)
		(target)	(sky)		($m < 14$)
Thomson	THX 7852	30μm	0$''$.40	90$''$	0.1
Thomson	THX 7863	23μm	0$''$.31	150$''$	0.3
Thomson	THX 1883	23μm	0$''$.31	250$''$	1.0
Thomson	512 × 512	20μm	0$''$.27	200$''$	0.6
Thomson	1000 × 1000	19μm	0$''$.26	380$''$	2.1
Tektronic	1000 × 1000	20μm	0$''$.27	400$''$	2.4
Tektronic	2000 × 2000	13μm	0$''$.18	560$''$	4.7
Loral	2000 × 2000	15μm	0$''$.20	580$''$	5.0
Loral	4000 × 4000	7.5μm	0$''$.10	580$''$	5.0

TABLE 1. The different CCD sizes and their corresponding fields.

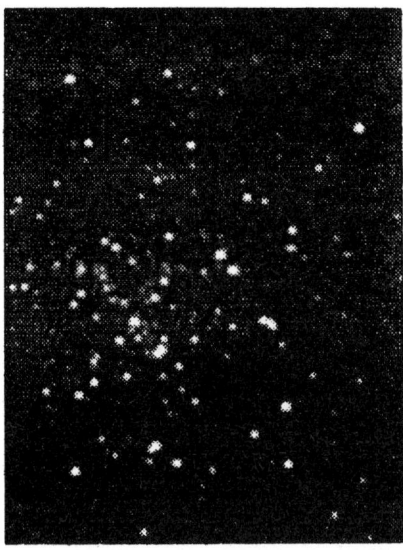

FIGURE 1. The centre of globular cluster M13 used for astrometric calibration.

as astrometric ones is very effective in the recording of events such as mutual occultations. Moreover, the presence of several objects and the background on the target allow us to obtain very reliable results for differential photometry as well as for astrometry.

2. Observational results using CCD cameras

At the Bureau des Longitudes we have performed observations of several satellite systems. Figure 3 shows images of the satellites of Mars, demonstrating the efficiency of anti-blooming in avoiding an excess of light from Mars and providing very measurable images of Phobos and Deimos even with Mars in the centre of the field (Colas & Arlot, 1991).

Figure 4 shows the accuracy of our series of observations of the Martian satellites. The

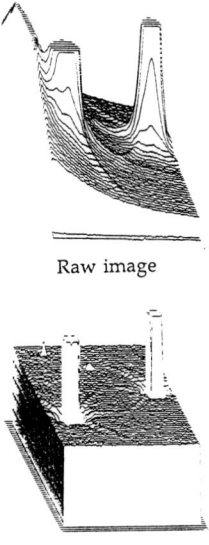

FIGURE 2. Numerical image of the Galilean satellites, before and after processing.

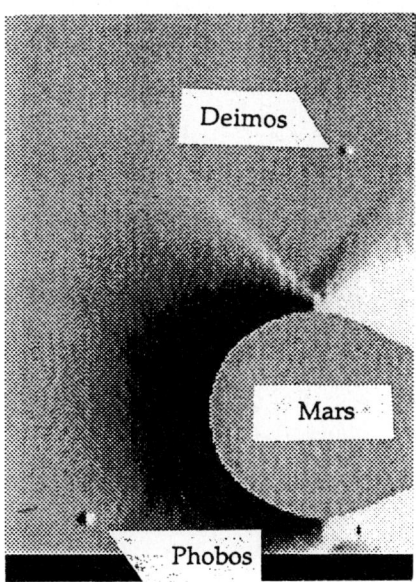

FIGURE 3. Mars, Phobos and Deimos.

accuracy depends on the Earth-Mars distance, being better at the time of the opposition (the satellites are brighter and the signal-to-noise ratio is better). The Galilean satellites of Jupiter are very easy to observe and in spite of the small size of the CCD it is not difficult to obtain several satellites on a single frame which allows the measurement of inter-satellite separations (cf. Figure 5).

The observation of faint satellites such as Amalthea or Thebe are much more difficult because they are very faint and close to their primary. Figure 6 shows an image of Thebe (magnitude 16) relative to the one of a Galilean satellites, despite the large magnitude

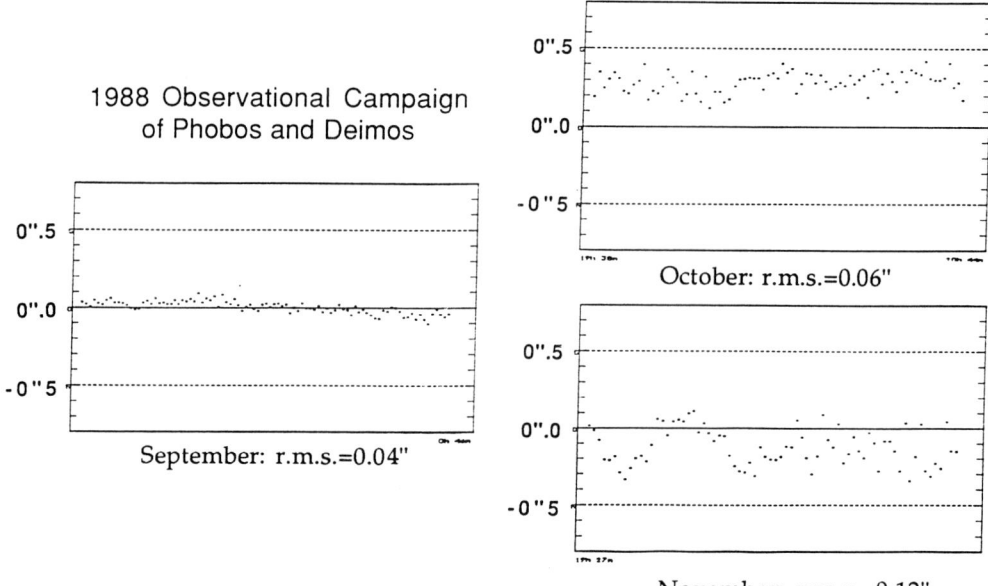

FIGURE 4. O−C of a series of observations of Phobos and Deimos.

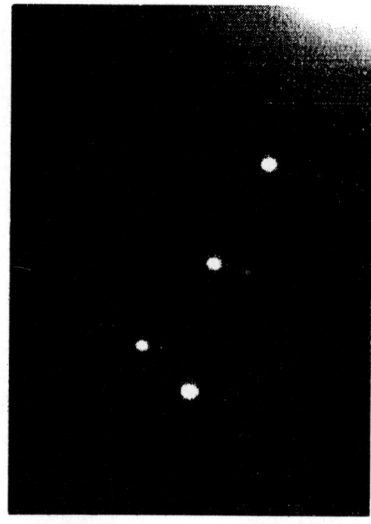

FIGURE 5. Observations of the Galilean satellites made at Observatoire de Haute-Provence (0.8 m, R filter).

difference. We obtained about 1000 positions of Thebe, allowing us to fit a new dynamical model.

Figure 7 shows images of the satellites of Saturn. Helene (magnitude 19) was observed at Pic du Midi under very good seeing conditions. Figure 8 presents an image of the Uranian system and Figure 9 an image of Neptune and Triton (magnitude 13.7) which is our only case of the measurement of a satellite relative to the primary.

Figure 10 shows the first ground-based observation of the satellite of Neptune, Proteus

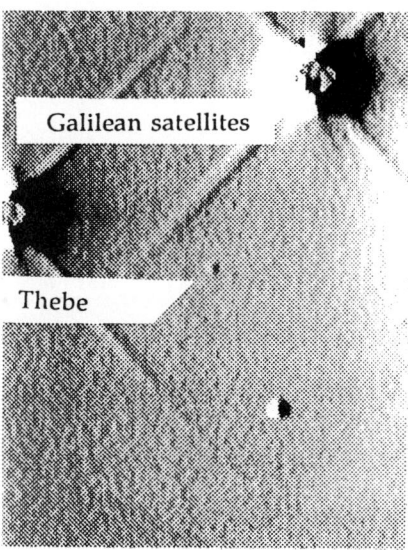

FIGURE 6. Image of Jupiter's satellite Thebe made at Pic du Midi.

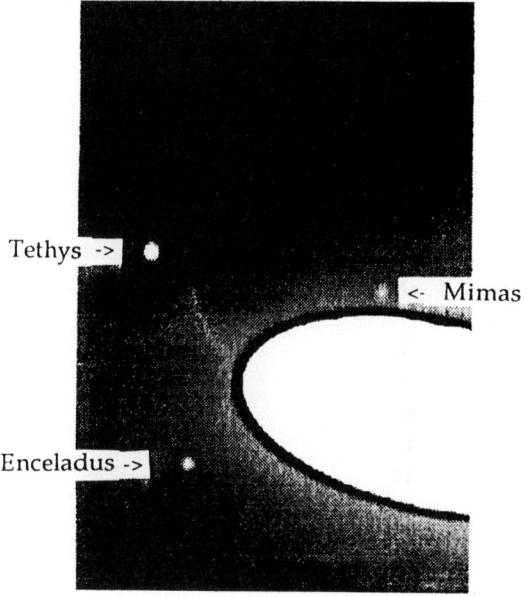

FIGURE 7. Satellites of Saturn (Pic du Midi, 1 m).

(magnitude 20.3) (Colas & Buil, 1992) which was discovered by the space probe Voyager. This image was obtained with our specially built CCD camera (made at Bordeaux Observatory), mounted on the 2.2-metre ESO telescope in Chile. The motion of Proteus has been compensated for, and images were accumulated during one hour and summed.

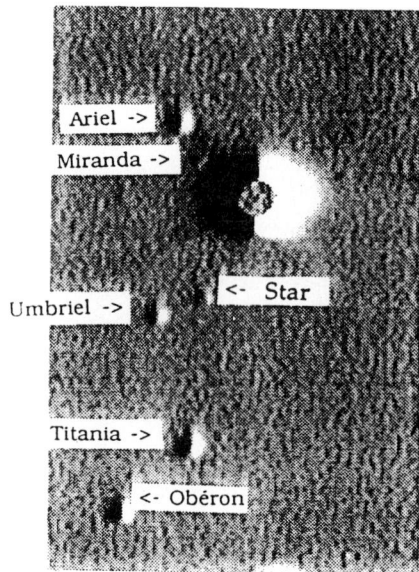

FIGURE 8. Uranus and its satellites (Pic du Midi, 1 m).

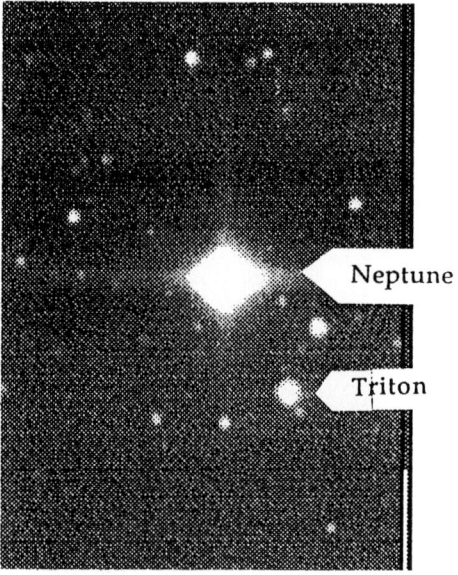

FIGURE 9. Neptune and Triton (Pic du Midi, 2 m).

3. Conclusion

In conclusion we would like to encourage observers to use such CCD cameras to make accurate series of observations of natural satellites; so that new dynamical models can be fitted to them and thereby develop high-precision ephemerides for these objects.

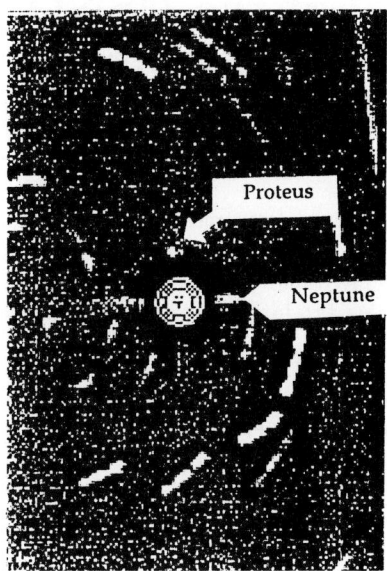

FIGURE 10. First ground-based observations of Proteus made at ESO, Chile (2.2 m).

REFERENCES

Colas, F. & Arlot, J.E., 1991, A&A, 252, 402

Colas, F. & Buil, C., 1992, A&A, 262, L13

Descamps, P., Arlot, J.E., Thuillot, W., Colas, F., Vu, D.T., Bouchet, P. & Hainaut, O., 1992, Icarus, 100, 235

Colas, F., 1991, Nouvelles observations CCD astrometriques pour l'etude dynamique des satellites des planetes, PhD thesis, Observatoire de Paris.

An appraisal of the USNO program for photographic astrometry of bright planetary satellites

By D. PASCU

U.S. Naval Observatory, 34th and Massachusetts Ave, Washington DC 20392, U.S.A

A program of photographic astrometry of the bright planetary satellites was started at the U.S. Naval Observatory 26 years ago in an effort to improve the aging satellite theories and to support the U.S. space program.

The distinguishing feature of the USNO program was the use of filters in the observations. The main function of these filters was to obtain a measureable image of the primary. While this has been largely successful, satellite observations relative to the planetary disk are not widely made or used.

Systematic effects, such as the phase effect and scale errors, are discussed, and methods for their resolution described. The accidental error in the observations is also analyzed in terms of signal-to-noise of the images, focal length of the instrument, and atmospheric effects. The most significant finding was that super accurate intersatellite observations can be obtained for separations less than 50″.

As a result of the new accurate satellite ephemerides, planetary positions are now obtained indirectly from observation of the satellites. When dense, high precision star catalogs become available, the Naval Observatory plate archive will be invaluable for producing high precision observations of Mars, Jupiter, Saturn over two orbits.

---

1. Motivations

The USNO program for photographic astrometry of the bright satellites was begun 26 years ago with observations of the Martian moons and the Galilean moons of Jupiter. The motivation was an effort to improve the aging theories of these satellites by the almanac-producing offices of the U.S. Naval Observatory, the Royal Greenwich Observatory and the Bureau des Longitudes (Pascu 1977, 1979). In 1973, the program was augmented by the inclusion of the Saturnian system in support of NASA's program of planetary reconnaissance (Seidelmann 1977, 1979). While NASA's space-navigation needs are still a major program driver, theoretical studies abroad have increased considerably and there is an increasing demand for these observations. Two separate programs for the fainter satellites were carried out at Flagstaff. A photographic program for observing the satellites of Uranus and Neptune was decribed by Walker et al. (1978) and Walker & Harrington (1988), and a CCD program for faint satellites was described by Pascu et al. (1983, 1987).

2. Technique

The long-focus photographic technique was used with the 26-inch refractor in Washington, its twin – the McCormick 26-inch at Charlottesville, and the 61-inch astrometric reflector at Flagstaff. Most plates were measured with an automatic measuring machine. Autocentering was performed on all images except those imbedded in the planetary halo, such as the Martian satellites and Mimas, or the planetary image itself. Those images were measured manually. Reduction of the plates was done by the trail/scale method because of its accuracy in most cases. If there were enough reference stars, and the right ascension and declination of the planet was the objective, a plate constants solution was made (Pascu & Schmidt 1990).

The distinguishing feature of the USNO program was the use of filters in the observations. Usually, small, partially transparent metallic Nichrome or Inconel spots were used to diminish the light of the planet. But for Jupiter's Galilean system, a larger, composite filter was used to filter the Galilean satellites themselves. The function of the filters is fourfold: (1) to reduce the reflection halo for close satellites, (2) to increase the exposure time for the Galilean moons (to reduce seeing anomalies), (3) for magnitude compensation between planet, satellites and stars and most important, (4) to get a measureable image of the planet. Of course, it is not necessary to connect the satellite position to the primary, provided that one had another satellite – not too far in distance or brightness – to which to refer it. Thus the position of the primary is important for only Triton, the faint outer satellites, and perhaps the Martian satellites. When Hermann Struve (1888) introduced the intersatellite method of orbital correction, it was very successful because systematic errors in the visual measurement of the planet were notoriously large. Struve understood, however, that the method had some serious flaws (Laves 1938, Pascu 1977) and recommended that planet/satellite measurements be continued concurrently with the intersatellite measurements (Struve 1903).

3. Systematic errors

3.1. *The phase effect*

The principal source of systematic error for observations made relative to the planet is the "phase effect"(de Vaucouleur 1963, Standish et al. 1976, Arlot 1980, Pascu & Schmidt 1990). It is also referred to as the "Phillips effect"(Cortesi 1978) or "phase exaggeration" (Smith & Reese 1968) in reference to its occurrence in Jovian studies. This effect occurs when there is an intensity difference between the east and west limbs, visible when the planet is observed away from opposition. Irradiation blooming at the brighter limb, and low light level at the terminator combine to systematically shift the measured planetary center toward the brighter limb by an amount several times that expected from the geometric phase displacement. This is illustrated in Figure 1; the planet-centered X-coordinates of the USNO observations of the Galilean satellites were shifted, as expected, in the manner of the geometric phase, but about three times as much!

Arlot (1975, 1980) managed to get around the phase problem in two ways. First, using the planet/satellite measurements, he introduced an empirical factor of the geometric phase into the conditional equations for the correction of the satellite orbits. His results were comparable to his intersatellite solutions. Second, he remeasured the plates using microdensitometry techniques and phase model algorithms to derive the center of figure for Jupiter. Although these results were better than the first, the amount of work involved was quite large and the microdensitometer measurements were not continued. Despite Arlot's efforts, only intersatellite observations are presently being used in the orbital adjustments.

Planet/satellite observations for the other moons have been more successful. Residuals for the Martian satellites indicate that the planet/satellite observations are generally more precise than the intersatellite observations. The reason for this appears to be that for an exposure long enough to obtain a black image of Deimos, Phobos was imbedded in the planetary halo. Alternately, when the exposure was short enough for a thin halo, the

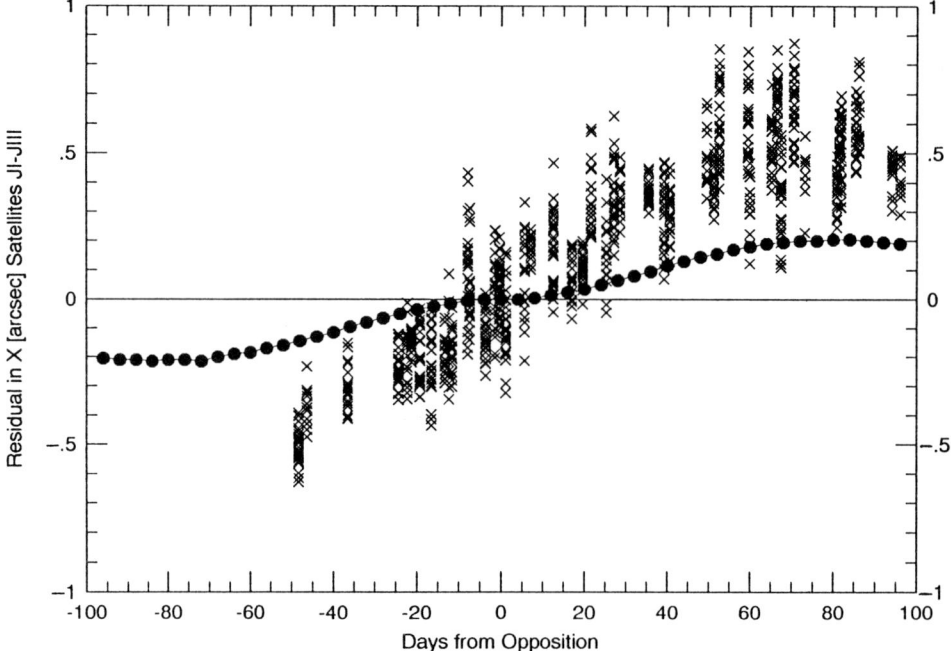

FIGURE 1. This figure demonstrates the phase effect for Jupiter. The heavy beaded line represents the expected X-displacement of a satellite, resulting from uncorrected geometric phase, when measured relative to the limbs of the planet. The crosses are the actual X-residuals for satellites JI-JIII for the apparitions of 1986-1989. The deviation of the observations from the expected geometric phase function is called the phase effect. The ephemerides were computed at the Bureau des Longitudes in Paris and are compared to observations made with the Naval Observatory 26-inch refractor.

image of Deimos was too gray. The phase effect was minimized by the thin atmosphere of the planet and our practice of observing the satellites when the defect was less than $0\rlap{.}''1$.

Planet/satellite observations for Saturn were also successful; the phase effect was not an issue since ring measurements were used to determine the center of the planet (Pascu & Schmidt 1990). The *rms* of the planet/satellite observations was $\pm 0\rlap{.}''14$ (Harper et al. 1988) and that of the intersatellite observations was $\pm 0\rlap{.}''10$. Although the precision of the intersatellite observations was greater, the issue is whether or not the orbital elements could be determined with greater precision from the less precise planet/satellite observations because of their greater leverage in the least-squares solution!

Walker et al. (1978, 1988) were quite successful in obtaining Neptune/Triton observations but seemed less confident about the planet/satellite observations in the Uranian system because of their method of measuring the disk of the planet. No phase effect has been reported for these planets.

3.2. Scale errors

For scale/trail reductions, scale errors are a major source of systematic error. Figure 2 demonstrates a scale error between the USNO observations of the Galilean moons and the ephemerides of the Bureau des Longitudes (BdL). The linear increase of separation residual with increasing separation is the hallmark of a scale disparity. The slope of this linear relation is the difference in scale between the observations and the ephemerides.

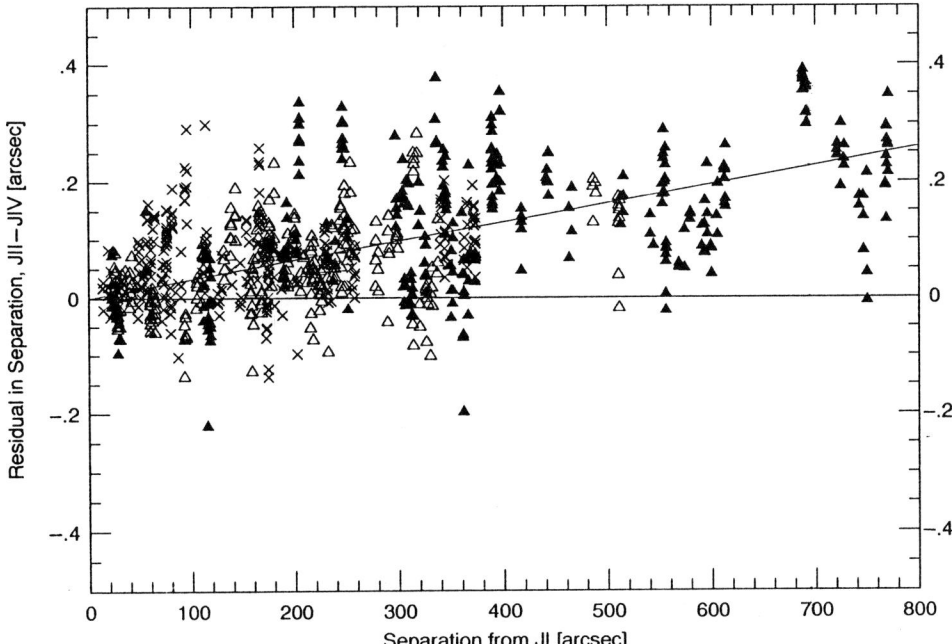

FIGURE 2. This figure demonstrates the effect of a difference in scale between the theory and the observations. The separation residuals increase with increasing separation of the satellites. Crosses, open triangles and filled triangles denote separation residuals of JII with respect to JI, JIII with respect to JI, and JIV with respect to JI respectively. The slope of the line represents the difference in scale. A scale of $20\rlap{.}''843$/mm was used to reduce the observations. The reason for the discrepancy is not yet understood. The observations are from the Naval Observatory, the ephemerides from Bureau des Longitudes.

A similar scale discrepancy exists between these observations and the JPL ephemerides. The instrumental scale, determined from stars on the satellite plates, is $20\rlap{.}''843$/mm, while Figure 3 indicates that the scale consistent with the BdL ephemeris is about $20\rlap{.}''839$/mm. Scale disagreements are due to instrumental scale errors, orbital scale errors in the theories, or errors in the correction of the observations to put them in the same coordinate system with the theory. All of these sources are being investigated.

4. Accidental errors

4.1. Atmospheric contributions

The accidental error of the Galilean satellite observations can be estimated from the scatter of points around the zero line in Figure 3 since the major systematic effects have been removed (or avoided). Root-mean-square errors were computed in strips $100''$ wide, every $50''$ in separation. In addition, the *rms* error for separations below $30''$ and $50''$ were computed. The data were grouped by satellite and by year. Although major differences were found in the errors for the satellites, the most significant results were obtained from the data grouped by year. It was found that results for 1986 were similar to those for 1987, so those years were grouped together. For the same reason the 1988 and 1989 observations were grouped together. The results are plotted in Figure 4.

The reason for the deviation of the 1988/89 set from the 1986/87 set is not clear. It is not due to a residual systematic scale error, but may be due to a random variation. If

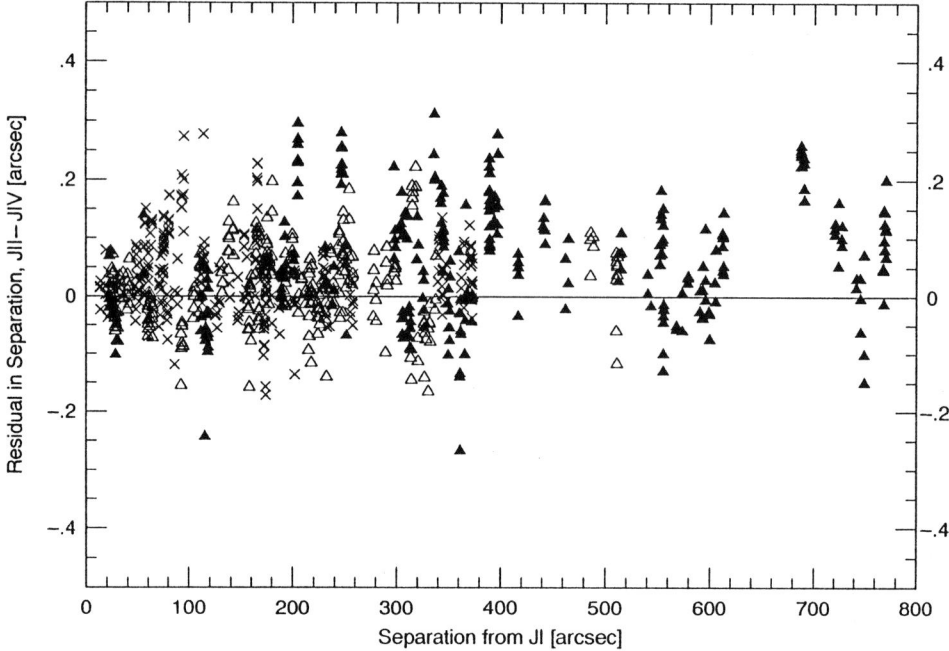

FIGURE 3. If one applies a scale of 20."839/mm to the observations, this figure results. The same result could be obtained by fitting the theory to these observations.

the deviation was due to atmospheric causes one would expect the 1988/89 set to have the smaller errors since the zenith distances were 20 degrees smaller on average than those for the 1986/87 observational set. The seeing, also, was at least as good for the 1988/89 set as for the 1986/87 set. However, there is one very interesting result for both data sets which can be attributed to the atmosphere, and that is the very small errors for separations of 50″ and below. Fifty arcsec appears to be the size of the seeing cell in which the seeing motions are correlated. The significance of this phenomenon to Galilean satellite observation is that one can produce normal points for these close separation observations which have an external precision of ±0."01. This makes them comparable to the mutual events in precision. But they also have the added advantage that 50″ separations occur much more frequently than the mutual events – averaging about 50 per month!

4.2. Relation to signal/noise(S/N)

The Saturnian system is a productive area for studying the effects of S/N (exposure) on astrometric precision. No other system covers the dynamic range of the photographic technique as well as the Saturnian system. A large set of observational data was recently fit to theory by Taylor & Shen (1988). Root-mean-square residuals were made available for analysis by Taylor(1985, 1986). These data gave errors by individual Titan-satellite pairs, and for several individual astrographs. The *rms* residuals are plotted, in Figure 5, against the satellite-Titan magnitude differences, as indicators of S/N. The data from six instruments are plotted.

While some of this data is puzzling, such as the McDonald data, it is not difficult to understand that a gray image does not represent a satellite's position on the plate as well as a black, well-exposed image. Or that a faint, gray image is not as accurately

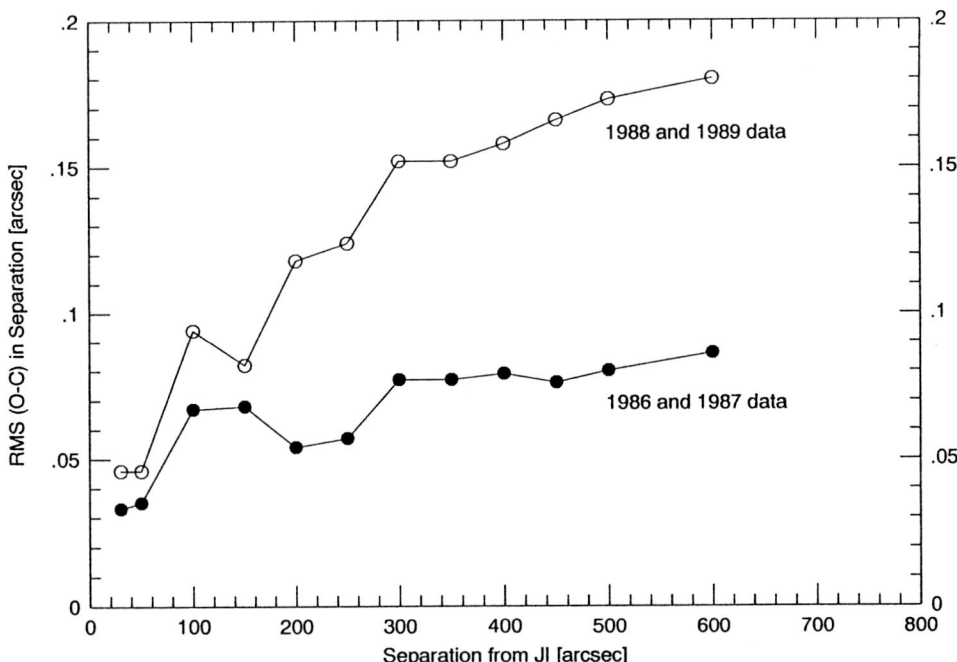

FIGURE 4. The accidental error is measured by the scatter about the horizontal zero line in Figure 3. Overlapping strips, 100" wide every 50", were used to compute the rms error as a function of distance(separation) between the satellites. The 1986 and 1987 data (filled circles) were found to be similar and combined. For the same reason, the 1988 and 1989 data (open circles) were combined. The first two points are the values for separations less than 30" and 50". It is not understood why the two sets should be so different from each other. See the text for further analysis.

measured as a larger, black image. For the long-focus refractors, the amount of the decline in external precision is 0pas033/magnitude. For the short-focus astrographs, the decline in precision is more rapid for the darkest images.

4.3. Relation to focal length

Last, Figure 5 illustrates the relationship of astrometric precision with focal length. Applied to satellite astrometry, this problem was most recently studied by Ferraz-Mello (1983). From a study of Galilean satellite observations taken with seven astrographs, ranging in focal length from 3.4 m to 10.9 m, he derived a relation between the external astrometric precision and instrumental focal length. Ferraz-Mello found that the minimum residual that could be obtained for the Galilean moons was 3 microns in the focal plane. This minimum applies to all separations since, apparently he was not aware of the precision premium at small separations. His limit is consistent with our Galilean satellite data. It is also consistent with the lower S/N images of Rhea in Figure 5. Although the rate of decline in astrometric precision with focal length is not linear, a mean value of 0!'017 per meter of focal length was determined from Ferraz-Mello's work. At Rhea's S/N, in Figure 5, a value of 0!'020 per meter of focal length was obtained. Figure 5 also indicates that the decline in precision for images of low S/N is more rapid than that for high S/N.

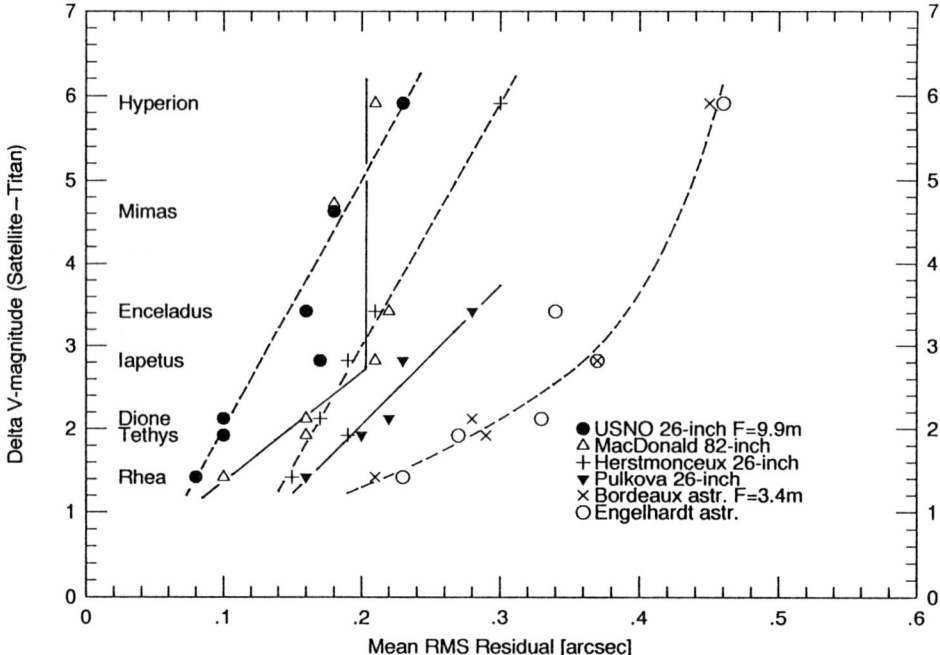

FIGURE 5. The Saturnian system is fertile ground for analyzing the effect of signal-to-noise on astrometric precision. Here we plotted rms residuals from fits to theory of Titan-satellite pairs. The data, which includes observations from several observatories, were provided by Taylor(private comm 1986) from work done at RGO. The rms residuals were plotted against the satellite-Titan magnitude difference as a measure of signal-to-noise.

5. Planetary positions

A major spin-off of accurate satellite ephemerides is improved positional observations of the planets (Pascu & Schmidt 1990, Taylor et al. 1991). The major planet observing methods have shifted from observing the planet directly, to observing the satellites relative to catalog reference stars and then inferring the planetary position indirectly by subtracting the satellite ephemeris coordinates from its observed position. The precision of these methods is about $\pm0''.2$, or a gain of 3 over conventional methods. In the long-focus photographic version of this method (Pascu & Schmidt 1990), the principal source of error is the lack of reference stars with good positions. When this problem is solved in the next decade, the long-focus focal plane (e.g. photographic) method will be able to produce observations of the planets with a precision of $\pm0''.05$. At that time, the USNO archive of photographic satellite plates can be used to obtain (recent) historical planetary positions to that precision. For Mars, observations from 1967 will give positions several times around its orbit, and for Jupiter, observations from 1967 will provide positions twice around its orbit. The archive contains plates of Saturn's system since 1974, and when combined with the Johannesburg series of 1926-1934, taken by Alden and O'Connell (also in the USNO archive), positions of Saturn, twice around its orbit, can be produced.

REFERENCES

Arlot, J.-E. 1975, Celest. Mech., 12, 39
Arlot, J.-E. 1980, A&A, 86, 55

Cortesi, S. 1978, Icarus, 33, 410

de Vaucouleurs, G. 1963, Mem. Soc. R. Sci. Liege, 7, 369

Ferraz-Mello, S. 1983, An. Acad. Brasil. Cienc., 55, 219 bibitemHT88 Harper, D., Taylor, D.B., Sinclair, A.T. 1988, A&A, 191, 381

Laves, K. 1938, Vierteljahrsschr. d. Astronom. Gesellschaft, 71, 310

Pascu, D. 1977, Astrometric Techniques for the Observation of Planetary Satellites, in *Planetary Satellites*, ed. J.A. Burns, University of Arizona Press, Tucson, p. 63

Pascu, D. 1979, The Naval Observatory Program for the Astrometric Observation of Planetary Satellites, in *Natural and Artificial Satellite Motion*, ed. P.E. Nacozy & S. Ferraz-Mello, University of Texas Press, Austin, p. 17

Pascu, D., Seidelmann, P.K., Baum, W.A. & Schmidt, R.E. 1983, Observations of Faint Planetary Satellites with a Charge- Coupled Device, in *The Motion of Planets and Natural and Artificial Satellites*, ed. S. Ferraz-Mello & P.E. Nacozy, University of Sao Paulo, Brazil, p. 253

Pascu, D., Seidelmann, P.K., Schmidt, R.E., Santoro, E.J. & Hershey, J.L. 1987, AJ, 93, 963

Pascu, D. & Schmidt, R.E. 1990, AJ, 99, 1974

Seidelmann, P.K. 1977, Tabulations of Satellite Positional Observations, and their Discussion, in *Planetary Satellites*, ed. J.A. Burns, University of Arizona Press, Tucson, p. 533

Seidelmann, P.K. 1979, Planetary Satellites, A Review of the Past and Assessment of the Future, in *Natural and Arttficial Satellite Motion*, ed. P.E. Nacozy & S. Ferraz-Mello, University of Texas Press, Austin, p. 3

Smith, B.A. & Reese, E.J. 1968, Observations of Io at Inferior Geocentric Conjunction, Contr. of *The Observatory*, New Mexico State Univ., vol. 1, no. 2

Standish, E.M., Jr., Keesey, M.S.W., & Newhall, X X 1976, NASA Technical Report No. 32-1603

Struve, H. 1888, "Beobachtungen der Saturnstrabanten", Obs. Pulkova, Suppl., 1, 1

Struve, H. 1903, PASP, 15, 183

Taylor, D.B. 1985, 1986, private communication

Taylor, D.B., & Shen, K.X. 1988, A&A, 200, 269

Taylor, D.B., Morrison, L.V. & Rapaport, M. 1991, A&A, 249, 569

Walker, R.L. & Harrington, R.S. 1988, AJ, 95, 1562

Walker, R.L., Christy, J.W. & Harrington, R.S. 1978, AJ, 83, 838.

CCD observations of Saturn's satellites

By KEVIN BEURLE

Astronomy Unit, School of Mathematical Sciences, Queen Mary and Westfield College, Mile End Road, London E1 4NS, U.K.

Improved elements and theories for the orbits of Saturn's satellites, particularly important given the forthcoming Cassini mission to the planet, depend on consistent and accurate observations.

A collaboration involving Queen Mary & Westfield (QMW) and the Royal Greenwich Observatory (RGO) is engaged in a continuing program of astrometry of the major Saturnian satellites. Observations have been made over a period of five years using CCD detectors on the Jacobus Kapteyn Telescope on La Palma. An accuracy of $0\rlap{.}''10$ is obtained for inter-satellite distances which corresponds to approximately 650 km at Saturn, and is comparable with the best photographic observations.

Calibration images, including double stars and star-trails, have been processed using a variety of techniques.

1. Introduction

The positions of Saturn's satellites were first measured systematically and accurately over a century ago. The records are, however, not continuous: the middle decades of this century saw few such observations, if any. In the case of Hyperion, the seventh major satellite, only one observation is recorded between 1922 and 1967, during which interval it made more than 770 orbits of Saturn. The coming of the space age renewed interest in satellite dynamics. Serious astrometric observations of Saturn's satellites recommenced in 1966, using photographic plates in contrast to the earlier technique of visual observations with a filar micrometer eyepiece. More recently observers have also used CCD imaging detectors.

Using these records, the orbits of the satellites are now better modelled than ever before - but not yet well enough. The forthcoming Cassini/Huygens mission to Saturn, due to arrive there early next century, will require still better ephemerides, accurate to within a few hundred kilometres. For example the position of the eighth major satellite, Iapetus, is at present known with an uncertainty of about 800 km – or just over one satellite radius. This is an impressive accuracy of about $0\rlap{.}''13$, or $\frac{1}{4000}$ of its distance from Saturn; Cassini, however, is planned to fly past at an altitude of just 950 km. For the later flybys of Cassini's tour, its own observations will have added greatly to our knowledge, but ground-based observations over longer timescales will still be important. This work will be particularly relevant in the very early stages of the mission when Cassini must pass within 1300 km of Titan's atmosphere, for even Titan's position has an uncertainty of ~600 km at present.

To improve ephemerides requires observations that are of high quality, and made regularly at each opposition. Ideally these should be made in a consistent manner to simplify and improve the combining of data from different years. The collaboration between Queen Mary and Westfield College (QMW) and the Royal Greenwich Observatory (RGO) has worked on such a program since 1987, including major observational campaigns in 1990 and 1991.

Satellite	N_{images}	Satellite	N_{images}
Mimas	16	Rhea	50
Enceladus	47	Titan	22
Tethys	54	Hyperion	14
Dione	46	Iapetus	0

Total = 249

TABLE 1. Number of CCD images of each satellite.

2. Observations

The observations of the QMW/RGO collaboration have been made using the 1 m Jacobus Kapteyn telescope in the Isaac Newton Group in La Palma. We obtained a full seven nights of data around each of the 1990 and 1991 oppositions, supplemented by a limited number of observations from 1987 to 1989 and in 1992. In all cases, images of Saturn, Uranus and Neptune and their satellites were taken at regular intervals throughout each night together with a range of images for calibration purposes.

The 1991 campaign from July 4 to July 11 was almost optimal given Saturn's opposition on July 27. A coated GEC CCD detector was used at the f/15 focus of the telescope. The pixel size of 22 microns corresponded to approximately $0''.3$, or about 1800 km at the distance of Saturn. The CCD size was 590 × 400 pixels, giving a field of view of about $180'' \times 120''$. Exposures of 4 s were generally used for the Saturnian system: Titan presented a V magnitude of 8.3, while Tethys, Dione and Rhea were in the range 10-11.

Saturn itself was overwhelmingly bright in comparison with its satellites, and this presented some problems. A Gunn Z filter, corresponding to a strong methane absorption band, was used to darken the globe of the planet somewhat. Nevertheless, the planet and its rings were always overexposed and bloomed and their positions could not accurately be measured. Often the innermost major satellite Mimas, faint and close to the planet, was unobservable in the glare.

In 1991 a total of 59 such observations of Saturn and its inner satellites were made. Excluding the more remote satellites Iapetus and Hyperion, 249 satellite positions were measurable (Table 1).

For astrometric calibration star-trail and double star images were captured along with standard bias and flat field amplitude calibration frames.

3. Reduction of observations

No centres could be measured for the saturated images of Saturn itself. The data were therefore reduced using the Cartesian separations between pairs of satellites: a similar technique to the micrometer observations of the last century! For each observation, expected positions of satellites within the field of view were calculated and presented as coordinate lists and finder diagrams (Figure 1).

The IRAF analysis package was used to scan the image for bright objects and report their approximate positions. Comparing images and coordinates, a list of preliminary satellite positions was assembled and submitted to the IRAF centroiding tool. Typical satellite images, with a FWHM of ~ 3 pixels gave centre coordinates with a standard error of ~ 0.23 pixels.

In all, the 249 measured satellite positions yielded 389 pairs as detailed in Table 2.

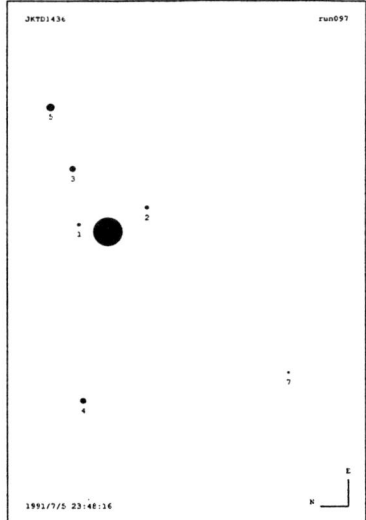

FIGURE 1. A JKT frame of Saturn and its satellites: 1=Mimas, 2=Enceladus, 3=Tethys, 4=Dione, 5=Rhea, 7=Hyperion.

N_{sats}	N_{images}	N_{pairs}	$N_{images} \times N_{pairs}$
2	5	1	5
3	12	3	36
4	23	6	138
5	12	10	120
6	6	15	90
		Total = 389	

TABLE 2. Number of satellites per frame.

The 389 resulting separation measurements were then analysed to determine both the internal consistency (Table 3) and the agreement with theory.

The observed relative positions were compared with predictions according to the theories in Harper & Taylor (1993). The residuals obtained when fitting to the theory were found to be comparable to the best photographic observations (Table 4).

4. Astrometric calibration

To translate the relative positions of satellites from CCD pixel coordinates to North-South and East-West arcseconds requires an orientation angle and a pixel scale factor. The object of the calibration procedures and images was to enable these quantities to be determined.

A variety of calibration techniques were used or investigated, starting with "internal calibration" of the pixel scale from the early stages of the data reduction.

The primary scale calibration was intended to be performed using observations of the wide double star 61 Cygni and at least one set of observations of this target was made

Datum	Used	Total	Mean	σ
Satellites 1-6				
$\Delta\alpha \cos\delta$	389	389	$-0\rlap{.}''042 \pm 0\rlap{.}''008$	$0\rlap{.}''159$
$\Delta\delta$	389	389	$-0\rlap{.}''002 \pm 0\rlap{.}''008$	$0\rlap{.}''150$
Satellites 3-6				
$\Delta\alpha \cos\delta$	185	185	$-0\rlap{.}''016 \pm 0\rlap{.}''007$	$0\rlap{.}''095$
$\Delta\delta$	185	185	$-0\rlap{.}''004 \pm 0\rlap{.}''008$	$0\rlap{.}''106$

TABLE 3. Statistics of O−C residuals from inter-satellite distances.

		Used	Total	Mean	σ
Alden(1929)	$\Delta\alpha \cos\delta$	108	111	$-0\rlap{.}''03$	$0\rlap{.}''09$
	$\Delta\delta$	109	111	$0\rlap{.}''00$	$0\rlap{.}''06$
Alden & O'Connell (1928)	$\Delta\alpha \cos\delta$	192	192	$-0\rlap{.}''04$	$0\rlap{.}''09$
	$\Delta\delta$	192	192	$0\rlap{.}''01$	$0\rlap{.}''07$
CCD data (this paper)	$\Delta\alpha \cos\delta$	185	185	$-0\rlap{.}''02$	$0\rlap{.}''10$
	$\Delta\delta$	185	185	$0\rlap{.}''00$	$0\rlap{.}''11$
Tolbin (1985)	$\Delta\alpha \cos\delta$	168	168	$0\rlap{.}''01$	$0\rlap{.}''08$
	$\Delta\delta$	167	168	$0\rlap{.}''00$	$0\rlap{.}''13$
Veillet & Dourneau (1992)	$\Delta\alpha \cos\delta$	434	434	$-0\rlap{.}''01$	$0\rlap{.}''13$
	$\Delta\delta$	434	434	$0\rlap{.}''01$	$0\rlap{.}''10$
Pascu (1982)	$\Delta\alpha \cos\delta$	762	770	$0\rlap{.}''00$	$0\rlap{.}''12$
	$\Delta\delta$	763	770	$0\rlap{.}''00$	$0\rlap{.}''10$

TABLE 4. Comparison of CCD residuals with the best photographic observations of Tethys, Dione, Rhea and Titan. The observations by Veillet & Dourneau are those made at Pic du Midi.

each night. Each frame was exposed five times for 0.1 s, at intervals as the stars drifted across the CCD field of view. Eleven such frames were obtained in all in 1991 giving a total of 55 separate images of the system. Fitting the observed separations to the well-determined orbit of 61 Cygni yielded both the E-W offset angle and the chip scale factor. The measured separations were not always consistent: 61 Cygni is too bright to be ideal, often saturating even at the 0.1 s minimum exposure. Calculating the scale factor from the median results of the 61 Cygni observations gives a scale factor of $0\rlap{.}''302 \pm 0\rlap{.}''001$ per pixel. This does not correspond exactly to the value of $0\rlap{.}''304$ which was found by minimising the residuals of the satellite observations. This small but possibly significant discrepancy is being investigated.

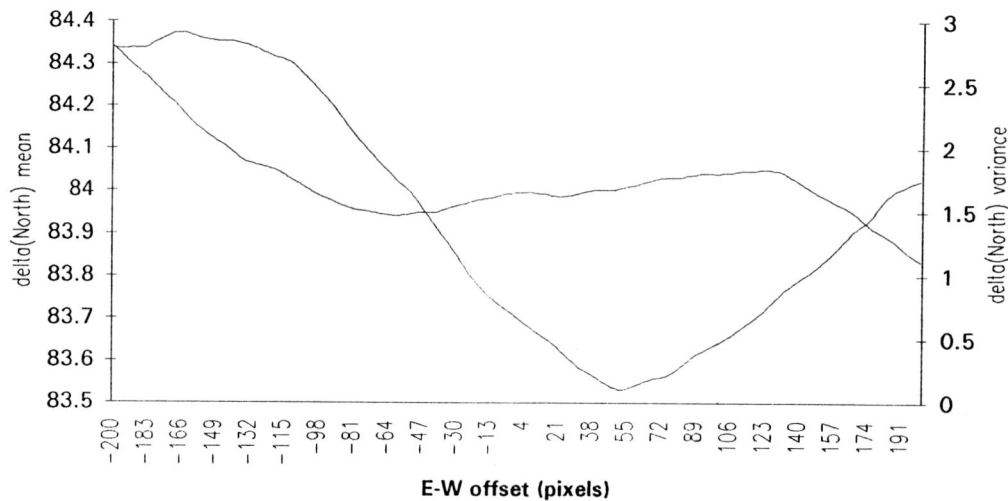

FIGURE 2. Correlation of star-trails for 61 Cygni (run 299).

A number of star-trail images were taken each night with the telescope fixed. These were intended to produce an E-W line across the CCD field, and thus the orientation. The results obtained were not consistent, and gave inferior angle measurements to the 61 Cygni observations. The problem may have been caused by pointing noise: the telescope design is such that it could not be physically clamped, but rather, had to be "driven stationary" by appropriate adjustment of the differential tracking rate.

A few of the star-trail frames were in fact of 61 Cygni, giving double trails. The correlation between the two trails was examined. A technique of determining E-W and N-S separations by finding the correlation peak was investigated (Figure 2). The method proved feasible, but again the excessive brightness impaired the accuracy: the results obtained were found to be inferior to those determined from the multiple exposures.

A further calibration method which we are using, which should give a much better value for the scale factor, is to compare CCD images of star-fields with photographic plates. Several deep exposures (256 s in R) had been made of the outermost Saturnian satellite Phoebe, fully 0.5 degrees, or 10 CCD widths, from the planet. No other satellites appeared in these images, so they were compared with photographic observations. To this end, a set of plates were taken with the JKT wide-field camera which has a field of view of 1.5 degrees diameter. Using these plates it proved possible to fix positions relative to the PPM star catalogue with a standard deviation of 0″.06. Appropriate stars near the CCD corners (713 pixels diagonal distance) will determine the scale factor ten times more accurately than the 61 Cygni observations.

5. Future programme

The QMW/RGO collaboration will continue to make regular CCD observations of the Saturnian satellites. A week of telescope time on the JKT has been scheduled for the opposition this year (1993) and for each subsequent opposition until 1996.

The CCD images are easy to capture, store and process; the detectors have good quantum efficiency and a favourable spectral range. When two or more satellites are in the field of view analysis is straightforward: for single satellite images we would expect to use photographic plates for calibration. The availability of larger CCD arrays will make them more useful still: already a larger chip on the JKT will allow us to observe Iapetus in the same frame as Titan.

We expect to use star-field images in comparison with wide-field photographic plates as the prime calibration for the CCD chip scale factor and alignment. We also intend to investigate alternative methods such as using the Uranian satellites as reference standards.

REFERENCES

Alden, H.L. 1929, AJ 40, 88

Alden, H.L. & O'Connell, W.C. 1928, AJ 38, 53

Harper, D. & Taylor, D.B. 1993, AA (in press)

Pascu, D. 1982, data held at the U.S. Naval Observatory, Washington D.C.

Tolbin, S.V. 1985, Glav. Astron. Obs. Navk. Leningrad, 14pp

Veillet, C. & Dourneau, G. 1992, A&AS **94**, 291.

Planetary satellites

By P. K. SEIDELMANN

U.S. Naval Observatory, 34th & Massachusetts NW, Washington DC 20392, U.S.A

Twenty-three years ago, a plan was initiated to improve the accuracy of the ephemerides of the planetary satellites for the Voyager mission. Now, post Voyager and Apollo, our knowledge of planetary satellites and their orbits is much better and the observational and theoretical requirements are different. The improvements have led to new opportunities and challenges. There is the continuing contest between accuracies of ephemerides and observations. Improvements in one lead to the capability or requirements for improvements in the other. The current status of ephemerides is described. The availability of CCDs and electronic and infrared detectors have changed the observing techniques, accuracies, and problems. The progress in satellite ephemerides, star catalogs, and observational techniques has led to the observations of satellites being the most accurate technique for planetary positions. The scientific purpose for satellite observations has also changed from planetary mass determinations to investigations of evolution, dynamics, and compositions. The question of completeness is now augmented by the problems of unobservable and unidentified satellites.

1. Introduction

The intent of this review paper is to give an overview of the status of both ephemeris computation and observation for the planetary satellites.

Twenty-three years ago, a group of us met at JPL, after an inauspicious flight, to consider the status of planetary satellites, specifically the possibilities of a spacecraft, Voyager, observing the satellites. The question was how accurately did we know the positions of the satellites, what was necessary to improve those orbits, and how accurately could we know the orbits of the satellites for an exploration mission. An informal satellite working group (Seidelmann, 1979) was established which eventually became the Satellite Working Group of IAU Commission 20. Now, Voyagers I and II have observed the planets and the satellites of Jupiter, Saturn, Uranus, and Neptune. Through the combination of ground-based observations and satellite observations over the past twenty-three years, the orbits of the satellites are much better known, and the number of known satellites has almost doubled (currently 61, vs. 32 in 1970).

As a result, the purpose, methods, and accuracies of observations and ephemerides of the satellites have changed significantly.

For the Moon, the change is even more significant than for the other satellites. Before the Apollo landing, observations of the Moon were restricted to optical and radar observations. Today, the preferred observing technique is lunar laser ranging, and the accuracies are at the centimeter level.

As is true in most cases in science, while the progress has been tremendous, that doesn't mean that all the problems have been solved. With improvements in accuracies of theories and observations come new and different problems. The intent of this review is not to emphasize the progress and discoveries of the past, but rather to consider the opportunities and challenges of the future.

2. Observations

Since the theories for the satellites were reviewed in a recent paper (Seidelmann, 1993), that material will not be repeated, but the observation status will be considered. There are certain cases where photographic observations continue to be an effective means of observing (Batrakov, 1991). These are primarily the brighter satellites. The program of Pascu, which extends over 25 years for Mars, Jupiter, and Saturn, is the prime example and an accurate source of observations for those satellites (Pascu, 1979; Pascu and Schmidt, 1990). The Japanese (Nakamura et al. 1991) and the University of Texas observing programs are other examples. With the availability of larger format CCDs, the use of photographic techniques can be expected to end, if some technique can be developed to overcome the fact that for bright satellites the exposures are so short that the atmospheric effects are not averaged. However, there are archives of photographic observations, such as the collection at the USNO, which would produce significantly improved positional data by reanalysis using new, more accurate star catalogs. In addition, using improved theories for the satellite motions with the new star catalogs, accurate observational positions of the planets can also be determined.

Mutual events are a continuing source of satellite observations providing in some cases sensitive tests of ephemerides (Soma, 1992). The mutual events provide accurate observations at limited, specified times, while observations taken when the satellites appear to be close together, thus minimizing atmospheric effects, may be equally accurate.

The observing technique of choice today, is the charge coupled device (CCD). This has been demonstrated primarily through the fainter satellites by Pascu et al. (1987). For the Galilean satellites of Jupiter, Monet has made a series of observations demonstrating the accuracy level that can be achieved, and the limitations imposed by an atmospheric effect. Colas and Arlot (1991) have a program for CCD observations of the satellites. Recently, Rohde and DeYoung, using a 24-inch telescope with a CCD camera in Washington, D.C., have successfully observed several of the outer satellites of Jupiter. This capability is made possible by the improved technology of the CCD, permitting much cleaner observations in a poor observing environment, and the availability of better star catalogs permitting the use of fewer reference stars. The USNO Astrographic Catalog Reference Stars (ACRS) should be used in place of the SAO. The HST Guide Star Catalog provides fainter star positions, at the risk of larger errors due to stars on the outer parts of the Schmidt plates. The anticipated Hipparcos catalog offers prospects for much better reference star positions in the near future. Thus, the previously accepted limitations of observing sites and the telescope requirements need to be reexamined based on the new technology.

Until recently, the co-orbital satellites of Saturn have only been observed from the ground during the ring plane crossings in 1966 and 1980, and during the two Voyager encounters. Recent observations using infrared detectors, when the northern portion of the rings is blocked by the planet, have provided a better understanding of the motions of these objects (Nicholson et al. 1992). IR detectors have also been used recently for observations of Jupiter XV and XVI, Adrastea and Metis. The librational satellites of Saturn have been observed since 1980 with CCDs, and data reduction analyses are needed on these satellites.

I should point out the coming passages of the Earth through the ring plane of Saturn on 22 May and 11 August 1995, and 12 February 1996. Saturn is at opposition on 14 Sept 1995, and 26 Sept 1996, and new Moons are 29 May, 27 July, and 26 Aug 1995, and 18 Feb 1996. USNO began CCD satellite observations for the 1980 ring plane passages, and the co-orbitals and librational satellites were discovered by ground-based CCD observations

NAME	Period	Dist (")	m_{app}	m_{bkg}	ΔV
Cordelia	08h 02m	4	24.1	22.3:	−1.8
Ophelia	09 02	4	23.8	22.3:	−1.5
Bianca	10 25	4	23.0	22.3:	−0.7
Cressida	11 07	5	22.2	22.9	+0.7
Desdemona	11 22	5	22.5	22.9	+0.4
Juliet	11 50	5	21.5	22.9:	+1.4
Portia	12 19	5	21.0	22.9:	+1.9
Rosalind	11 54	5	22.5	22.9:	+0.4
Belinda	14 57	6	22.1	24.3:	+2.2
Puck	18 17	7	20.2	24.5:	+4.3
Miranda	33 56	10	16.3		
Ariel	60 30	14	14.2		

TABLE 1. Inner Satellites of Uranus

at the time. Thus, the 1995 dates offer another search opportunity which will be pursued both from the ground and by the Hubble Space Telescope.

The Voyager spacecraft has provided a wealth of data concerning the satellites; a long list of inner satellites have been discovered around Jupiter, Saturn, Uranus, and Neptune, along with unique rings around each of the planets. In addition, the Voyager spacecraft have determined the values of the masses of the planets, so that there is no longer the justification of observing the satellites to determine the planetary masses. Voyager has also presented the problem of how to continue to observe these inner satellites.

There is a possibility of observing some of these satellites with the Hubble Space Telescope, once it has corrected optics. This presents a challenging opportunity. Table 1 indicates the difficulty of making the observations due to the magnitude of the satellites (m_{app}) compared to the expected background magnitude (m_{bkg}) from the planet. However, as with other astrometric observing programs, it is difficult to get time on a telescope like the Hubble Space Telescope for an astrometry program.

Colas and Arlot (1991) are observing Martian satellites and fainter satellites. Radar and VLA observations have been made of the Galilean satellites and Titan (Muhleman et al. 1986). However, these are isolated observations and, due to the complexity of arrangements, cannot be expected on a regular basis. In the future, regular observations from these sources may be possible.

So, what are the future prospects for new observational capabilities for the satellites? Interferometers offer the most accurate technique for positional observations in the future. An optical interferometer has basically two restrictions for observing satellites. First is a magnitude limitation. Presently, the optical interferometer planned for Anderson Mesa will observe down to 10th magnitude. There are prospects for it to observe as faint as 14th magnitude. Also, the optical interferometer can resolve sources of 50-milliarcsecond diameter, so, it prefers sources of 20 milliarcseconds or smaller in diameter. Table 2 is a listing of selected satellites, including the smaller and brighter ones. It shows that all

	Satellite	Mag	Radius	Opposition Distance	Opposition Apparent Diameter
		V_e	km	au	"
MARS					
I	Phobos	11.3	11.1	0.523	0.059
II	Deimos	12.4	6.2	0.523	0.033
JUPITER					
II	Europa	5.29	1565	4.2	1.03
V	Amalthea	14.1	86.2	4.2	0.056
VI	Himalia	14.84	85	4.2	0.056
VII	Elara	16.7	40	4.2	0.026
VIII	Pasiphae	17.0	25	4.2	0.016
XV	Thebe	15.7	50	4.2	0.032
SATURN					
I	Mimas	12.9	199	8.5	0.064
II	Enceladus	11.7	250	8.5	0.081
III	Tethys	10.2	530	8.5	0.172
VI	Titan	8.28	2575	8.5	0.835
VII	Hyperion	14.19	141	8.5	0.046
VIII	Iapetus	11.1	718	8.5	0.233
IX	Phoebe	16.45	110	8.5	0.035
URANUS					
I	Ariel	14.16	579	18.2	0.087
IV	Oberon	13.94	762	18.2	0.115
V	Miranda	16.3	236	18.2	0.036
NEPTUNE					
I	Triton	13.47	1353	29.1	0.128
II	Nereid	18.7	170	29.1	0.016

Magnitude > 10; 14 in future
Diameter ≈ $0''.02$

TABLE 2. Optical Interferometry Observations

the satellites that are bright enough can be resolved. There are a few that are close to meeting the requirements, and might be future candidates. Deimos is the best candidate. The use of optical interferometry appears to be limited for satellite observations.

The radio wavelength would be a highly desirable observation, since it would present an opportunity to observe the satellites directly in the extragalactic reference frame. The possibilities are radar, VLA or VLBA observations of black body radiation from the satellites, or VLBI observations of radio sources from probes landed on the satellites. There is a proposal under consideration to land radio sources on Venus in order to study the internal activity of Venus. Possibly, there will be future plans to land radio sources on selected satellites.

The Carlsberg Meridian Circle has observed some of the brighter satellites, primarily as a means of determining the positions of the planets. With the availability of accurate satellite ephemerides and accurate star catalogs, the observations of the satellites offer a more accurate means of determining the positions of planets than the direct observation of the planets themselves. Also, it is not possible to observe the extended disks of the planets with some electronic detectors. Thus, one of the principal reasons for observing satellites in the future will be to obtain accurate planetary positions, because our observational techniques and our accuracy regimes have moved beyond the capability of accurately, directly observing the positions of the planets. This will require accurate knowledge of the orbits of the satellites.

3. Discrepancies, Uncertainties, Challenges

There is the continuing accuracy game between theory and observation. With the use of CCD technology and the availability of more accurate star catalogs, the observations are more accurate. In addition, their use for determining planetary positions requires improved ephemerides for the satellites. As the Martian satellite theories have been greatly improved (Ivanov et al., 1988; Shor, 1991), other satellite theories will require improvement.

In a number of cases, the inner satellites, which were discovered by Voyager, cannot be observed by any current techniques, except by another spacecraft to the planet. Thus, the limited accuracy of the mean motions of the satellites will result in an increasing inaccuracy in the knowledge of the position of the satellite until it will be essentially unknown. In some cases, due to resonances, there are interactions between the satellites and the rings. The limited accuracy of the knowledge of the motions of the satellites can be a significant factor in the uncertainty concerning the dynamics in the inner satellite systems. The use of the Hubble Space Telescope, ground-based CCD's, IR detectors, and other clever techniques needs to be considered for these unobservable, or very difficult-to-observe, inner satellites.

The Galilean satellites have a long, good history of observations, and have been the subject of an intense accurate, theoretical development. As a result, there are theories for the Galilean satellites (Lieske, 1977, 1980; Vu 1977; and Thuillot and Vu, 1983). However, recent accurate observations by Pascu (1992) indicate a scale discrepancy between the theory and observations. This discrepancy could be due to observational reduction inaccuracies, a system mass error, or some other hypothetical source.

There are also the undiscovered satellites; or, more particularly, cases where individual images were obtained of what were thought to be satellites, e.g. Charles Kowal's image which was temporarily labelled Jupiter XIV. The designation was later re-used for a satellite with sufficient observations to determine its orbit. There are also a number of images obtained by Voyager, particularly at Saturn, which indicate satellites, but sufficient observations were not obtained to determine a satellite orbit.

Voyager did a marvelous job of discovering satellites, but is our knowledge of planetary satellites now complete? Could there be more satellites, close in, or in rings? Could there be outer satellites or more librational satellites of Saturn?

4. Scientific Interests

Other than positional data, is there any scientific reason for observing planetary satellites? In the past, one of the purposes for observing the satellites was to determine the planetary masses. Clearly, this is no longer necessary. When you look back at past determinations, with the knowledge of the true value of the mass as determined from the Voyager mission, there does not appear to be any definitive lessons to be learned. There is no pattern that mass determinations by one method, or one type of data, gave values closer to the true value than other methods or data. It may be possible, with more accurate observations of the satellites, to determine improved values of the masses of some of the other satellites. Certainly, there is knowledge to be gained concerning the interaction between the satellites and the rings, and the resonant motions among the different satellites.

Voyager observations provided information concerning the composition and photometry of the satellites, in addition to positional data (Helfenstein et al., 1988; Thomas and Veverka, 1991). In many cases the surfaces are not uniform, and the variations indicate a complex activity involving neutral and ionized particles in the field around the planets, and the resulting surface colors, or darkness. The distribution of dark material in the outer solar system, in the rings, and on the satellites is a mystery which raises questions concerning the composition and origin of the satellites and their material.

The dynamics of the satellite systems leads to studies and investigations concerning the stability and presence of chaos in their orbital and rotational motions. The lack of understanding of the interactions fits with the question of the completeness of the discoveries of the satellites. So, in the area of dynamics within the solar system, the satellites offer a fertile ground for investigations.

5. Conclusion

This is a time of change for planetary satellites. Observational accuracies have changed due to the use of CCD technology and improved star catalogs. The use of new detectors means the satellites can be observed more accurately than the planets themselves. Both of these developments require improved theories for the motions of the satellites.

The use of computer-based almanacs means that accurate predictions of the satellites can be readily available to the observers. Besides the continued need for positional observations, there is also scientific interest in the dynamics and photometry of the satellites and the rings.

REFERENCES

Batrakov, Yu. V. (1991) "On the Determination of Orbits of Satellites of the Distant Planets and Some Requirements to their Positional Observations," First Spain - USSR Workshop on Positional Astronomy and Celestial Mechanics, edited by A. Lopez Garcia, R.F. Lopez Machi and A. G. Sokolsky, Observatorio Astronomico, Universitat de Valencia 51.

Colas, F. and Arlot, J.E. (1991) AA, 252, 402.

Helfenstein, P., Veverka, J., and Thomas, P.C. (1988) Icarus, 74, 231.

Ivanov, N. M., Kolyuka, Yu. F, Kudryavtsev, S.M. and Tikhonov, V.F.(1988) Pis'ma Astron. Zh. 14, 956-960= Sov. Astron. Lilt, 14, 5.

Lieske, J.H. (1977) AA, 56, 333-352.

Lieske, J.H. (1980) AA, 82, 340-348.

Muhleman, D.O., Berge, G.L., Rudy, D.J. et al. (1986) in Fundamental Astronomy and Solar System Dynamics, edited by R.L. Duncombe, J.H. Lieske, and P.K.Seidelmann, D. Reidel Publishing Co., Dordrecht, 329.

Nakamura, T., Kinoshita, H. and Kozai, H. (1991) AJ, 101, 290.

Nicholson, P.D., Hamilton, D.P., Mathews, K. et al. Icarus (in press).

Pascu, D. (1992) private communication.

Pascu, D. (1979) in Natural and Artificial Satellite Motion, edited by P. E. Nacozy and S. Ferraz-Mello, University of Texas Press, Austin, 17

Pascu, D., Seidelmann, P.K., Schmidt, R.E. et al. AJ 93, 963-967.

Pascu, D., and Schmidt, R.E. (1990) AJ 99, 1974-1984.

Seidelmann, P.K. (1979) in Natural and Artificial Satellite Motion, edited by P.E. Nacozy and S. Ferraz-Mello. University of Texas Press, Austin, 3-12.

Seidelmann, P.K. (1993) Celestial Mechanics and Dynamical Astronomy 56, 1-12.

Shor, V.A. (1991) in First Spain-USSR Workshop on Positional Astronomy and Celestial Mechanics, edited by A. Lopez Garcia, R.F. Lopez Machi and A.G. Sokolsky, Observatorio Astronomico, Universitat de Valencia 55-61.

Soma, M. (1992) AA 265, L21.

Thomas, P. and Veverka, J. J. of Geophys. Res. 96, Supplement, 19261.

Thuillot, W. and Vu, D.T. (1983) in The Motion of Planets and Natural and Artificial Satellites, edited by S. Ferraz-Mello and P.E. Nacozy, Universidade de Sao Paulo, 273-294.

Vu, D.T. (1977) AAS, 30, 361-367.

Satellite astrometry with a long-focus astrograph

By A. A. KISSELEV

Pulkovo Observatory, 196140 St Petersburg, Russia.

The problem of determining the relative positions of satellites and planets by means of photographs obtained with a long-focus astrograph is considered. Two astrometric plate reduction techniques are analysed: (1) a traditional method using some reference stars and (2) so-called "scale-trail" method using the accurate value of the scale of the telescope. The author proposes rigorous generalized formulae which allow for differential refraction in the scale and initial orientation of measured coordinates on the plate and enable the calculation of accurate, angular, tangential coordinates of the objects. It is demonstrated that if the measured distances do not exceed $600''$, the "scale-trail" technique provides a higher accuracy of reduction than the classical one. The values of the rms are as follows: $\pm\ 0''.05$ to $0''.10$ ("scale-trail" technique), $\pm\ 0''.15$ to $0''.30$ (plate constants technique using six well-distributed reference stars).

1. Introduction

Astronomical photographs obtained with a long-focus telescope still provide the best observational base to determine the relative positions of satellites and planets in the solar system. There are two astrometric plate reduction techniques to transform the measured coordinates of photographic images on the plate into precisely-defined, angular, tangential coordinates. The first one is the traditional plate-constant method using reference stars. The second is the so-called "scale-trail" method using the accurate value of the scale of the telescope. The rigorous analysis of these two methods is considered in this paper.

2. Comparison of methods

We start with the following assumptions:
(i) The focal-length (f) of the astrograph is about 10 m, hence the scale (M) is
$$M = 20''/\text{mm}.$$
(ii) The measured (working) field of the astrograph (2ρ) does not exceed $40'$:
$$2\rho \leq 40'.$$
(iii) The tangential point of the astrograph is near its centre, and so the optical aberrations – coma and distortion – may be neglected.
(iv) The errors of the measured coordinates of star-like objects on the plate are about $\pm 2\mu$m ($0''.04$). Hence the mean error of a measured distance is about $\pm 0''.06$, and the corresponding relative errors of a measured arc σ ($\sigma < 20'$) and the error of the measured angle (θ) are:
$$\frac{\varepsilon_m}{\sigma} \geq 5 \times 10^{-5}; \quad \varepsilon_\theta \geq 5 \times 10^{-5}.$$
(v) The errors of positions ($\varepsilon_{\alpha,\delta}$) of the great majority of reference stars contained in astrographic catalogues AGK3 or PPM corresponding to the present time of observations (1990) are as follows:
$$\varepsilon_{\alpha,\delta} = \pm(0''.3 - 0''.5) \quad AGK3$$

$$\varepsilon_{\alpha,\delta} = \pm(0\rlap{.}''2 - 0\rlap{.}''3) \quad PPM.$$

(vi) The number of reference stars in the field of a long-focus astrograph may be 2-9 (the mean is 4), but the distribution of these stars is usually far from uniform.

2.1. Classical method (reference stars)

The statements above allow us to estimate the accuracy of the relative positions of star-like objects using the classical technique with reference stars. We assume that the plate constants have been computed by the least-squares method, with N uniformly distributed reference stars. Then the relative error of the empirical scales M_x and M_y is determined from (Kisselev, 1989),

$$\frac{\varepsilon(M)}{M} = \frac{2\varepsilon_1}{\rho\sqrt{N}} \left(= 5.6\frac{\varepsilon_1''}{\rho°\sqrt{N}}10^{-4}\right), \tag{2.1}$$

where $\varepsilon_1'' = \sqrt{\varepsilon_m^2 + \varepsilon_{\alpha,\delta}^2}$ means that the unit error of the least-square solution is expressed in arcsec, and $\rho°$ denotes the mean radius of area containing reference stars expressed in degrees. Equation (2.1) allows us to estimate the errors of orientation θ (2.2) and distance σ (2.3) due to the reduction technique:

$$\varepsilon_\theta = \frac{\varepsilon(M)}{M} \quad (= 2.5 \times 10^{-4}), \tag{2.2}$$

$$\varepsilon_\sigma = 2\frac{\varepsilon_1}{\sqrt{N}}\frac{\sigma}{\rho} = 0\rlap{.}''30). \tag{2.3}$$

These numerical results are obtained using the following data:

$$\varepsilon_1 = 0\rlap{.}''3, \quad \rho = 20'(\rho° = 0.33), \quad \sigma = \rho, \quad N = 4.$$

2.2. "Scale-trail" method

Now we consider the accuracy of the "scale-trail" technique working on the same problem. We assume that the geometrical scale (M) of the astrograph is known to high accuracy:

$$\varepsilon_M = \pm 0\rlap{.}''001/\text{mm}; \quad \frac{\varepsilon(M)}{M} = 5 \times 10^{-5}.$$

To apply the "scale-trail" technique we must take into account certain known geometrical and physical phenomenon which distort the true configuration of the stars. These are:
(i) Difference between radial and transverse scales (M_r and M_t) in the central projection. The amount of this effect is estimated from:

$$\frac{M_t - M_r}{M_0} \simeq \frac{1}{2}\rho^2; \tag{2.4}$$

$$\Delta\theta \leq \frac{\Delta M}{M} \leq 1.7 \times 10^{-5}; \quad \rho < 20'.$$

(ii) Differential refraction. The first and second order effects of this phenomenon are evaluated from (2.5) and (2.6):

$$\left|\frac{\Delta\sigma}{\sigma}\right|_{IR} \leq \beta_\zeta \sec^2\zeta. \tag{2.5}$$

$$\left|\frac{\Delta\sigma}{\sigma}\right|_{IIR} \leq \beta_\zeta \sigma \tan\zeta \sec^2\zeta. \tag{2.6}$$

Here β_ζ is the photographic refraction coefficient,

$$\beta_\zeta = \beta_0 - \beta'\tan^2\zeta; \quad \beta_0 \simeq 29.6 \times 10^{-5}; \quad \beta' = 0.042 \times 10^{-5},$$

ζ	$\left|\frac{\Delta\sigma}{\sigma}\right|_{IR}$	$\left|\frac{\Delta\sigma}{\sigma}\right|_{IIR}$	ρ_A	$\left|\frac{\Delta\sigma}{\sigma}\right|_{IA}$
0°	29.1	0.0	0°	9.8
15°	31.2	0.1	15°	9.6
30°	38.8	0.1	30°	8.6
45°	58.2	0.3	45°	7.0
60°	164.4	1.2	60°	5.0
70°	248.7	4.0	75°	2.6
75°	434.2	9.4	90°	0.0

TABLE 1. The maximum refraction and aberration differential effects (in units of 10^{-5}), assuming $\sigma = 20'(0.0058\,\text{rad.})$

where ζ is the the zenith distance of the object.
(iii) Differential aberration. The first order effect of this phenomenon is given by:

$$\left|\frac{\Delta\sigma}{\sigma}\right|_{IA} \leq \kappa \cos \rho_A, \tag{2.7}$$

where κ is the annual coefficient of aberration, 9.8×10^{-5} rad ($20''\!\!.5$), and ρ_A is the spherical distance between the object and the apex of the annual motion of the Earth. The maximum differential values of the refraction and aberration effects are listed in Table 1.

The data above show that the main effect which must be taken into account in using the "scale-trail" technique is the first order of differential refraction; the other distorting phenomenon (see 2.4, 2.6 and 2.7) may be neglected compared to the errors of measurement (5×10^{-5}). The rigorous formula for transforming the measured coordinates into tangential ones have to allow for the foreshortening of the spherical distance ($\Delta\sigma$) and the alteration of the true positional angle $\Delta\theta$ relative to zenith caused by refraction. The required formulae are as follows:

$$\Delta\sigma = -\sigma\,\beta_\zeta\,(1 + \tan^2\zeta\,\cos^2\theta), \tag{2.8}$$

$$\Delta\theta = \beta_\zeta \tan^2\zeta \sin\theta \cos\theta. \tag{2.9}$$

Taking these effects into account, the generalised relationship between the standard coordinates (ξ, η) and the measured coordinates (x, y) of the objects (satellites) relative to the centre (planet) is (Kisselev, 1989):

$$\xi = M_0[1 + \beta_\zeta(1 + K_1^2)]\,x + M_0\beta_\zeta[2K_1K_2\sin^2\psi_0 + (K_2^2 - K_1^2)\sin\psi_0\cos\psi_0]\,y, \tag{2.10}$$

$$\eta = M_0[1 + \beta_\zeta(1 + K_2^2)]\,y + M_0\beta_\zeta[2K_1K_2\cos^2\psi_0 - (K_2^2 - K_1^2)\sin\psi_0\cos\psi_0]\,x, \tag{2.11}$$

where
M_0 is the geometrical scale of the astrograph,
K_1, K_2 is the tangential coordinates of the zenith,
ψ_0 is the positional angle of the measured coordinates relative to the true pole.
We assume that the measured coordinates of the astrograph are orientated in such a way, that they satisfy:

$$\frac{x_0}{y_0} = \frac{\xi_0}{\eta_0} = \tan\psi_0.$$

Here ψ_0 is the theoretical positional angle of a star relative to the planet. If we use the diurnal trail as the orientation reference, we have $\psi_0 = 90°$ and the formulae reduce to the simple well-known form:

$$\xi = M_0[1 + \beta_\zeta(1 + K_1^2)]\, x + 2M_0\beta_\zeta K_1 K_2 y \tag{2.12}$$

$$\eta = M_0[1 + \beta_\zeta(1 + K_2^2)]\, y \tag{2.13}$$

The formulae 10-13 provide the astrometric reduction results which maintain the accuracy of the initial measured coordinates i.e. $\sim \pm(0\rlap{.}''04$ to $0\rlap{.}''06)$ in measured distances, where the working field used is sufficiently small ($\rho < 20'$) and the true refraction does not depart far from the mean value used.

3. Conclusion

Thus we come to the conclusion that the "scale-trail" technique provides better accuracy for the astrometric reduction than the classical one. To illustrate this result, we apply it to the photographic observations of the Galilean satellites obtained with the 26-inch refractor at Pulkovo (Kisseleva, 1987). The observations were carried out during two periods: 1974–1981 and 1986–1992. More than 300 plates containing 10 exposures and two diurnal trails were obtained. The "scale-trail" reduction technique was used. The accuracy of the relative positions of the satellites and Jupiter were derived from comparisons with modern theories G-5 (Arlot). The *rms* errors were as follows:

(1): satellite to satellite: $\pm\ (0\rlap{.}''05 - 0\rlap{.}''15)$

(2): satellite to Jupiter: $\pm\ (0\rlap{.}''10 - 0\rlap{.}''20)$.

The largest errors occurred for measured distances exceeding $600''$(30mm).

REFERENCES

Kisselev, A.A., 1989, *Theoretical foundations of photographic astrometry*. Moscow, Nauka, (in Russian)

Kisseleva, T.P., 1987, A determination of precise coordinates of Galilean satellites of Jupiter with the Pulkovo 26" refractor. *IZVESTIA GAO* (Proceedings of the Main Astronomical Observatory), **204**, 57-64, (in Russian).

Some results obtained during the 1991 campaign of observation of the mutual events of the Galilean satellites

By B. MORANDO AND P. DESCAMPS

Bureau des Longitudes, 77, Avenue Denfert-Rochereau, Paris, F-75014, France

The observations of their mutual phenomena, which occur every six years, give the most accurate positions of the Galilean satellites of Jupiter. Examples are given showing that simple observations made from bad sites (Paris Observatory), as well as observations using refined modern techniques (infrared observations), are equally useful.

---

1. The mutual phenomena of the Galilean satellites

Every six years the equatorial plane of Jupiter, with which the orbital planes of the Galilean satellites nearly coincide, is in such a situation relative to the Sun and the Earth that mutual eclipses and occultations of the satellites are possible. Since 1973 these events have been predicted systematically, and campaigns of observation were organised by Bureau des Longitudes in 1979, 1985 and 1991 (Arlot et al. 1982; Arlot 1987; Arlot & Thuillot 1988). Many observations obtained in 1979 and 1985 have been published (Froeshle et al. 1988; Thuillot et al. 1991; Arlot et al. 1989; Souchay et al. 1992). The reason for this is the great accuracy with which the instant of the maximum of an eclipse or occultation can be determined, which leads to very accurate positions of the satellites. These are needed to build good ephemerides of these bodies.

Table 1 shows the progress made in the precision of the positions using different techniques. The second column of the Table gives the precision of the position of a satellite on the celestial sphere as seen from the Earth, and the third column the accuracy in space. Before the mutual phenomena were observed, the best observations were the photographic observations carried out by D. Pascu at the U.S. Naval Observatory. Such observations are necessary to fill the gaps between mutual phenomena campaigns and are still being made. Yet it is seen in Table 1 that the very first and crude observations of the mutual phenomena give a precision three times better. More refinements in the models used for the comparison of the theory with observation have now improved the precision by 20 times that of photographic observations.

Bureau des Longitudes organized international campaigns of observation starting in 1979. It was realised that as many instruments, sites and observers as possible have to be involved in order to minimise the consequences of bad weather and obtain a large number of observations. As the Galilean satellites are bright objects and the photometry involved is simply the determination of the minimum of a light-curve, the observations may be carried out in many places that are not supposed to be fit anymore for astronomical purposes. Moreover, clever amateurs may obtain valuable results. Many of them, who are well-equipped and own photometric photometers, have been considered as professional observers in Table 2, which shows where the 311 good light-curves of the campaign were obtained. The amateurs mentioned at the bottom of the first column of Table 2 are those who used visual or photographic methods.

Type of observation	Precision in angle (arcsec)	Precision in space (km)
Photographic	0.083	300
Mutual phenomena (first models)	0.028	100
Mutual phenomena (with phase models)	0.008	30
Mutual phenomena (with phase and surface properties)	0.004	15

TABLE 1. Precision of observations of the Galilean satellites.

Place	Type	No.	Place	Type	No.
Paris Obs.	PM	8	Jungfrau Obs.	PM	10
Meudon Obs.	2D	13	Belgrade	PM	1
Pic du Midi Obs.	2D	13	Chile	PM-IR	18
Haute Prov. Obs.	2D-IR	36	Brasopolis	PM	2
Bordeaux Obs.	2D	16	Rio de Janeiro	PM	2
Cote d'Azur Obs.	PM	39	India	PM	16
Italy	PM-2D	42	Japan	PM-IR	14
Barcelona	PM	20	Canada	PM	1
Canary Islands	PM	10	USA	PM-2D	7
Romania	PM	6	Hawaii	PM	3
Germany	PM	11	Australia	PM	4
Bulgaria	PM	2	Amateurs	Visual & Photo.	26
Benelux	PM-2D	11			

Total No. 311

TABLE 2. Mutual phenomena obtained during the 1991 campaign organized by the Bureau des Longitudes.

2. Results obtained at Paris Observatory

As an example of what was done during the 1991 campaign some results obtained at Paris Observatory – now in the heart of a big city – are given. The telescope is a 38 cm refractor with a 8.70 m focal length established on the terrace of the old seventeenth century building in 1854. The object glass was remade by the Henry brothers in 1882. A photoelectric photometer made of an RCA 4840 photomultiplier was built for the purpose by F. Sevre and R. Vitry at the Institut d'Astrophysique in Paris.

Figure 1 shows the assembly diagram of the instrumentation. The signal is received by an Amstrad 512 micro computer. The results obtained will be published in detail elsewhere. We just show here one of the light-curves obtained (Figure 2) and the O–C of the instants of the maximum (Table 3). In the second column of Table 3, for example **2 O 1** means that satellite J2 (Europa) Occults satellite J1 (Io): the letter E stands for eclipse.

The O–Cs seem to be systematically positive by about 18 s on the average when both Io and Europa are involved. In a paper published in *Icarus*, Mallama (1992) finds the same result, observing different phenomena from a different place.

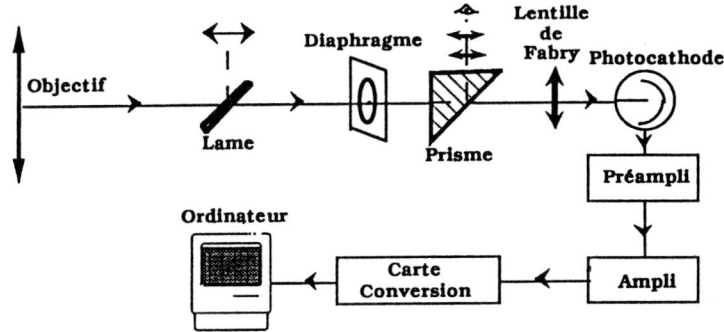

FIGURE 1. Assembly diagram of the instrumentation used at Paris Observatory.

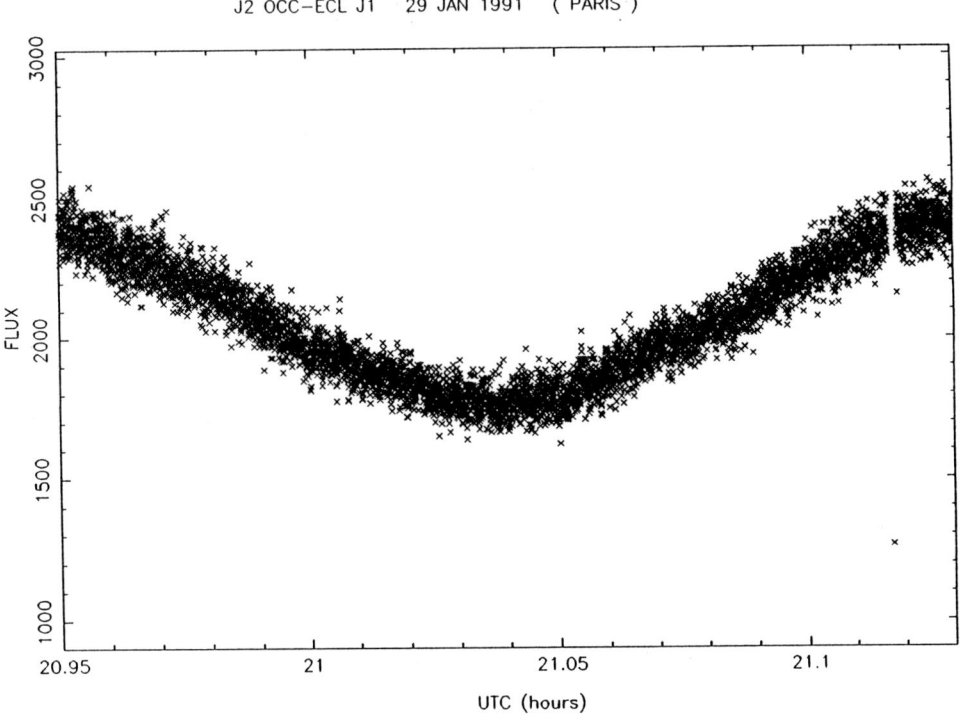

FIGURE 2. Light-curve of a mutual event recorded at Paris Observatory.

3. Observations of the volcanoes on Io

Another interesting result is shown in Figure 3. The first satellite Io is known to possess very large and active volcanoes, the most important being called Loki. The hot lava expelled by the volcanoes radiates in the infrared band of the spectrum at μm wavelengths.

At the bottom of Figure 3 the geometry is explained of an occultation of Io by Europa observed at that wavelength. As the light-curve proceeds, the light is not altered at first because, apart from the volcanoes, the surface of Io does not radiate much at $3.8\,\mu$m. But a sudden drop in the light-curve is observed when Loki is occulted (top of Figure 3), followed by a similar drop when another less active volcano, Pele, is occulted. As the

20 Février 1991 (ESO) - 3.8 µm

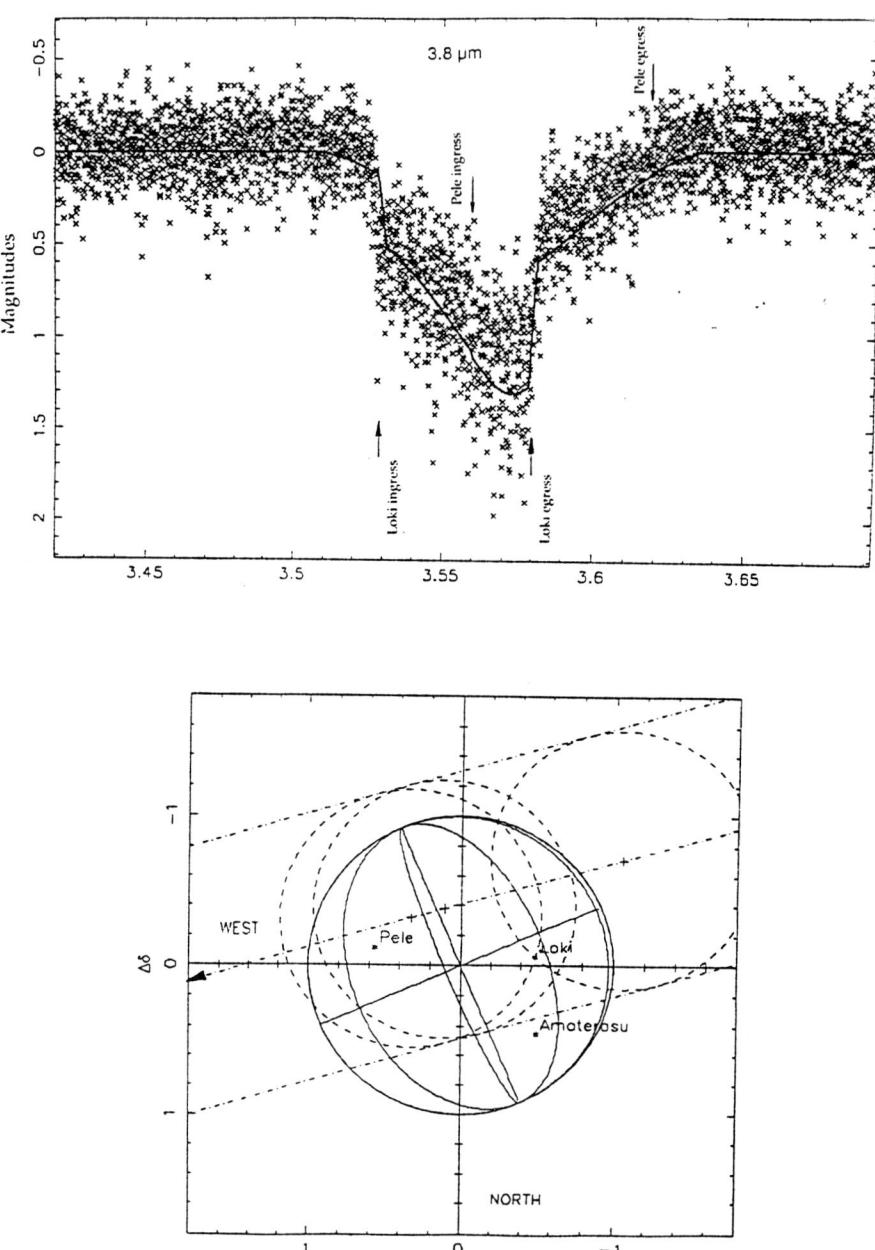

FIGURE 3. Observation in the infrared of an occultation of Io by Europa.

Date	Phenomenon	UTC of minimum (h m s)	O−C (s)
21/1/91	2 O 1 & 2 E 1	21 02 19	11.9
5/2/91	2 O 1	23 13 35	28.7
5/2/91	2 E 1	23 36 24	22.4
9/3/91	2 O 1	20 58 39	15.8
9/3/91	2 E 1	22 26 33	16.2
10/3/91	2 E 4	03 43 34	−0.7
1/4/91	3 E 2	22 56 41	16.5
10/4/91	2 E 1	20 49 19	31.7

TABLE 3. Results obtained with the Paris Observatory 38 cm refractor.

light-curve at visual wavelength is simultaneously recorded, the time when the apparent disc of Europa comes into contact with the apparent disc of Io is determined accurately and compared with the time corresponding to the drop in the infrared light-curve. From the known diameters of the satellites and the theory of their motions, the geometry of the problem is known and, thus, the position of Loki on the surface of Io can be precisely determined. More detailed results of this observation can be found in Descamps et al. 1992.

4. Conclusion

There is a big demand from space missions, as well as from Earth-based astronomers for more and more precise ephemerides of the satellites of the planets in general, and of the Galilean satellites of Jupiter in particular. The mutual phenomena ensure the best precision available for that purpose, with the models used for fitting the observations becoming more and more sophisticated. On top of this, the benefit to planetology appears as an important by-product of these observations, as shown by the observations of the volcanoes of Io.

But the mutual phenomena of the Galilean satellites suffer from an important drawback: they only take place for a few months every six years. However, new techniques, such as CCD camera observations, now well understood and already used during the 1991 campaign, will lead to a new type of observation between mutual phenomena periods. Close approaches of satellites can be recorded in two dimensions and will yield a dense set of continuous data.

REFERENCES

Arlot, J.-E., Bernard, A., Bouchet, P., Daguillon, J., Dourneau, G., Figer, A., Helmer, G., Lecacheux, J., Merlin, Ph., Meyer, C., Mianes, P., Morando, B., Naves, D., Rousseau, J., Soulie, G., Terzan, A., Thuillot, W., Vapillon, L., Wlerick, G., 1982, Les resultats de la campagne d'observation PHEMU79 des phenomenes mutuels des satellites galileens de Jupiter en 1979, A&A, 111,151

Arlot, J.-E., 1987a, Amateurs' contributions to the PHEMU85 campaign of observation of the mutual phenomena of Jupiter's satellites, in IAU Coll., 98, Kluwer, Dordrecht, p.198

Arlot, J.-E., 1987b, Editor of the Proceedings of the Journees PHEMU85, Suppl. Ann. Phys., 12, 1

Arlot, J.-E. & Thuillo, W., 1988, The coordination of the observations during the PHEMU85 campaign, in *Coordination of Observational Projects in Astronomy*, ed. Jaschek C. & Sterken C., Cambridge University Press, p.171

Arlot, J.-E., Bouchet, P., Ciuiffes, C.H., Schmider, F.X., Thuillot, W., 1989, Mutual events of the Galilean satellites. An analysis of the observations made in 1985 at ESO, AJ, 98, 1890

Arlot, J.-E., Thuillot, W., Barroso Jr, J., Bergeal, L., Blanco, C., Boninsegn, R., Bouchet P., Bourgeois, J., Briot, D., Bulder, H., Burchi, R., Cano, J.A., Colas, F., D'Ambrosio, V., Di Paolantonio, A., Dourneau, G., Dumont, M., Ferrand, S., Figer, A., Francou, G., Froeschl, M., Gomez-Forellad, J.M., Gouiffes, C., Helmer, G., Jablonsky, F.J., Laques, P., Le Campion, J.F., Lecacheux, J., Lecontel, J.M., Manfroid, J., Meyer, C., Morando, B., Quast, G.R., Rmis, J., Renauddineau, J., Rouan, D., Ruatti, C., Sareyan, J.P., Schmieder, F.X., Svre, F., Souchay, J., Valtier, J.C, Vu, D.T., Wahiche, J.D., 1992, A catalogue of the observations of the mutual phenomena of the Galilean satellites of Jupiter made in 1985 during the PHEMU85 Campaign, A&AS, 92,151

Descamps, P., Arlot, J.-E., Thuillot, W., Colas, F., Vu, D.T., 1992, Observations of the volcanoes of Io, Loki and Pele made in 1991 at ESO during an occultation by Europa, Icarus, 100, 235

Froeschle, M., Helmer, G., Meyer, C., 1988, Observations of mutual phenomena of the Galilean satellites of Jupiter made at CERGA, A&A, 189, 277

Souchay, J., Hiromoto, N., Takami, T., Nakamura, T., Descamps, P., Aruga, T., 1992, Galilean satellite observations, A&A, 264, 314

Thuillot, W., Arlot, J.-E., Fettig, S. & Colas, F., 1991, Video techniques applied to astrometry, Ap&SS, 177, 321.

Observation and analysis of mutual satellite events

By K. AKSNES

Institute of Theoretical Astrophysics University of Oslo, Box 1029 Blindern, 0315 Oslo 3, Norway

Photoelectric observations have been made worldwide of the mutual occultations and eclipses of the Galilean satellites in 1973, 1979, 1985, and 1991, yielding a total of more than 500 light curves. From the 1973 observations, accurate satellite radii and relative positions were obtained. The radii agreed to within 35 km with the radii measured on Voyager TV images in 1979, while the positions have *rms* errrors of about 100 km when Sampson's theory is fitted to them.The mutual events technique is probably still superior to any other known technique for determining satellite positions.

In 1979-80 13 photoelectric light curves were made for the first time also of the four satellites Saturn 2-5. The midtimes of those light curves show an *rms* deviation of about 30 s from those predicted with the latest satellite ephemerides by Dourneau. In 1995 and 1996, some 200 mutual events will occur between Saturn 1-7.

1. Introduction

Twice during the revolution periods of Jupiter and Saturn; i.e., about every 6 and 15 years, Earth and Sun will pass through the nearly coplanar orbital planes of the inner satellites of those planets. The satellites will then occult and eclipse each other. Each of these mutual satellite events can last from a few seconds up to a couple of hours and they occur in series lasting a year or more. This will happen next in 1995-96 for Saturn's satellites and in 1997 for Jupiter's satellites.

Currently, the main reason for observing the mutual events is to obtain precise positional data on the satellites for ephemeris improvement, in support of the Galileo mission en route to Jupiter and the upcoming Cassini mission to Saturn, as well as for better theoretical insight into the dynamics of the satellites. Of special interest is the possibility to detect secular terms in Io's motion due to tidal interaction with Jupiter.

The theory of motion for the Galilean satellites by Sampson (1921), and updated by Lieske (1980), has been based mainly on timings of eclipses of the satellites in Jupiter's shadow. The timings are affected by considerable statistical uncertainties and likely systematic effects due to the variable conditions in Jupiter's upper atmosphere. Despite the fact that progress has been made recently in modeling these eclipses , the attainable accuracy will be considerably less than for mutual satellite eclipses and occultations, for the simple reason that the satellites have no appreciable atmospheres. The orbital theories for Saturn's satellites due to Struve (1924-33), and most recently Dourneau (1993), are based largely on micrometric and photographic observations which cannot compete in accuracy with the mutual events either. Considerable progress has been made through the introduction of CCD imaging, but Pascu (1993) reports unexplained systematic errors in orientation and scale increasing with distance from the center of the frame.

Although relatively few in number, the mutual events appear to be well worth observing also in the future.

2. Prediction and observation

The mutual satellite events have since 1931 been predicted regularly in the *Handbook of the British Astronomical Association*, based on methods published by Levin (1931) and Comrie (1931) for Jupiter's and Saturn's satellites, respectively.

It is rather surprising that the mutual events had received very little attention from professional astronomers until the mutual events of the Galilean satellites in 1973-74. On the initiative of R. T. Brinkmann of the Lunar Science Institute and R. L. Millis at Lowell Observatory, a worldwide observing campaign with modern photoelectric equipment was organized for the mutual events that year. More detailed predictions than those available in the BAA Handbook, including estimates of the light losses and the geometry of the events, had been distributed to many observers by Brinkmann (1973) and Aksnes (1974). Similar predictions have been computed by Aksnes and Franklin (1978,1984,1990) and by Arlot (1978,1984) for the mutual events of the Galilean satellites in 1979, 1985-86 and 1990-92, and for Saturn's satellites in 1979-80. Based on improved ephemerides of Saturn's satellites by Dourneau (1993), detailed predictions of about 200 mutual events among those satellites in 1995 and 1996 have been computed by Arlot and Thuillot (1993) and Aksnes and Dourneau (1993).

The 1973-74 campaign resulted in about 100 light curves of very high photometric quality, the best ones coming from observers in the southern hemisphere, since Jupiter had a high southern declination. Many events were favorable for observation because Jupiter reached opposition near the middle of the event series. The next series of mutual events, in 1979, with Jupiter at a high northern declination, were much less favorable, since most occurred near Jupiter conjunction. Only about a dozen observations were made of the 1979 events. The 1985-86 events were again favorably, but the 1990-92 events rather unfavorably, placed for observation. In total, about 500 observations of the mutual events of the Galilean satellites have been observed and analyzed so far.

Under good conditions, a photometric accuracy of 1 to 2 percent has been achieved, the limiting factor being normally the variable sky background due to Jupiter's glare. This forces the use of small photometer diaphragms and the risk of spillage of light outside the diaphragm because of inaccurate centering on the target satellite, smear due to poor seeing or high airmass, or telescope shaking in gusts of wind.

The use of CCD photometry is becoming a promising alternative (Mallama 1992) to photomultipliers. The CCD images can also be used for making accurate astrometric measurements of satellite separations before or after a mutual event or when satellites only have close approaches to one another.

Although at least two hundred mutual events were predicted to occur among Saturn's satellites in 1979-80, only about a dozen were reported observed. Because of the small sizes of most of the satellites and their proximity to Saturn, the predictions are very sensitive to satellite ephemeris errors, and the events will be difficult to observe with a photometer. While an eclipsed Galilean satellite can usually be isolated in the diaphragm, a Saturnian satellite undergoing a mutual eclipse will often have to be observed together with one or even two other satellites. During a mutual occultation the measured light contribution will, of course, always involve at least two satellites. This will reduce the depths of the light curves and therefore their sensitivities to the light variations of the satellite undergoing change. In order to isolate those light variations, accurate intensity readings will be required for all the satellites in the diaphragm, preferably by direct observation just before and after the event or otherwise from published rotation light curves of the satellites. Most observations have been performed with a standard V-filter, sometimes with simultaneous recording also in U,B or even R filters.

3. What has been learned from the mutual events?

As a basis for the later discussion of what has actually been learned from observations of mutual satellite events, we shall first briefly describe the light curve models involved. Reference is made to Aksnes and Franklin (1976) for further details.

The light variation during a mutual occultation can be expressed simply as

$$I_{oc} = 1 - \frac{\beta A}{1+\gamma} \tag{3.1}$$

Here I_{oc} is the combined intensity of the two satellites, normalized to unity outside the event, γ is the ratio of the full-disk intensity of the occulting satellite to that of the occulted satellite, and A is the occulted area of average surface brightness β. The units are such that for a total occultation, $\beta = A = 1$. If β varies considerably across the disk of the occulted satellite, βA can be replaced by the sum,

$$\beta A = \sum_i \beta_i A_i \tag{3.2}$$

It is much more involved to model a mutual eclipse. If the eclipsed satellite can be observed alone, its intensity variation can be expressed as

$$I_{ec} = 1 - A_u + \int_0^{A_p} (1 - I_x) dA_p \tag{3.3}$$

where A_u and A_p denote the fractions of the satellite's disk covered by the umbra and penumbra, respectively. At a given point in the penumbra, the light intensity I_x will be proportional to the unobstructed portion of the Sun's disk as seen from that point, corrected for limb darkening. Since I_x must itself be obtained through an integration, a double integration is implied in this equation. The situation is further complicated by albedo variations across the satellite or limb darkening if it has an appreciable atmosphere. No attempts have been made to model the last two effects on the satellites. The areas A, A_u, and A_p in the above equations can be calculated by means of geocentric ephemerides of Jupiter and Saturn and planetocentric ephemerides of their satellites(Aksnes et al. 1986).

If we assume that the light curve parameters most in need of correction are the radii R_1 and R_2 of the two satellites involved, their minimum apparent separation, x in longitude and z in latitude, and the albedo coefficients β_i of the occulted or eclipsed satellite, we may formally combine Eqs. (1) and (3) into the single equation,

$$I = I(R_1, R_2, x, z, \beta_i) \tag{3.4}$$

This theoretical model is then fitted to the observed light curves by making least squares adjustments in these parameters.

In the seventies, before the Voyager spacecraft had made precise astrometric and photometric measurements of the Galilean satellites, the main interest in the mutual events was the possibility of obtaining albedo maps from the occultations and better values for the satellites' sizes, which were then uncertain by hundreds of kilometers. Satellite radii accurate to 35 km were derived for Europa, Ganymede, and Callisto from the mutual events in 1973 (Aksnes and Franklin 1976), but this result was superseded already by the Voyager result in 1979. The Voyager passage of Saturn's satellites in 1980 and 1981 preempted the need for deriving radii for those satellites from the mutual events in 1979-80.

Attempts to derive albedo maps of Europa in the V-band on the basis of a series of occultations by Io were disappointing. No obvious improvement was obtained by subdividing the surface of Europa into a few subareas according to Eq. (2) and solve for the β-values, since a single β did just as well. This was perhaps partly due to the fact that Europa has a rather uniform surface brightness, as later revealed by Voyager images. On the other hand, IR observations of mutual occultations of Io by other satellites have revealed light curve features associated with hot spots on Io (Goguen et. al 1988). This has made it possible, in a crude way, to monitor the activity of volcanos on Io.

Even on the assumption of a uniform visual satellite surface, there is an important photometric correction which needs to be applied to the light curve models derived above. For several years there was an unexplained bias between the observed midtimes of the 1973 occultations of Europa by Io and the midtimes observed for the correspoding eclipses, occurring just before or after the occultations. The cause of this bias, which amounted to a longitude shift of up to 150 km, was eventually (Aksnes et al. 1986) traced to the phase defect on Europa. Although the observed geometric center of Europa is shifted by at most 8 km, the effect on the *light center* is actually ten times larger; furthermore, the shift is in opposite directions for the occultations and the eclipses.

In the mutual events analyses by Aksnes and Franklin (1976), Aksnes et al. (1984,1986), Franklin et al. (1991) and Kaas et al. (1993), a longitude correction Δx and latitude correction Δz to Sampson's theory at midevent are obtained for each event. Corresponding theory-independent satellite separations, $\Delta\alpha cos\delta$ in right ascension and $\Delta\delta$ in declination, are also computed. A combined analysis of all longitude and latitude corrections then yields least squares corrections to the proper elements of the Galilean satellites; i.e., the mean longitudes, perijoves, eccentricities, nodes, and inclinations. A fit to all the mutual events data between 1973 and 1991 gives rms residuals of about 100 km ($\sim$ $0''.02$) both in longitude and latitude. Similar residuals have been obtained for mutual events data analyzed by Lieske (1980), Arlot (1982),and Arlot et al. (1989). Most of these residuals are probably due to model deficiencies, since multiple observed mutual events frequently are internally consistent to around 10 km.

By adjusting the most important constants in Struve's theories in the same way, Aksnes et al. (1984) obtained rms residuals of about 80 km in the relative satellite positions derived from 14 light curves of Saturnian mutual events in 1980. The observed midtimes agree to within 30 s with the midtimes predicted with Dourneau's ephemerides.

REFERENCES

Aksnes, K. (1974): Icarus **21**, 100.

Aksnes, K. & Dourneau, G. (1993): personal comm.

Aksnes, K. & Franklin, F.A. (1976): Astron. J. **81**, 464.

Aksnes, K. & Franklin, F.A. (1978): Icarus **34**, 194.

Aksnes, K. & Franklin, F.A. (1990): Icarus **84**, 542.

Aksnes, K., Franklin, F. & Magnusson, P. (1986): Astron. J. **92**, 1436.

Aksnes, K., Franklin, F., Millis, R., Birch, P., Blanco, C., Catalano, S. & Piironen,J. (1984): Astron. J. **89** ,280.

Arlot, J.E. (1978): Astron. & Astrophys. Suppl. **34**, 195.

Arlot, J.E. (1982): Astron. & Astrophys. **111**, 151.

Arlot, J.E. (1984): Astron. & Astrophys. **138**, 113.

Arlot, J.E., Bouchet, P., Couiffes, Ch., Schmides, F.X. & Thuillot,W. (1989): Astron. J. **98**, 1890.

Arlot, J.E. & Thuillot, W. (1993): Icarus, in press.

Brinkmann, R.T. (1973): Icarus **19**, 15.

Comrie, L.J. (1931): Mem. Brit. Astron. Assoc. **30**, 97.

Dourneau, G. (1993): Astron. & Astrophys. **267**, 292.

Franklin, F.A. et al. (1991): Astron. J. **102**, 806.

Goguen, J., Sinton, W.M., Matson, D.L., Howell, R.R., Dyck, H.M., Johnson, T.V., Brown, R.H., Veeder, G.J., Lane, A.L., Nelson, R.M. & Mc Laren, R.A. (1988): Icarus **76**, 465.

Kaas, A.A., Franklin, F.A., Aksnes, K. & Lieske, J.H. (1993): Astron. J., (submitted).

Levin, A.E. (1931): Mem. Brit. Astron. Assoc. **30**, 149.

Lieske, J.H. (1980): Astron. & Astrophys. **82**, 340.

Mallama, A. (1992): Icarus **95**, 309.

Pascu, D. (1994): Proceedings of this Workshop.

Sampson, R.A. (1921): Mem. Roy. Astron. Soc. **63**, 1.

Struve, G. (1924-33): Veroff. Univ. Sternw. Berlin-Babelsberg **6**, Parts 1,4, & 5.